Elegant Circuits
Simple Chaotic Oscillators

To Jun:

With best wishes!!

May 25 2022

Elegant Circuits
Simple Chaotic Oscillators

Julien Clinton Sprott
University of Wisconsin-Madison, USA

Wesley Joo-Chen Thio
University of Michigan, USA

World Scientific

NEW JERSEY · LONDON · SINGAPORE · BEIJING · SHANGHAI · HONG KONG · TAIPEI · CHENNAI · TOKYO

Published by

World Scientific Publishing Co. Pte. Ltd.
5 Toh Tuck Link, Singapore 596224
USA office: 27 Warren Street, Suite 401-402, Hackensack, NJ 07601
UK office: 57 Shelton Street, Covent Garden, London WC2H 9HE

Library of Congress Control Number: 2021055560

British Library Cataloguing-in-Publication Data
A catalogue record for this book is available from the British Library.

ELEGANT CIRCUITS
Simple Chaotic Oscillators

Copyright © 2022 by World Scientific Publishing Co. Pte. Ltd.

All rights reserved. This book, or parts thereof, may not be reproduced in any form or by any means, electronic or mechanical, including photocopying, recording or any information storage and retrieval system now known or to be invented, without written permission from the publisher.

For photocopying of material in this volume, please pay a copying fee through the Copyright Clearance Center, Inc., 222 Rosewood Drive, Danvers, MA 01923, USA. In this case permission to photocopy is not required from the publisher.

ISBN 978-981-123-999-1 (hardcover)
ISBN 978-981-124-000-3 (ebook for institutions)
ISBN 978-981-124-001-0 (ebook for individuals)

For any available supplementary material, please visit
https://www.worldscientific.com/worldscibooks/10.1142/12362#t=suppl

Dedicated to Leon O. Chua

Leon O. Chua, 1936–

Photo: https://www2.eecs.berkeley.edu/Faculty/Photos/Fullsize/chua.jpg

Preface

Chaos in electrical circuits has been known since at least 1927 when van der Pol and van der Mark (1927) observed it in a relaxation oscillator but dismissed it as a 'subsidiary phenomenon.' Yoshisuke Ueda also observed 'randomly transitional phenomena' in 1961 with an analog computer, but the observation was so contentious that its publication was suppressed for many years [Abraham and Ueda (2000)]. The modern interest in chaotic circuits is due largely to the work of Leon Chua, who rightly deserves the title of the 'father of chaotic circuits.' His role parallels and complements the role that Edward Lorenz played as the 'father of chaos' and that Benoit Mandelbrot played as the 'father of fractals.' By now hundreds of chaotic circuits have been designed and studied, and they serve as one of the most important practical applications of chaotic dynamics for basic studies and for secure communications.

One of us (JCS) began building electrical circuits as a teenage amateur radio operator in the 1950s when many of my circuits exhibited strange behaviors, but it was not until the 1980s that I learned about chaos and chaotic circuits. Over the years I have repeatedly returned to the challenge of finding especially simple chaotic circuits to complement my passion for finding the simplest systems of ordinary differential equations whose solutions are chaotic and that resulted in the companion book, 'Elegant Chaos: Algebraically Simple Chaotic Flows' [Sprott (2010)]. In that book the elegance is in the form of the equations that produce the chaos. Here, the elegance refers to the simplicity of the circuit as indicated by the number and type of components that it contains. Most of these circuits were designed by others, but there are a few to which I lay claim and some that are presented here for the first time.

The other one of us (WJT) also built circuits for fun as a teenager, which led me to begin my undergraduate studies in electrical engineering at The Ohio State University. While perusing the library, I stumbled across the 'Elegant Chaos' book and thought it would be an enjoyable exercise to start building circuits for the chaotic systems that it described. I eventually contacted Professor Sprott to report my progress, and he immediately recruited me to design and study some new systems and circuits that we subsequently published. Thus began a collaboration that culminated in the writing of this book while I continued my studies as a graduate student at the University of Michigan. We have worked closely together, albeit mostly by email, where I mainly handled the experimental side of our projects.

This book should be of interest to researchers and hobbyists looking for a simple way to produce a chaotic signal. It should also be useful to students and their instructors as an engaging way to learn about chaotic dynamics and electronic circuits. The book assumes only an elementary knowledge of calculus and the ability to understand a schematic diagram and the components that it contains.

You will get the most out of this book if you can construct the circuits for yourself. There is no substitute for the thrill and insight of seeing the output of a circuit you built unfold as the trajectory wanders in real time across your oscilloscope screen. A goal of this book is to inspire and delight as well as to teach. We hope you will enjoy reading and studying it as much as we did writing it.

Many people have contributed to the ideas in this book, and they are referenced in the description of the individual circuits. We particularly thank the colleagues with whom we have coauthored papers that include circuit designs based on new chaotic systems that we were studying for other reasons, especially Chunbiao Li who keeps us busy with his unrelenting stream of clever ideas and suggestions, the group at the University of Catania who are world leaders in the construction and analysis of chaotic circuits, and Ludovico Minati and Buncha Munmuangsaen who read an early manuscript and provided many helpful comments and suggestions.

J. C. Sprott
Madison, Wisconsin

W. J. Thio
Ann Arbor, Michigan

September, 2021

Contents

Preface   vii

1. Introduction   1
   - 1.1 Electronic Oscillators 1
   - 1.2 Relaxation Oscillators 4
   - 1.3 van der Pol Oscillator 8
   - 1.4 Sinusoidally Forced van der Pol Oscillator 10
   - 1.5 Primer on Chaos 13
   - 1.6 Basins of Attraction and Robustness 17
   - 1.7 Early Chaotic Oscillators 19
   - 1.8 Circuit Dynamical Equations 21
   - 1.9 Theoretical Analysis 25
   - 1.10 Parameter Scaling 26
   - 1.11 Construction and Analysis 28
   - 1.12 Circuit Elegance 38

2. Conventional Diode Circuits   41
   - 2.1 Diode Characteristics 41
     - 2.1.1 Ideal diode 41
     - 2.1.2 PN junction diode at equilibrium 42
     - 2.1.3 PN junction with an applied voltage 44
     - 2.1.4 I-V characteristic of the PN junction 45
     - 2.1.5 Capacitance of the PN junction 48
   - 2.2 Forced Diode Resonator 49
   - 2.3 Vilnius Oscillator 51
   - 2.4 Banlue–Rattikarn Circuit 57

	2.5	Banlue–Buncha Diode Circuit	61
	2.6	Chaotic Wien Bridge Oscillator	67
	2.7	Elwakil–Kennedy Diode Oscillator	72
	2.8	Saito Family Diode Circuit	77
	2.9	Diode Jerk Circuit	84

3. Transistor Circuits   89

	3.1	Transistor Characteristics	89
		3.1.1 Bipolar junction transistor (BJT)	89
		3.1.2 BJT I-V characteristic	90
		3.1.3 Field effect transistor	94
		3.1.4 FET I-V characteristic	96
	3.2	Chaotic Colpitts Oscillator	97
	3.3	Minati Circuit	101
	3.4	Minati–Frasca Double-scroll Circuit	108
	3.5	Minati–Frasca Spiking Circuit	111
	3.6	Chaotic BJT Switch	117
	3.7	Lindberg–Murali–Tamasevicius Circuit	125
	3.8	Chaotic Hartley Oscillator	130
	3.9	JFET-based Wien Bridge Oscillator	133
	3.10	Chaotic MOS Amplifier	140

4. Tunnel Diode Circuits   149

	4.1	Tunnel Diode Junction	149
		4.1.1 Tunnel diode I-V characteristic	150
		4.1.2 Tunnel diode emulator	151
	4.2	Forced Relaxation Oscillator	153
	4.3	Autonomous Relaxation Oscillator	158
	4.4	Chua Tunnel Diode Oscillator	162
	4.5	Coupled Relaxation Oscillator	165

5. Thyristor Circuits   173

	5.1	Thyristor Characteristics	173
		5.1.1 Silicon controlled rectifier	173
		5.1.2 Silicon bilateral switch	175
		5.1.3 Thyristor I-V characteristic	177
	5.2	Forced Thyristor Circuit	179
	5.3	van der Pol Relaxation Oscillator	184

	5.4	Autonomous Relaxation Oscillator	187
	5.5	Coupled Relaxation Oscillators	191
	5.6	Many Coupled Oscillators	195
	5.7	Saito Family Thyristor Circuit	199
6.	Saturating Amplifier Circuits	205
	6.1	Operational Amplifiers	205
		6.1.1 Operational amplifier transfer characteristic	206
		6.1.2 Comparators	207
	6.2	Saturating Wien Bridge Oscillator	208
	6.3	Murali–Lakshmanan–Chua Circuit	212
	6.4	Wang–Zhang–Bao Circuit	216
	6.5	Coupled RC Circuits	217
	6.6	Ketthong–Banlue Circuit	221
	6.7	Saito Family Hysteresis Circuit	224
	6.8	Saito Family Switch Circuit	228
	6.9	Simplified Piper–Sprott Circuit	233
7.	Analog Multiplier Circuits	241
	7.1	Analog Multipliers	241
		7.1.1 Analog computers	241
		7.1.2 AD633 multiplier	242
	7.2	Lorenz System	243
	7.3	Rössler Prototype-4 System	245
	7.4	Original Ueda System	249
	7.5	Simple Jerk System	253
	7.6	Petrzela–Polak Circuit	260
	7.7	Dissipative Nosé–Hoover System	261
	7.8	Signum Thermostat	268
8.	Nonlinear Inductor Circuits	275
	8.1	Ferromagnetism	275
		8.1.1 Magnetic properties of ferromagnets	275
		8.1.2 Saturating inductor model	277
		8.1.3 Ferrite-core inductor construction	278
		8.1.4 Ferrite-core inductor measurement	279
	8.2	Forced Ferroresonant Circuit	281
	8.3	Saito Family Inductor Circuit	285

	8.4	Minimal 3D Autonomous Inductor Circuit	292
9.	Memristor Circuits		297
	9.1	Memristors	297
		9.1.1 Memristor I-V characteristic	298
		9.1.2 Ag-chalcogenide memristor	301
		9.1.3 Memristor measurement and model	303
	9.2	Forced Memristor Circuit	305
	9.3	Saito Family Memristor Circuit	310
	9.4	Memristive Wien Bridge Oscillator	313
	9.5	Elwakil–Kennedy Memristor Oscillator	317
	9.6	Senani–Singh Memristor Oscillator	321

Bibliography	329
Index	339
About the Authors	343

Chapter 1

Introduction

This chapter contains the motivation and background for the book, a brief history of chaotic circuits, what we mean by a circuit being 'elegant,' and details that are common to all the circuits in the book, their construction, and their analysis. The subsequent chapters are organized according to the essential nonlinear component that is responsible for the chaos.

1.1 Electronic Oscillators

We are surrounded by electrical circuits, the most obvious of which are the radios, televisions, and computers we use every day. Circuits are often miniaturized and hidden, such as the hearing aids that many people wear or the pacemakers that regulate their heartbeats. These circuits were engineered by humans, but nature has also produced circuits as evidenced by the electrical signals measured with electrocardiograms and electroencephalograms, as well as those arriving at our radio telescopes from outer space.

Perhaps the most common and useful electronic circuit is the *oscillator*. Oscillators control the frequency of radio transmitters, the timing of clocks and logic operations in computers, the sounds produced by music synthesizers, and much else. They provide a means for converting a constant *direct current* (*DC*) into an *alternating current* (*AC*).

Most electronic oscillators are periodic with a well-defined fundamental frequency that is typically in the range of a few cycles per second to billions of cycles per second. They can produce waveforms that are *sinusoidal* (a single frequency) or ones that are rich in harmonics such as those arising from square waves or from a regular train of identical pulses.

There are several factors that determine the frequency of an oscillator. For example, a pendulum will swing at a fixed frequency called the *resonant frequency* that is dependent on the length of its string. Older electronic oscillators controlled their frequency using an inductor (L) and a capacitor (C) where their resonant frequency is determined by $\omega_0 = 1/\sqrt{LC}$.

Throughout this book, we will use the Greek symbol ω to denote the angular frequency in radians per second with ω_0 being the natural resonant frequency. The regular or linear frequency f is the number of complete oscillations in a given unit of time and is given by $f = \omega/2\pi$ and measured in units of *Hertz* (Hz).

All resonant circuits have losses, mostly in the form of resistance (R) especially in the inductor, and those losses cause the oscillations to decay unless energy in some form is continually added to the circuit. The rate at which energy is lost from a circuit is determined by its *quality factor* (Q), which is defined as 2π times the energy stored divided by the energy dissipated per cycle, or equivalently, the number of radians required for the electrical energy to decrease to $1/e \approx 37\%$ of its initial value. (The quantity e is *Euler's number* given by $e = 2.71828....$) For the series RLC circuit in Fig. 1.1, the Q is given by $Q = \omega_0 L/R = \sqrt{L/C}/R$. The oscillating voltages and currents everywhere in this circuit are *damped* as shown in Fig. 1.2 with $Q = 10$, and decay toward zero over time.

Practical oscillators require an *active component* such as a transistor or operational amplifier (colloquially called an 'op amp') and an external source of electrical power to produce sustained oscillations. Two classical circuits of this type are the *Hartley oscillator* in Fig. 1.3 and the *Colpitts oscillator* in Fig. 1.4. Both oscillators involve an LC circuit whose resistive loss is compensated by feeding back a small part of the signal, amplified by an operational amplifier in this case. If the Q of the circuit is sufficiently high ($Q \gg 1$), the oscillations are nearly sinusoidal and produce a voltage V at the output of the operational amplifier that varies in time t according to $V = V_0 \sin \omega_0 t$ as shown in Fig. 1.5. The resulting signal is periodic with period $T = 1/f = 2\pi/\omega_0$ and *amplitude* V_0.

An alternate way to display the behavior of a circuit is to plot a voltage on one axis and the time derivative of that voltage on the other axis. For example, the current in the capacitor of an LC circuit is related to the capacitor voltage through $I = C\dot{V}$, where the overdot in $\dot{V}$ is the symbol we will use for the time derivative of a quantity ($\dot{V} = \frac{dV}{dt}$). Such a plot is called a *phase space plot*, an example of which for a sinusoidal oscillator is shown in Fig. 1.6.

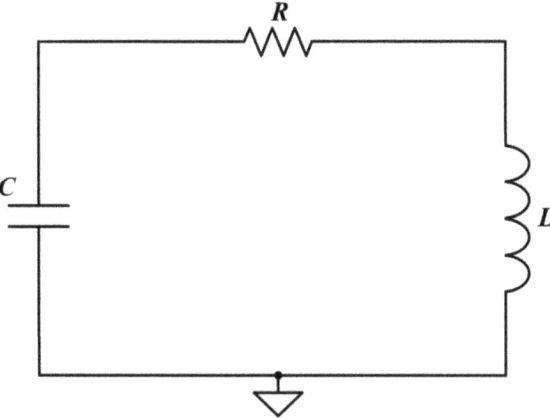

Fig. 1.1 *RLC* circuit.

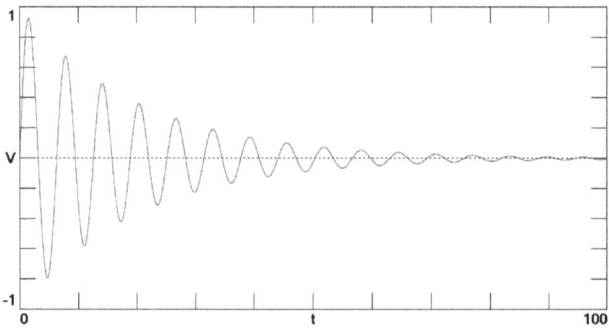

Fig. 1.2 Damped voltage versus time for an *RLC* circuit ($Q = 10$).

In this case, the resulting curve is an ellipse that you can make into a circle through a linear rescaling of the current. If the capacitor voltage is $V_0 \sin \omega_0 t$, the capacitor current is $I = \omega_0 C V_0 \cos \omega_0 t$, giving a curve that is a simple circle for $\omega_0 C = 1$ since $\sin^2 \omega_0 t + \cos^2 \omega_0 t = 1$ and a current that is 90° out of phase with the voltage. The voltage obeys a second-order differential equation given by $\ddot{V} = -\omega_0^2 V$, as does the current. The important conclusion is that any periodic oscillation will have a closed curve in phase space. In fact, any two circuit quantities that are not strictly proportional (not 'in phase') will have such a closed curve if the oscillations are periodic.

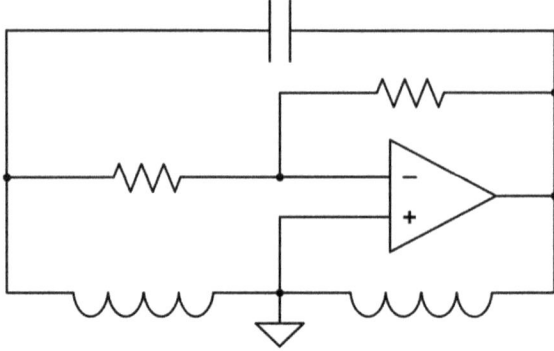

Fig. 1.3 Hartley oscillator using an operational amplifier.

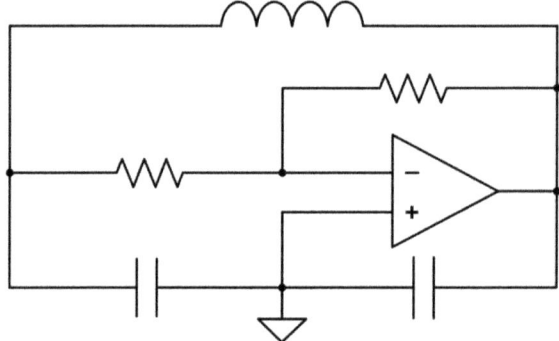

Fig. 1.4 Colpitts oscillator using an operational amplifier.

1.2 Relaxation Oscillators

Oscillators can also be constructed without using resonant circuits, and it is often desirable to avoid inductors since they are hard to miniaturize and adjust over a wide range of values. Electrical circuits that exhibit a periodic abrupt switching when a voltage or current reaches some threshold value are called *relaxation oscillators*. The term was coined by Balthasar van der Pol (1926) by analogy with a deformable medium that periodically 'relaxes' in the presence of an induced stress. The resulting waveform typically resembles a sawtooth or triangle.

One example of an electronic relaxation oscillator is shown in Fig. 1.7. The operational amplifier acts as a comparator, abruptly switching its

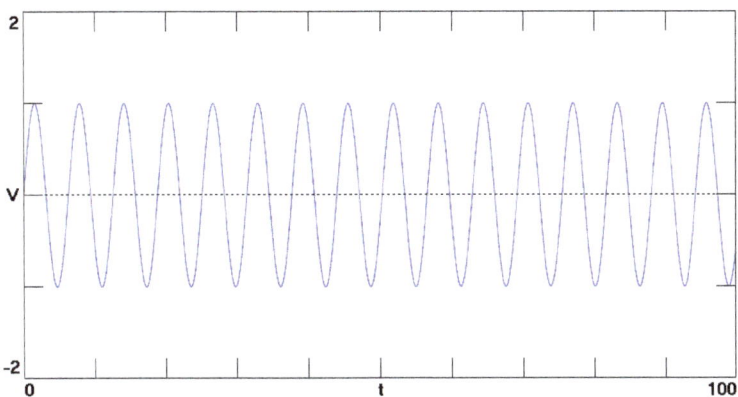

Fig. 1.5 Voltage versus time for a sinusoidal oscillator.

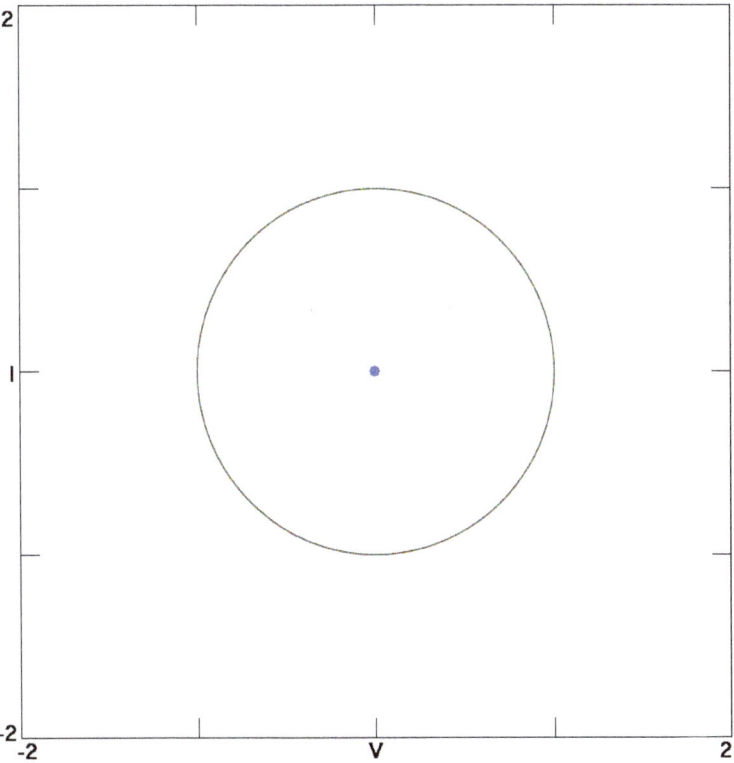

Fig. 1.6 Phase space plot for a sinusoidal oscillator.

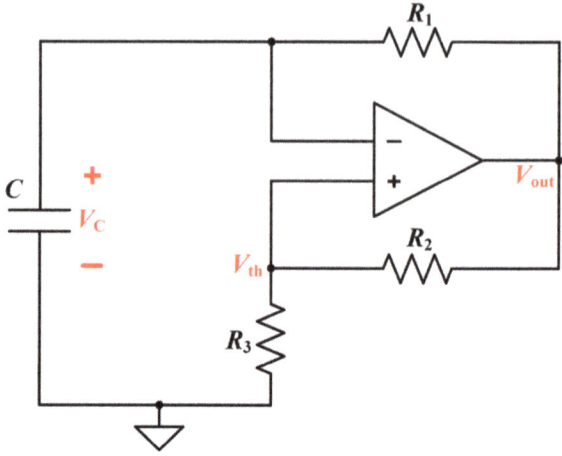

Fig. 1.7 Relaxation oscillator described by Eq. (1.1).

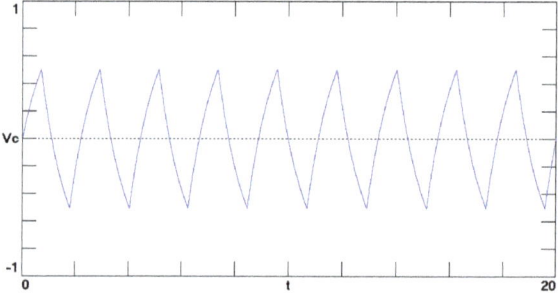

Fig. 1.8 Capacitor voltage for the relaxation oscillator in Fig. 1.7.

output from a large positive voltage (e.g., $V_{out} = 1$ V) to a large negative voltage ($V_{out} = -1$ V) when the capacitor charging through R_1 reaches a threshold voltage V_{th} determined by the voltage divider made up of R_2 and R_3 ($V_{th} = 0.5$ V if $R_2 = R_3$).

This circuit is also called an *astable multivibrator*. The output voltage is a square wave, but the voltage across the capacitor is approximately triangular as shown in Fig. 1.8 for the case where all the components have unit value. The oscillation is periodic but far from sinusoidal. In fact, it is rich in odd harmonics, $3\omega_0, 5\omega_0, ...,$ where $\omega_0 = 2\pi/T = 0.9\pi$ in this case.

The capacitor voltage obeys the differential equation $\dot{V}_C = (V_{out} - V_C)/R_1C$, where $V_{out} = 1$ V until V_C exceeds 0.5 V, whereupon it abruptly

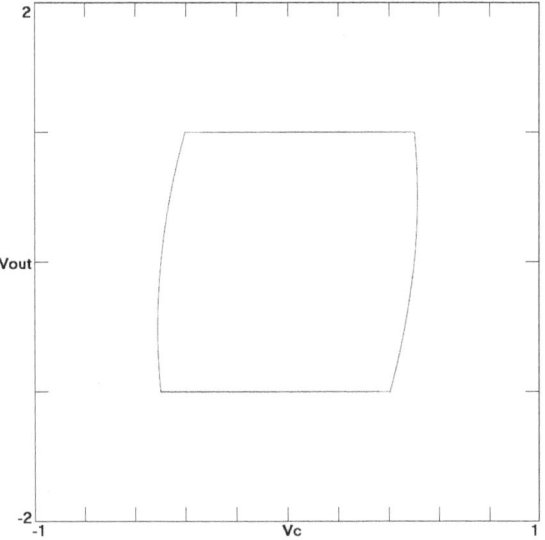

Fig. 1.9 Phase space plot for the relaxation oscillator in Fig. 1.7 as given by Eq. (1.1) with $S = 10$.

switches to $V_{out} = -1$ V until V_C drops below -0.5 V, whereupon it abruptly switches back to $V_{out} = +1$ V, and so forth for each successive cycle.

In a real circuit, V_{out} does not switch instantly, but rather its time derivative is limited to $|\dot{V}_{out}| = S$, where S is the *slew rate* of the amplifier, typically on the order of a few volts per microsecond. This allows us to describe the circuit by a pair of differential equations,

$$\dot{V}_C = (V_{out} - V_C)/R_1 C$$
$$\dot{V}_{out} = \begin{cases} S & V_C < V_{out}/2 \text{ and } V_{out} < 1 \\ -S & V_C > V_{out}/2 \text{ and } V_{out} > -1 \\ 0 & \text{otherwise,} \end{cases} \quad (1.1)$$

from which a phase space plot as shown in Fig. 1.9 with $S = 10$ can be constructed. The closed curve approaches a perfect rectangle for an ideal comparator with $S \to \infty$, but in that limit the equations cannot be solved using the usual method since $|\dot{V}_{out}|$ is infinite.

This is an example of *hysteresis*, which is a kind of 'stickiness' in which the circuit retains a memory of its past state and changes that state only reluctantly. The curve in Fig. 1.9 is called a *hysteresis loop*. Many chaotic circuits rely on hysteresis for their operation.

Relaxation oscillations typically involve two time scales, an interval during which voltages change slowly, alternating with intervals of rapid change. Such systems are said to be *stiff*, and you will need to pay special attention to the numerical integration method. To navigate the sharp corners in the curve requires a tiny time step, while a larger time step is needed to efficiently follow the orbit for more than a few cycles. Thus we strongly recommend a numerical integrator with an adaptive step size such as the adaptive fourth-order Runge–Kutta algorithm described by Press *et al.* (2007) for such cases. Once you have implemented such an integrator, it will be useful and sufficient for the analysis of all the circuits in this book.

1.3 van der Pol Oscillator

To model relaxation oscillations as well as resonant circuits in which significant electrical energy is lost and regained during a single cycle (low Q), van der Pol (1920) proposed a simple second-order differential equation given by $\ddot{x} - b(1-x^2)\dot{x} + x = 0$, which can be written in terms of two first-order differential equations as

$$\begin{aligned} \dot{x} &= y \\ \dot{y} &= b(1-x^2)y - x, \end{aligned} \qquad (1.2)$$

where x can represent the voltage or current at some point in the circuit and y is its time derivative.

In general, throughout this book, we will use lower-case symbols ($x, y, b, ...$) for dimensionless model quantities and upper-case symbols ($V, I, R, L, C, ...$) for actual circuit values, although usually measured in units such as milliamperes, kilo-ohms, henries, microfarads, and milliseconds. However, we will use lower case t for the time in seconds (or more usually milliseconds) and Greek τ for the dimensionless time.

Without the x^2 term and with $b < 0$, Eq. (1.2) is just the equation for a *damped harmonic oscillator* such as the series *RLC* circuit in Fig. 1.1. However, for $b > 0$ the circuit has *antidamping* as might occur in the presence of positive feedback so that small oscillations grow in amplitude. Eventually, x^2 becomes sufficiently large that the damping averages to zero, and an oscillation of constant amplitude results. However, this oscillation is not sinusoidal except in the limit $b \to 0$, but rather it looks as shown in Fig. 1.10 for $b = 2$. The corresponding phase space plot in Fig. 1.11 is a closed loop, but it is certainly neither a circle nor an ellipse nor a rectangle.

This curve is an example of a *limit cycle*, and it is the periodic analog of a stable equilibrium. It is an *attractor* in the sense that any non-zero

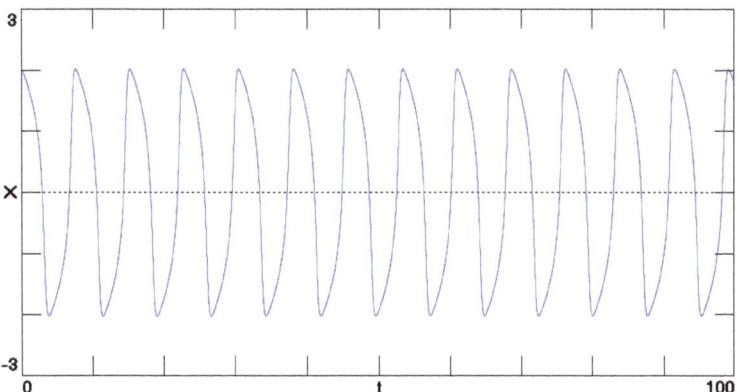

Fig. 1.10 Solution of the van der Pol equation in Eq. (1.2) with $b = 2$.

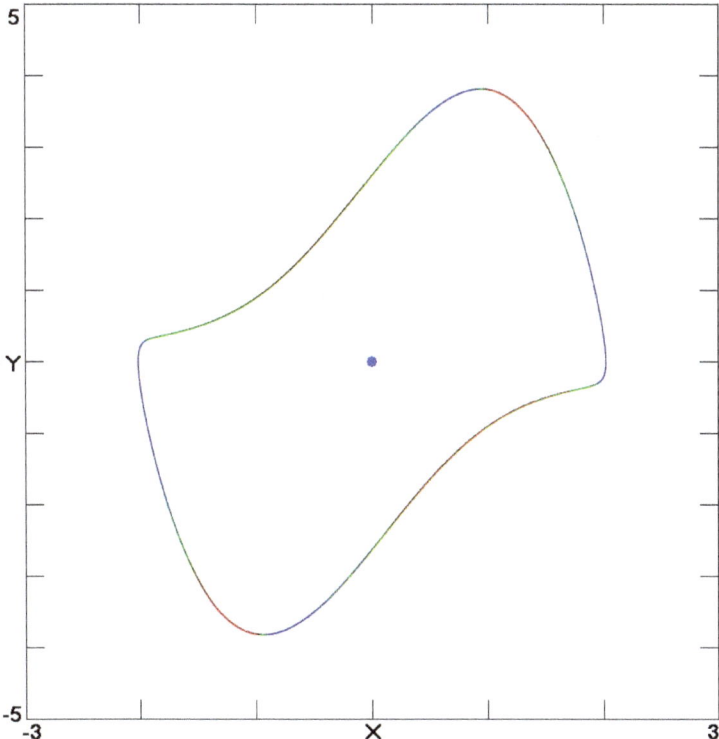

Fig. 1.11 Phase space plot for the van der Pol equation in Eq. (1.2) with $b = 2$.

initial condition for x and y approaches the indicated curve typically within a few cycles of the oscillation, just as the point at the origin (indicated by a dot at $x = y = 0$ in Fig. 1.11) is an attractor for the damped harmonic oscillator. The remainder of this book is concerned with attractors that are considerably more complicated and interesting ('strange attractors') than either a stable equilibrium or a limit cycle.

1.4 Sinusoidally Forced van der Pol Oscillator

If we add to the van der Pol oscillator an externally applied sinusoidal forcing $a \sin \omega \tau$, Eq. (1.2) takes the form

$$\begin{aligned} \dot{x} &= y \\ \dot{y} &= b(1 - x^2)y - x + a \sin \omega \tau, \end{aligned} \quad (1.3)$$

where a is the amplitude of the forcing function and ω is its (angular) frequency, which is generally different from the natural frequency of oscillation for the unforced oscillator, denoted by ω_0. For $b = 2$, the natural frequency of the van der Pol oscillator in Eq. (1.2) is $\omega_0 \approx 0.8235$, corresponding to a period of $T = 2\pi/\omega_0 \approx 7.630$.

When a is sufficiently small, the resulting $x(t)$ consists of the superposition of two components at respective frequencies ω and ω_0 as shown in Fig. 1.12 for $a = 2$ and $\omega = 2$. Whenever the two frequencies are *incommensurate* (their ratio is not a rational number), the result is not periodic. The corresponding phase space plot as shown in Fig. 1.13 is no longer a closed curve, but it winds endlessly around the surface of a *torus* (a 'doughnut'), never intersecting itself. We will explain the meaning of the colors later. This is an example of *quasiperiodicity*, and the torus is a two-dimensional attractor embedded in a three-dimensional space in which the third dimension is the phase of the forcing function ($z = \omega \tau$).

In fact, Eq. (1.3), which is *nonautonomous* because it involves τ explicitly on the right-hand side, can be written in *autonomous* form as

$$\begin{aligned} \dot{x} &= y \\ \dot{y} &= b(1 - x^2)y - x + a \sin z \\ \dot{z} &= \omega. \end{aligned} \quad (1.4)$$

Thus the phase space (more generally called *state space*) is three-dimensional, which allows the trajectory to fill a two-dimensional region (the surface of the torus) without ever intersecting itself since such an intersection would violate the condition for a *deterministic dynamical system* that every point in state space has a unique future.

Introduction 11

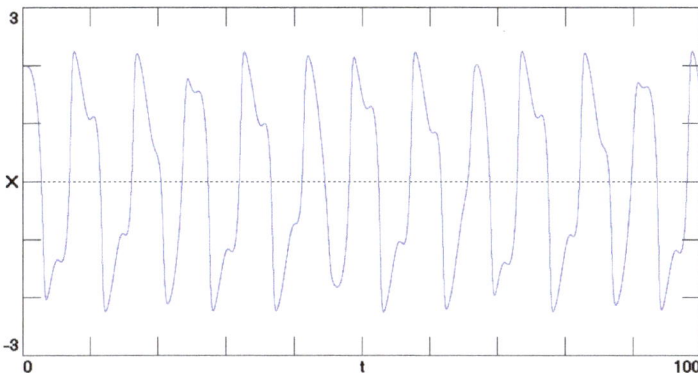

Fig. 1.12 Solution for the forced van der Pol oscillator in Eq. (1.3) with $a = b = \omega = 2$ showing quasiperiodicity.

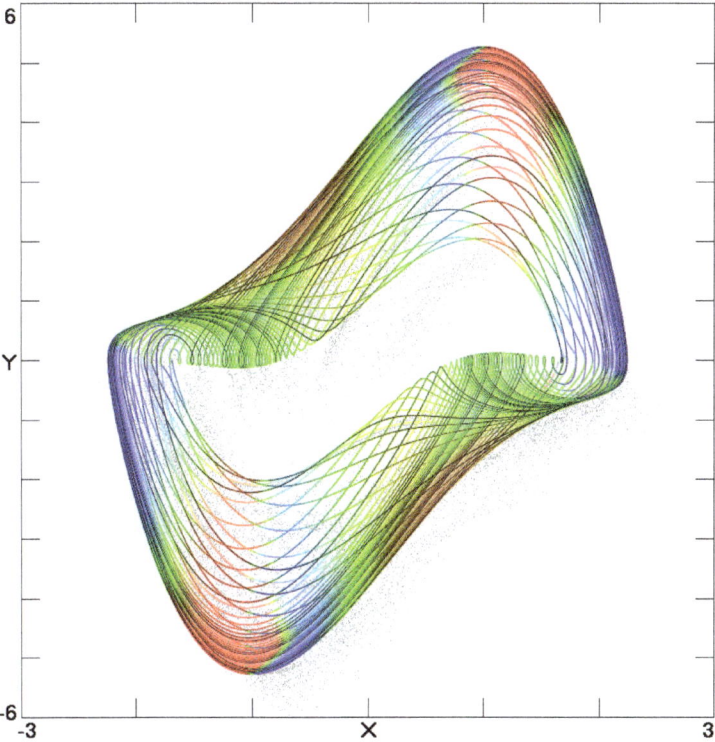

Fig. 1.13 Phase space plot for the forced van der Pol oscillator in Eq. (1.3) with $a = b = \omega = 2$ showing quasiperiodicity with an orbit that lies on the surface of a torus.

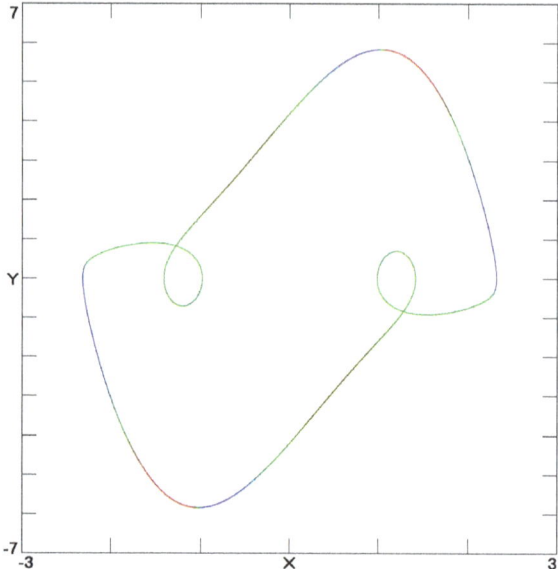

Fig. 1.14 Phase space plot for the forced van der Pol oscillator in Eq. (1.3) with $a = \omega = 2$ and $b = 3$ showing a limit cycle.

However, not all combinations of the parameters produce quasiperiodicity. For example, increasing a to 3.0 gives the limit cycle shown in Fig. 1.14, which is a closed loop but now embedded in three-dimensional space so that its projection onto the xy-plane appears to intersect but actually passes beneath itself. This is an example of *frequency-locking* or *entrainment*, with the circuit oscillating at the frequency of the forcing function, which is $\omega = 2$ in this case, but with higher harmonics (integer multiples of ω).

A different kind of behavior is also possible that is neither periodic nor quasiperiodic, but rather is *aperiodic*, an example of which for $a = 1$, $b = 2$, and $\omega = 0.24$ is shown in Fig. 1.15. Although its attractor looks similar to the one in Fig. 1.13, it is neither a closed curve nor a torus, but something more complicated called a *strange attractor*, which is a *fractal* [Mandelbrot (1982); Sprott (2019)] with a fractional dimension in this case (by one way of calculating) of about 2.0134. It consists of a continuous spectrum of frequencies rather than the discrete frequencies in the previous cases. It is an example of *chaos*, to which the remainder of this book is devoted and to which we now turn.

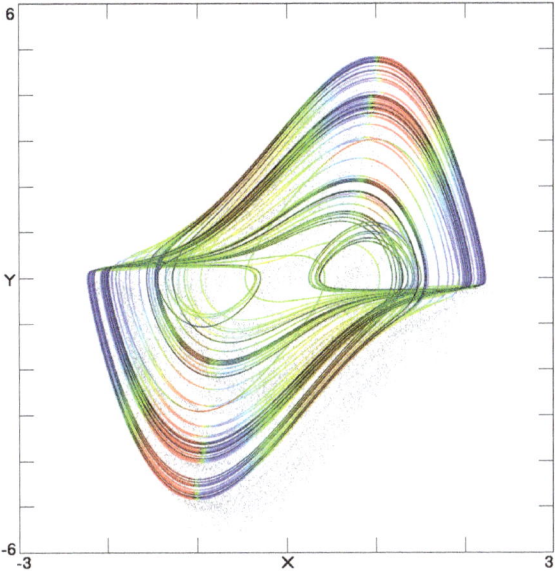

Fig. 1.15 Phase space plot for the forced van der Pol oscillator in Eq. (1.4) with $a = 1$, $b = 2$, and $\omega = 0.24$ showing aperiodicity (chaos).

1.5 Primer on Chaos

In addition to being aperiodic and having an attractor that is a fractal with a non-integer dimension, the most salient characteristic of a chaotic system is its *sensitive dependence on initial conditions*. If you repeat the calculation with a slightly different initial condition, the difference between the two solutions will grow exponentially on average until they eventually lie in totally different regions of the attractor. As a practical matter, that means that long-term prediction of a chaotic system is impossible, except perhaps in a statistical sense. Lorenz (1993) called this the *butterfly effect* since a butterfly flapping its wings in Brazil can cause a tornado in Texas several weeks later because the atmosphere is chaotic.

Crucially important is that the error grows at an *exponential* rate since there are non-chaotic systems such as an inaccurate clock in which the error grows linearly in time or even as some power of time. Conversely, exponential growth of errors can occur in systems that are not chaotic such as those governed by the simple equation $\dot{x} = x$ where x increases exponentially without bound according to $x = x_0 e^\tau$. The increase in the value of a financial asset due to compounding of interest is an example of nonchaotic exponential growth. Two investors who invest slightly different amounts will have very different wealth after a sufficiently long time.

Thus we must also insist that the solution remain bounded, which requires a nonlinear term in the equations to bring the trajectory in state space back to the vicinity of its starting point. In fact, it will eventually come arbitrarily close to where it started on the attractor, but never exactly since that would imply periodicity. Arbitrarily close to every such orbit is one that *does* return exactly to where it started, but those are a *set of measure zero*, which means that you will never accidentally stumble across one. Most such periodic orbits have very long periods. Even if you managed to find one, even the slightest perturbation would take you away from it onto an orbit that never repeats. We say that the attractor is *dense in unstable periodic orbits*, and this is another defining characteristic of chaos.

You can also view the sensitive dependence as a *stretching* of the attractor, while the nonlinearity causes a *folding*. This process may remind you of a taffy-making machine, and the resulting *mixing* that results is yet another characteristic of chaos. In an electronic circuit, boundedness is rarely a problem because voltages typically cannot exceed the power supply voltage, although unbounded solutions of the governing equations will usually manifest in your amplifiers saturating and becoming unresponsive to their inputs.

A consequence of this requirement is that a circuit that contains only linear elements (ideal resistors, capacitors, inductors, and linear amplifiers) can never exhibit chaos no matter how complicated you make the circuit, but rather a chaotic circuit must contain at least one nonlinear element. We have organized the subsequent chapters in this book according to the essential nonlinear device that is responsible for the chaos, although some circuits contain more than one nonlinear component. Transistors and saturating amplifiers are nonlinear devices and thus can produce chaos as discussed in Chapters 3 and 6, respectively.

The exponent (power of *e*) describing the average rate at which errors grow is called the *Lyapunov exponent* (LE), and it provides a means to establish that a system is chaotic if its value is positive and to quantify how chaotic it is. However, its calculation requires averaging the *local* Lyapunov exponent along the orbit, and the orbit usually passes through regions of the attractor where the value is strongly negative as well as where it is strongly positive.

In fact, the positive and negative values nearly cancel for typical cases, and so the calculated Lyapunov exponent converges slowly and requires you to follow the orbit for a long time to obtain accurate values. As a rule of thumb, to get four-digit accuracy (1 part in 10^4), you must follow the orbit for about 10^8 cycles of the dominant frequency, and hardly anyone does that.

Fortunately, it is easy for you to assess the precision of the calculated Lyapunov exponent for a chaotic system by simply repeating the calculation with a slightly altered initial condition, making use of sensitive dependence to provide a new orbit essentially uncorrelated with the original one but whose Lyapunov exponent should be identical. Do not make the common mistake of calculating the Lyapunov exponent for a short time with a single initial condition and quoting a value with extraneous insignificant digits! Most of the Lyapunov exponent values that we give in this book have been calculated for a time of at least 10^7 (in units of the natural period of oscillation).

Furthermore, there are systems that exhibit long-duration *chaotic transients*, in which the solution appears chaotic with a well-converged positive Lyapunov exponent, but that eventually settle onto a limit cycle or some other non-chaotic orbit. Such behavior is often restricted to a narrow range of parameters and can be revealed by changing the parameters slightly. However, transients are rarely a problem in actual chaotic circuits since their time scale is usually much faster than the time required to perform the corresponding numerical calculation. Nature can solve the equations much faster than can a digital computer!

In the state space plots such as Figs. 1.13 and 1.15, the colors show the local Lyapunov exponent, with red indicating positive values and blue indicating negative values, while intermediate values follow the colors of the rainbow. Thus green indicates values near zero. You can think of red as denoting regions of poor short-term predictability and blue as regions of good predictability. Note that even the quasiperiodic case in Fig. 1.13 has regions of positive local Lyapunov exponents, but these values are exactly cancelled by the regions of negative Lyapunov exponents. Thus periodic and quasiperiodic systems have Lyapunov exponents that are exactly zero.

Since it is difficult to distinguish zero from a small positive value in a numerical calculation, it is important that any claim of chaos based on a positive Lyapunov exponent include an estimate of the precision of the calculation. All the LE values quoted in this book have an uncertainty only in the least significant quoted digit. Furthermore, it is very difficult to determine the Lyapunov exponent from an experimental measurement of an electronic circuit, and so the evidence that a circuit is exhibiting chaos usually comes only from observing its aperiodicity and showing that the governing equations have a positive Lyapunov exponent, which likely annoys the mathematicians.

A three-dimensional system such as Eq. (1.4) actually has three Lyapunov exponents (λ_1, λ_2, λ_3), which by convention are ordered according to $\lambda_1 \geq \lambda_2 \geq \lambda_3$, and it is the largest of them (λ_1) that determines whether the system is chaotic. The other exponents are useful for determining the dimension of the attractor and the *entropy*.

Unless the system has a stable equilibrium, one of the exponents will always be zero. For example, the forced van der Pol oscillator in Eq. (1.3) with $a = 1$, $b = 2$, and $\omega = 0.24$ has an *LE spectrum* of (0.0269, 0, −2.0105). It is common for the negative exponent(s) to be much larger in magnitude than the positive exponent(s). The negative exponents denote how fast nearby initial conditions approach the attractor.

Chaos cannot occur in an autonomous system with fewer than three first-order differential equations (three dimensions) because the orbit would otherwise intersect itself, leading to periodicity. This fact is formalized in the *Poincaré–Bendixson theorem* [Hirsch et al. (2004)], which says that the bounded state space trajectory of a two-dimensional system can only approach a stable equilibrium or a limit cycle. As a consequence, oscillator circuits (without external forcing) generally require at least three reactive components (capacitors and inductors) to exhibit chaos. A circuit with a periodic external forcing needs only two such reactive components.

For a stable equilibrium (a damped oscillation) in three dimensions, the exponents have signs $(-, -, -)$, for a limit cycle they are $(0, -, -)$, for an attracting torus they are $(0, 0, -)$, and for a strange (chaotic) attractor they are $(+, 0, -)$, with respective attractor dimensions of 0, 1, 2, and $2 - \lambda_1/\lambda_3$, the latter according to the Kaplan and Yorke (1979) conjecture. The sum of the exponents is equal to the trace of the Jacobian matrix averaged along the orbit, $\lambda_1 + \lambda_2 + \lambda_3 = \langle \frac{\partial \dot{x}}{\partial x} + \frac{\partial \dot{y}}{\partial y} + \frac{\partial \dot{z}}{\partial z} \rangle$ and is less than or equal to zero for a bounded system. In fact, it will be strictly less than zero for the *dissipative systems* that dominate this book. Calculation of the LE spectrum is slightly technical and beyond the scope of the present discussion but is discussed in many sources such as Wolf et al. (1985).

While chaos requires at least three coupled autonomous ordinary differential equations and most of the elegant circuits that follow can be so modeled, there are some circuits that require higher dimensional models, in which case there will be more than three Lyapunov exponents, but at least one must be positive if the circuit is chaotic, one must be zero (corresponding to the direction parallel to the orbit), and the sum of the exponents cannot be positive if the orbit is bounded. Systems with more than one positive Lyapunov exponent are called *hyperchaotic*, and you will see some such cases later. Systems in which the exponents sum to zero are called *conservative* (as opposed to dissipative), and they are very hard to construct.

1.6 Basins of Attraction and Robustness

Any dissipative dynamical system with a bounded solution, such as a chaotic oscillator, will have an attractor, and that attractor is surrounded by a *basin of attraction* that represents those initial conditions that approach the attractor during the initial transient phase. It is important to identify the basin of attraction since that will determine the required initial voltages and currents when the circuit is turned on, and it will foretell any difficulty in making the circuit oscillate.

Most of the circuits we describe in this book have an unstable equilibrium at or near the origin, and so the circuits will properly start when the capacitors are initially discharged since there is always a small amount of thermal noise voltage present to get the orbit off the point attractor. We will explicitly note the occasional exceptions. However, many of the attractors have relatively small basins of attraction, which means that they may not oscillate if the capacitors have a significant initial voltage. In principle, the inductors could have a nonzero initial current, but in practice, that is rarely the case and hence not an issue.

Furthermore, some of the circuits have no equilibrium points, which implies that they will either oscillate for any initial condition or they will be unbounded, meaning that the voltages grow to the point where the circuit equations are no longer valid, typically saturating the amplifiers. It is also possible for a circuit to have multiple equilibrium points. Initial conditions in the vicinity of one equilibrium may lead to the desired chaotic attractor, while the other does not. We say that such systems are *multistable*. If there is no equilibrium or if none of the equilibria lie within or adjacent to the basin of attraction, the attractor is said to be *hidden*; otherwise it is *self-excited* [Leonov and Kuznetsov (2013)].

Basins can be categorized into four classes depending on the probability that an initial condition far from the attractor is in its basin [Sprott and Xiong (2015)]. Generally, the probability has the form $P = P_0/r^\gamma$ for large r, where r is the distance from the 'center of mass' of the attractor in units of the 'size' of the attractor. For a system with dimension D, the four classes represent a hierarchy of basin sizes from the largest to the smallest as follows:

Class	P_0	γ	Description
1a	1	0	global except for a set of zero measure
1b	1	0	global except for a set of finite measure
2	<1	0	basin is a fixed fraction P_0 of the space
3	P_0	$<D$	basin is infinite but is a zero fraction of the space
4	P_0	D	basin has a finite size of order $r \sim \sqrt[D]{P_0}$

For each of the circuit models in the following chapters, we give the basin class along with the values of P_0 and γ as appropriate, and we will note if there are coexisting attractors. However, the basin for the actual circuit may differ from the model since the model will usually ignore amplifier saturation and other effects that are not relevant for the desired attractor. Systems with a very small basin of attraction are said to be *delicate* since even small initial voltages can cause them not to operate correctly.

A system can also be delicate if a small change in the values of the parameters (the component values) cause it to fail. This idea can be quantified in the mathematical model by choosing random values of the normalized parameters within a hypersphere centered on the nominal values and determining by what percentage they can vary before the probability of chaos drops to 50%. This value is then a measure of the *robustness* of the system and will be given for those cases that are especially delicate (not robust). As a standard of comparison, the classic Lorenz (1963) system is about 25% robust (a 25% change in parameters is likely to destroy the chaos).

In some cases, the predicted robustness from the model does not reflect the observed robustness of the circuit. It is common for a system to be very sensitive to some parameters and very insensitive to others. Also the model and circuit may be in different regions of parameter space, or the model is just not a very good description of the circuit. We do not mean to imply that you must replicate our component values to this level of precision, but only that you should be prepared to make correspondingly small adjustments to your parameters to observe the chaos.

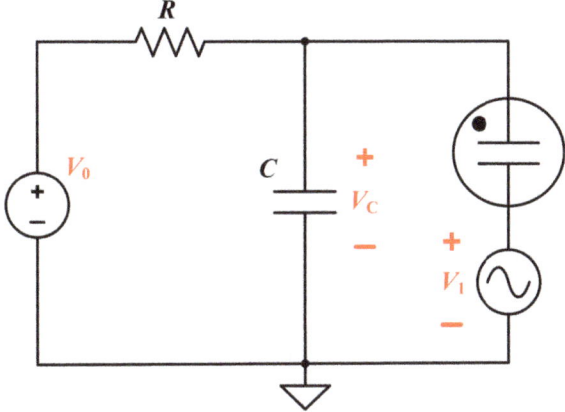

Fig. 1.16 van der Pol's periodically-forced relaxation oscillator circuit.

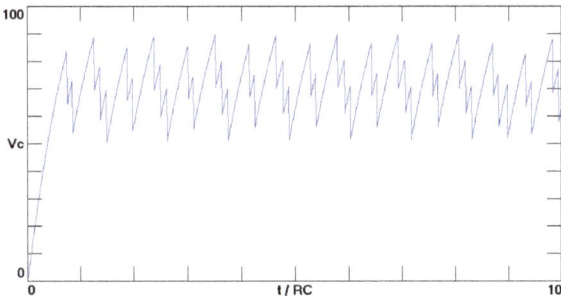

Fig. 1.17 Typical waveform for the capacitor voltage in van der Pol's periodically-forced relaxation oscillator ($V_0 = 160$, $V_1 = 10\sin(11t/RC)$).

1.7 Early Chaotic Oscillators

The earliest recorded observation of chaos in electrical circuits came from a brief comment published in *Nature* by van der Pol and van der Mark (1927), who described a circuit that could generate submultiples of an input sinusoidal voltage. However, between each frequency jump as they increased the frequency of the input signal, they would hear an 'irregular noise' in the telephone receiver they used to detect the frequency changes.

Their periodically-forced relaxation oscillator circuit as shown in Fig. 1.16 consisted of an input sinusoidal voltage source, a neon lamp, and a variable capacitor. A typical chaotic waveform for the voltage across the capacitor is shown in Fig. 1.17.

At low voltages, the neon lamp is an open circuit, which allows the capacitor to charge from the DC voltage source through the resistor. When the voltage reaches a sufficiently high value (typically about 80 volts), the neon lamp goes into conduction and abruptly discharges the capacitor to some lower voltage (typically about 60 volts) where the lamp ceases conduction, and the process repeats. The exact values of the capacitor voltage where the lamp turns on and off depend on the amplitude and phase of the sinusoidal voltage source.

Although this circuit is not very well modeled by Eq. (1.3), we now know that such periodically-forced nonlinear oscillators can exhibit chaos, but van der Pol and van der Mark dismissed the noise as a 'subsidiary phenomenon' and did not investigate further. A modern implementation of the circuit would use a low-voltage diac or breakover diode in place of the neon lamp, eliminating the possibility of electrocution with the large voltages required to activate the neon lamp.

Developments in computing technology a few decades after van der Pol and van der Mark's work provided a quick way to obtain the solutions of differential equations. In 1961, a Japanese graduate student at Kyoto University named Yoshisuke Ueda used an analog computer to solve a periodically-forced nonlinear ordinary differential equation (ODE). While plotting the recorded output data on graph paper, he noticed an unexpected behavior that he called 'randomly transitional phenomena' [Abraham and Ueda (2000)]. However, his advisor was unconvinced and prevented him from publishing his results until 1973 [Ueda *et al.* (1973)]. Thus the modern discovery of chaos is credited instead to Edward Norton Lorenz, whose 1963 publication [Lorenz (1963)] describes the solution of a simple model of atmospheric convection using a Royal McBee LGP-30 digital computer [Gleick (1987)].

Despite these developments, engineers continued to think of chaos as a pathological phenomenon that only existed in abstract mathematical equations. To demonstrate the practical application of chaos, Professor Takashi Matsumoto of Waesada University constructed an electrical circuit in 1983 to model the Lorenz equations [Chua (1992)]. This circuit often failed to operate due to the unreliability of early analog multipliers. Today, such circuits can be easily constructed [Strogatz (1994)], and we consider such examples in Chapter 7.

After a visit to Matsumoto's laboratory in 1987 and witnessing a failed demonstration of the Lorenz circuit, Leon O. Chua was inspired to design a more robust chaotic circuit for a different chaotic system. His circuit contained only a few passive components with an active piecewise nonlinearity called *Chua's diode* that can be built with operational amplifiers, circumventing the need for analog multipliers. This well-known circuit is now called *Chua's circuit* [Matsumoto (1984)], and it is the basis for some of the circuits in this book.

Chaotic circuits have been proposed for computing [Munakata et al. (2002)] and for secure communications [Cuomo and Oppenheim (1993)], but such applications have not progressed beyond research. One of the difficulties in finding practical applications is the little improvement they offer over other technologies such as digital random number generators or stronger encryption methods. Although time will tell whether chaotic circuits will find an important application, the ubiquity of chaos throughout the advances of technology from the earliest vacuum tube oscillators to innovative devices for artificial intelligence likely portends new applications beyond our imagination.

1.8 Circuit Dynamical Equations

We now provide an overview of the basic theory required to understand the exciting world of chaotic circuits. The hands-on experiment described here will illustrate the method for determining the differential equations governing the circuits in this book and will help develop the skills needed to build them.

We consider the example of a simple oscillator circuit shown in Fig. 1.18 [Sprott (1981)] and analyze its behavior in detail. This is a simple circuit that can generate nearly sinusoidal signals, but not chaos. The circuit has only a single inductor, capacitor, and resistor, along with an operational amplifier to sustain the oscillations. Thus, in some sense, it is the most elegant example of an electrical harmonic oscillator.

The first step in determining the equations that model a circuit is to use *Kirchhoff's current law*. This is an intuitive law that treats electrical conductors like water pipes. If water enters a pipe, there must be at least one path for the water to exit, and furthermore, the total flow of water out of the pipe must exactly equal the flow into it. The electrical analogy is that whenever an electric current flows into an electrical conductor, that same current must exit the conductor somewhere else. You can confirm that this is the case in Fig. 1.18 by noting the direction of the blue arrows. The number of currents entering and exiting a *node* (one of the black dots where the conductors form a junction) does not matter as long as there is

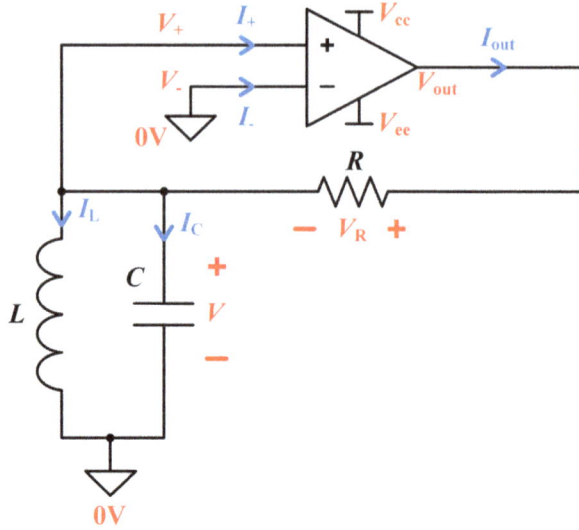

Fig. 1.18 Simple oscillator circuit described by Eq. (1.8).

at least one of each. At the basic physics level, Kirchhoff's current law is just the law of conservation of electric charge. Charge can move from place to place, but it cannot be created or destroyed.

There are a few conventions that we adopt when labeling the currents, which are also intuitive and easy to follow. Water naturally flows from a high place to a low place because of gravity. Likewise, electric current usually flows from a high voltage to a low voltage (except in a voltage source). The high and low voltages are represented by a red plus (+) and minus sign (−), respectively, since positive is greater than negative. Zero voltages are represented by a *ground* (the downward pointing triangle symbol). In fact, points labeled by the ground should be connected to each other by a conductor (typically a copper wire) as well as to the low voltage side of the power supply (discussed later).

Of course the voltages and currents are generally time dependent, so that at a given instant, their values may be either positive or negative relative to the way the +, −, and arrows are drawn. It is only necessary that the arrows point from the labeled + to the − for each passive component, whether it is a resistor, capacitor, inductor, or one of the nonlinear elements such as the diodes to be discussed later. Applying Kirchhoff's current law to this circuit gives

$$I_{out} = I_C + I_L + I_+. \tag{1.5}$$

The next step is to apply *Kirchhoff's voltage law*, which states that the voltage drops around any loop in a circuit must sum to zero. There are two loops in the circuit in Fig. 1.18, one involving L and C that gives the trivial result that the inductor and capacitor have the same voltage V, and the other involving the operational amplifier and resistor that gives

$$V_{out} - V_R = V = V_+. \tag{1.6}$$

At the basic physics level, Kirchhoff's voltage law is just the law of conservation of electrical energy. A charge moving around a loop in a circuit must return to its starting point with the same electrical potential energy with which it started.

The final step is to apply the equations that describe the components in the circuit, collected here for convenience:

$$\begin{aligned} V &= RI \quad \text{(resistor)} \\ V &= L\dot{I} \quad \text{(inductor)} \\ I &= C\dot{V} \quad \text{(capacitor)} \\ V_{out} &= A(V_+ - V_-) \quad \text{(op amp)}. \end{aligned} \tag{1.7}$$

The operational amplifier is a remarkable device that we describe in more detail in later chapters for the many circuits in which it is employed, but for the present it suffices to say that the output voltage V_{out} is an amplified version of the difference between its two inputs $V_+ - V_-$ with a large amplification factor A that is typically 10^5 to 10^6 but that decreases with increasing frequency. As a result, even a tiny difference between the input voltages causes the output to saturate at $\pm V_{sat}$ where the magnitude of V_{out} approaches the power supply voltage of V_{cc} or V_{ee}. Typically $V_{sat} \approx V_{cc} = -V_{ee} \approx 15$ volts.

Thus the operational amplifier in this application acts as a comparator, switching from $-V_{sat}$ to $+V_{sat}$ whenever the input voltage difference changes from negative to positive, conveniently written as $V_{out} = V_{sat} \, \text{sgn}(V_+ - V_-)$. The function sgn x, called the *signum function*, is -1 if x is negative, 0 if $x = 0$ and $+1$ if x is positive. It can also be written as sgn $x = x/|x|$ or as sgn $x = |x|/x$ if you prefer.

For simplicity, we will omit the voltages V_{cc} and V_{ee} in subsequent circuit schematics, but they are always required to power the operational amplifiers. The other important property of an operational amplifier is that it has a very high input resistance (usually many megohms), so that the input currents I_+ and I_- are essentially zero and can be neglected for most purposes.

Combining Eq. (1.7) with Eq. (1.5) and Eq. (1.6) leads to a pair of differential equations that model the circuit:

$$\begin{aligned} \dot{V} &= [(V_{sat}\,\text{sgn}\,V - V)/R - I_L]/C \\ \dot{I}_L &= V/L. \end{aligned} \tag{1.8}$$

Inductors always have some internal series resistance R_L that often cannot be neglected. This resistance arises from the wire used to construct the inductor and accumulates as the wire gets longer. Furthermore, there is an additional AC resistance due to *skin effect* (the current flows mainly near the surface of the conductor) as well as hysteresis and *eddy current* losses if the inductor uses an iron core. This AC resistance increases with frequency and can easily exceed the DC resistance even at modest frequencies.

This resistance 'steals' a voltage $\Delta V = I_L R_L$ from the inductor so that the inductive voltage is reduced to $V_L = V - I_L R_L$, where V is the voltage across the real inductor. To include this effect in the model equations, you would simply change the second equation in Eq. (1.8) to $\dot{I}_L = (V - I_L R_L)/L$. The same AC effects causes the inductance to change slightly with frequency. A real inductor also has some inherent parallel capacitance due to the proximity of the winding, and this capacitance is distributed throughout the inductor but is usually unimportant at the low frequencies used for the circuits in this book.

These are examples of *parasitic effects* since they are usually undesirable but unavoidable. They can generally be ignored but sometimes used to advantage. Any resistance in series with an inductor will be considered parasitic and part of a single electrical component since you could in principle construct an inductor with the desired resistance. This resistance will usually not be shown to simplify the schematic.

All electrical components have such parasitic effects, which can often be modeled by a series resistance and inductance and by a parallel capacitance, especially as the frequency increases. In fact, at sufficiently high frequencies, inductors behave more like capacitors, and capacitors behave more like inductors. Generally, the frequencies employed in this book are sufficiently low that the parasitic resistance in the inductors is the most problematic effect, although you will see circuits in the following chapters where parasitic capacitance is important and useful. Circuit simulation software takes some of the parasitic components into consideration, but their neglect in the mathematical models is the main reason the circuits do not behave exactly as predicted.

1.9 Theoretical Analysis

Having reduced a circuit to a set of coupled ordinary differential equations, the next step is to find appropriate parameters of the model and analyze its behavior. For the system in Eq. (1.8), the parameters are R, L, C, and V_{sat}. Generally, we will seek parameters that give chaotic solutions and that are simple (such as small integers) using the method described in the companion book *Elegant Chaos: Algebraically Simple Chaotic Flows* [Sprott (2010)].

However, the system in Eq. (1.8) does not have chaotic solutions because there are only two state variables (V and I_L) and two reactive components (C and L), but it does have periodic solutions provided $R^2C/L > 1/4$. Thus the elegant values of $R = L = C = 1$ give a limit cycle solution with a nearly sinusoidal oscillation of the variables at a frequency of $\omega_0 \approx 1/\sqrt{LC} = 1$ rad/s and a 90° phase shift between them and with an amplitude determined by the value of V_{sat}. In the limit $R^2C/L \to \infty$, the oscillation is perfectly sinusoidal with an amplitude of $\sqrt{2}V_{sat}$, but parasitic resistance in the inductor does not allow oscillation in this limit since the positive feedback provided by R is insufficient to overcome the resistive losses in the LC circuit.

To continue the theoretical analysis, it is helpful to simplify and reduce the equations to dimensionless form by defining new variables, $x = V/V_{sat}$, $y = I_L R/V_{sat}$, and $\tau = t/RC$. Then Eq. (1.8) becomes

$$\begin{aligned}\dot{x} &= a\,\text{sgn}(x) - x - y \\ \dot{y} &= bx,\end{aligned} \quad (1.9)$$

where the new dimensionless parameters are a and $b = R^2C/L$, and the overdot now denotes a derivative with respect to the dimensionless time τ. The parameter a is called an *amplitude parameter* since it proportionally controls the amplitude of the voltages and currents at every point in the circuit and thus can be taken as $a = 1$ without loss of generality.

The parameter b is called a *bifurcation parameter* since it controls the dynamic behavior of the system. In fact, in this case, it is nothing more than the square of the Q for the oscillator (defined reciprocally from the earlier case because this is a parallel rather than a series RLC circuit). This is only one of several ways to transform the equations, and the bifurcation parameter could just as well have been chosen as the coefficient of one of the other terms. In any case, this system has a single bifurcation parameter whose variation allows complete characterization of its possible dynamics.

As already mentioned, for $b < 0.25$ the circuit does not oscillate, but rather any initial voltages and currents rapidly decay to zero because the origin ($x = y = 0$) is a stable equilibrium (a point attractor). As b increases, this equilibrium loses its stability at $b = 0.25$ in a *Hopf bifurcation* [Marsden and McCracken (1976)], and the stable periodic limit cycle is formed.

Note that to have a stable oscillation (a limit cycle) requires a nonlinearity, in this case provided by the signum $\text{sgn}\, x$ function. An oscillator constructed with purely linear components will have oscillations that either decay to zero or grow without bound. Fortunately, all real circuits have nonlinearities that eventually limit the magnitude of the output signal.

We contend that it is better to eliminate the tedious algebra that led to Eq. (1.9) and perform the parameter optimization directly on Eq. (1.8) using the circuit component values as the parameters without introducing any numerical values. This may cause the dimensionless equations to be slightly less elegant, but it automatically gives elegant circuit component values, which generally means as many as possible are set to unity with the remaining ones having rounded digits. In this case, it makes no difference since all the parameters are unity.

The parameters $a = 1$ and $b = 1$ give the limit cycle attractor in Fig. 1.19 and a time series as shown in Fig. 1.20. The Lyapunov exponents are $(0, -0.5)$, and the Kaplan–Yorke dimension is $D_{ky} = 1.0$ as expected for a limit cycle. The attractor has a global Class 1a basin which means that all initial conditions except the one exactly at $(0,0)$ end up on the limit cycle. The equilibrium at $(0,0)$ is an unstable node with eigenvalues $(\infty, 0)$ as a consequence of the discontinuity in $\dot{x}$ at $x = 0$.

1.10 Parameter Scaling

If the circuit in Fig. 1.18 were built with $R = 1\,\Omega$, $C = 1$ F, and $L = 1$ H, it would in principle work and would oscillate at a frequency of 1 radian/second. However, such values are impractical and inconvenient. Fortunately, it is a simple matter to rescale the component values provided R^2C/L remains constant, and this works for any circuit, not just the one described here. For example, if R is in k$\Omega = 10^3\,\Omega$, C is in μF $= 10^{-6}$ F, and L is in henrys, then the dimensionless quantity R^2C/L is unity, and the

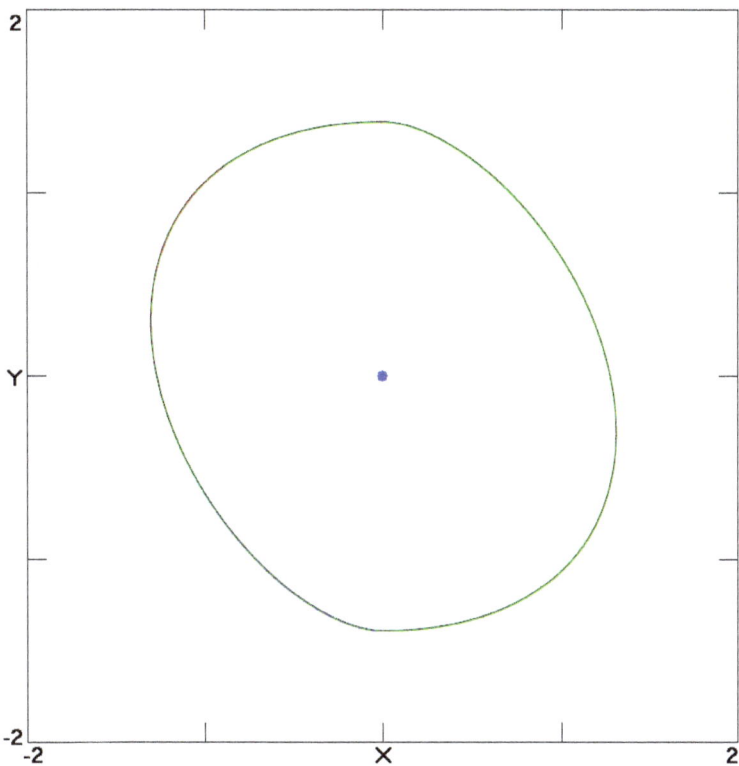

Fig. 1.19 Numerical solution of Eq. (1.9) with $a = 1$ and $b = 1$.

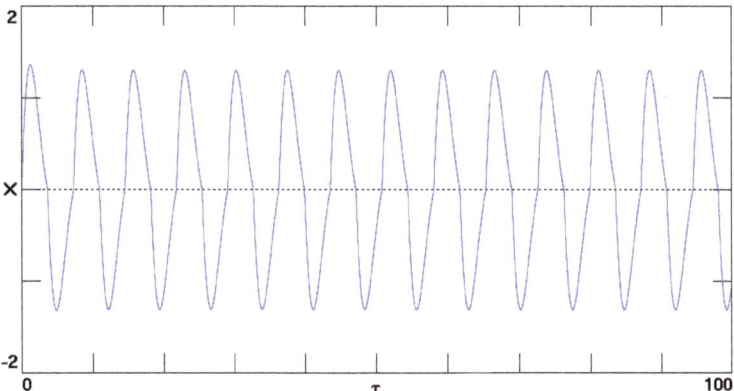

Fig. 1.20 Numerical waveform for $x(\tau)$ from Eq. (1.9) with $a = 1$ and $b = 1$.

time scale is shortened to $\tau = RC = L/R = \sqrt{LC} = 10^{-3}$ s = 1 millisecond. In addition, the ratio of current to voltage is $I/V = 1/R = 10^{-3}$, so that if voltages are measured in volts, then the currents will be in milliamperes (mA).

These are the default units that we will use throughout the book, although other choices are sometimes used, especially to avoid inductances as large as 1 H. If the circuit has no inductors, then it is only necessary to scale all the capacitors by the same factor or all the resistors by a perhaps different factor. Such a rescaling may be necessary if some of the parameters are beyond your control such as parasitic resistance or capacitance.

This choice of parameters makes the characteristic time scale in the circuits of order one millisecond and the corresponding characteristic frequency equal to one radian per millisecond or $1/2\pi$ cycles per millisecond which is about 159 Hz. Since a chaotic system will typically have most of its power at frequencies lower than the characteristic frequency (all the way down to zero since the period of a chaotic oscillation is infinite), it is often better to shorten the time scale in the circuits.

For example, increasing the characteristic frequency by two orders of magnitude to about 16 kHz puts most of the power in the audio range of frequencies. You can accomplish this by reducing the inductances and capacitances by a factor of one hundred while keeping the resistances constant or by reducing the inductances by a factor of ten thousand and the resistances by one hundred while keeping the capacitances constant or any other convenient combination that preserves R^2C/L.

Similarly, the circuits should work if you rescale all the voltages and currents by the same constant factor while keeping the component values constant as might be necessary if one of the voltages is beyond your control and determined, for example, by the saturation voltage of an operational amplifier or the forward voltage drop in a particular diode.

1.11 Construction and Analysis

One of the pleasures of studying chaotic circuits is their accessibility to electronic hobbyists since most of the circuits can be built with common low-cost components. The component values we use throughout this book scale the time to the order of a millisecond, the dominant frequency to the order of a kilohertz, and the voltages in the range ±15 V unless otherwise noted. This choice of the time scale has the additional advantage that it is in the audio range so that the chaos can be heard (with an appropriate amplifier and speaker) as well as seen on an oscilloscope. The low voltage ensures that the circuits do not pose a danger of electrocution except as explicitly noted in the text, but you should take appropriate precautions if you scale them up to higher voltages or currents.

The most common components used throughout this book are resistors, capacitors, inductors, and operational amplifiers. You may wish to acquire several 1 μF class 1 ceramic capacitors, 1 kΩ and 10 kΩ carbon composition (not wire wound) resistors (a $\pm 10\%$ tolerance is usually sufficient with some circuits requiring $\pm 1\%$), and operational amplifiers. High-precision operational amplifiers are not required, and we mainly use a type LM741, commonly known as a 'jellybean' because of its low cost. Additionally, you should have several potentiometers that can be adjusted from zero to at least 10 kΩ to allow you to vary the parameters of the circuit.

Inductors are especially problematic since they typically have significant unavoidable internal resistance including skin effect and eddy current losses that are difficult to anticipate and control, motivating the study of inductorless chaotic circuits. Since the 1 H inductor used in many of the circuits is large and expensive, you may want to purchase only one and reuse it for each chaotic circuit. You should also acquire a few 0.1 H (100 mH) inductors.

Inductors larger than a few mH typically use an iron core, which means that the inductance can vary considerably with the magnitude of the current flowing through them because of the nonlinear permeability of the iron. In fact, at a sufficiently high current, the iron will saturate, and the inductance will drop by orders of magnitude. If you buy an inductor rated for 1 H at a specific current such as 200 mA, it is likely intended as a *choke* in a power supply filter, and you may find that the inductance is closer to 2 H when the current through it is just a few milliamperes. However, such chokes can still be used to observe chaos in the circuits presented in this book provided the actual inductance is taken into account, with a possible difference in the frequency. Standard inductors with values up to about 10 H, including inductor decade boxes that maintain their inductance to within a few percent are available, but they are quite expensive. We use a standard 1 H inductor (General Radio Type 1481-K) that is accurate but expensive and no longer in production.

We recommend that you prototype your circuits on a breadboard prior to implementing a permanent design such as a printed circuit board. Chaotic circuits are sensitive to nonideal effects in the components which can change the behavior of the circuit, and it can be difficult to adjust

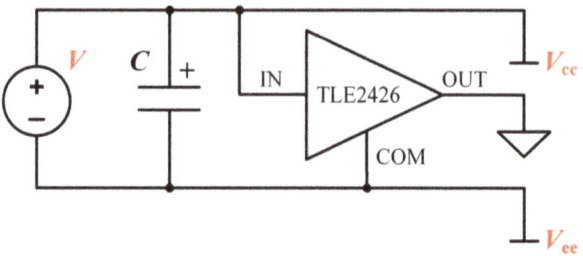

Fig. 1.21 Dual supply circuit (rail splitter).

them afterwards if you have committed to a design. Be aware that operating the circuits at frequencies higher than about 1 MHz may require you to use specialized components or a different circuit layout to reduce the stray inductances and capacitances.

You will need to power the operational amplifier using a *dual-voltage supply* which has three terminals, one *positive* (V_{cc}), one *negative* (V_{ee}), and a third terminal for the ground. This type of power supply can be provided using batteries, but we discourage that because batteries will discharge over time, leading to irreproducible behavior since the values of certain parameters in the circuit are dependent on the power supply voltage. It is better to invest in or construct a stable (voltage-regulated) power supply that can supply at least 30 V. The current rating need not be large (1 A is sufficient) since most of these circuits only draw a few milliamperes of current.

If your power supply only provides a single polarity, you can construct an additional circuit called a *rail splitter* that takes a DC voltage V and converts it into a dual voltage supply. The rail splitter, as schematically shown in Fig. 1.21, requires a TLE2426 integrated circuit and a polarized 220 μF electrolytic capacitor and is constructed at the top of the breadboard in Fig. 1.22.

To check the dual supply circuit, measure the voltage between the positive (red) lead and the ground (green) lead. If your voltage supply is 30 V, this should read around +15 V. Then measure the voltage between the negative (blue) lead and ground, which should read around −15 V. If your connections are correct but you read very different values, it is likely that the integrated circuit is damaged and needs to be replaced.

You will also need another type of power supply called a *function generator*, which we will use to generate sine-wave and square-wave signals for circuits requiring external forcing. Your function generator needs to produce signals at frequencies in the kilohertz range with an amplitude of at least 10 V.

Some chaotic circuits will also require you to generate a square wave signal that ranges from 0 V to 5 V. These signals are suitable for digital logic, also called *transistor-transistor logic* (TTL), where 5 V represents a 'high' logic state, and 0 V represents a 'low' logic state. Some function generators have a specialized output that will generate this signal, but you can also accomplish this by adjusting the amplitude and the DC offset of a square wave.

The LM741 operational amplifier has eight terminals (also called 'pins') that are numbered counter-clockwise with the first pin starting at the top left. The power supply pins are the positive supply terminal (pin 7) which goes to V_{cc} (+15 V) and the negative power supply terminal (pin 4) which goes to V_{ee} (−15 V). The other pins are the inverting input terminal (pin 2), the noninverting input terminal (pin 3), and the output (pin 6). You can connect the inverting input terminal directly to ground (green wire) for this oscillator and the output to R which is replaced with a potentiometer so that it can be varied for further study. The remaining unused pins 1, 5 and 8 can be left disconnected. This operational amplifier is constructed on the breadboard below the rail splitter in Fig. 1.22.

The LC circuit that sustains the oscillations in Fig. 1.18 consists of the blue-colored ceramic capacitor in Fig. 1.22 and the large inductor to the left of the breadboard. Both the capacitor and inductor are nonpolarized, and you can operate them in either direction for the circuits described in this book.

One important parameter to know is the DC resistance (R_L) of the inductor which can be determined by connecting an ohmmeter across the inductor terminals. The parasitic resistance in this inductor should be low in this circuit (ours was 40 Ω), but other circuits may require you to add a resistor in series with the inductor to supplement the parasitic resistance.

You will need an oscilloscope to view the time and phase-space plots, and to diagnose problems with your circuit. Most cost effective is a computer oscilloscope of which any model would suffice. In fact, it is possible to use the (stereo) sound card in the computer along with appropriate software. Try to obtain an analog oscilloscope if you can find one since they have no sampling-rate limitation, and the traces appear more attractive.

Your oscilloscope will also need at least two oscilloscope probes to measure the signals. The tip of the probe measures the signal, while the *ground*

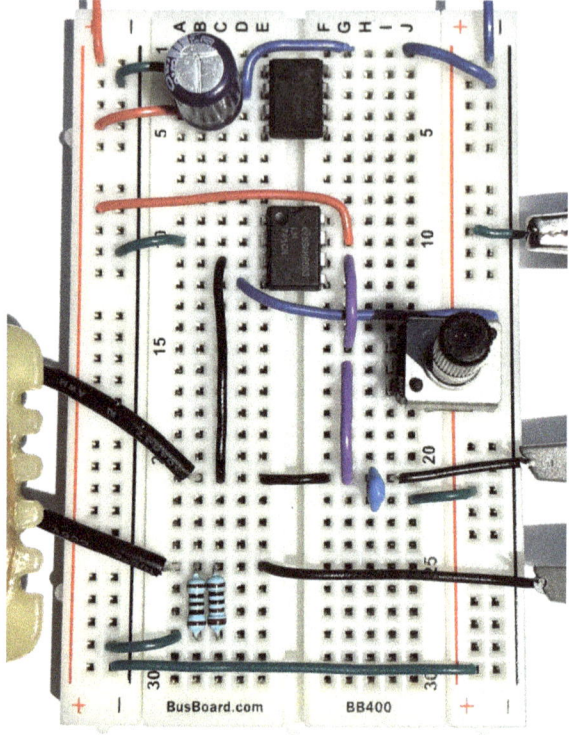

Fig. 1.22 Simple oscillator constructed on a breadboard.

lead (usually a black alligator clip) should be connected to the ground (green wire) of zero volts. If you connect the tip of the probe to V and the ground lead to ground for the circuit in Fig. 1.18, you should see an approximate sinusoidal waveform as shown in Fig. 1.27. Note that the frequency of oscillation is approximately $1/\sqrt{4LC/3} \approx 0.866$ rad/ms (138 Hz) because the Q of the parallel RLC circuit is near unity rather than infinite.

Sometimes a component will be located in the middle of the circuit with neither end connected to ground. To view the voltage across such a component, you can place the two oscilloscope probes on the two terminals of the component and take the difference between the voltages. If you only have one oscilloscope probe available, you can measure the voltage using the circuit in Fig. 1.23, called an *instrumentation amplifier*, which takes the difference between its inputs V_1 and V_2 and outputs the voltage $V_{out} = V_2 - V_1$ when $R_1 = R_2 = R_3 = R_4 = 1$ kΩ. You can then measure

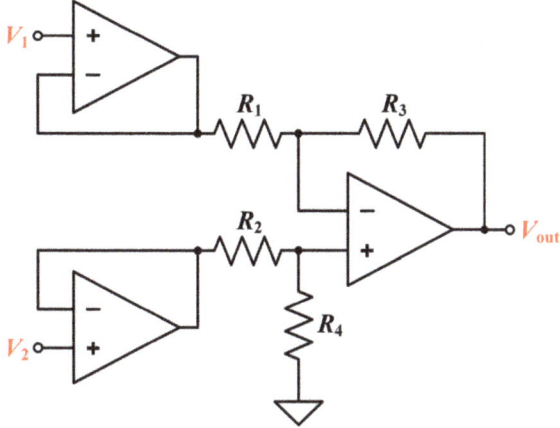

Fig. 1.23 Instrumentation amplifier.

the voltage of the capacitor by connecting V_1 and V_2 across it, which should not affect the circuit since the instrumentation amplifier has a *high input impedance*, meaning that it draws negligible current from the circuit being measured.

Measuring the inductor current I_L requires adding a small ($\lesssim 50 \, \Omega$) 'sensing' resistor in series with the inductor to convert the current into a voltage. This is done in Fig. 1.22 with the two parallel 100 Ω resistors between the inductor and ground. You then calculate the current from $I_L = V_{sense}/50 \, \Omega$ where V_{sense} is the voltage across the small resistor. This method is simple but may be troublesome for certain circuits with small currents.

If the inductor is grounded or connected to a DC voltage, you can use a *current-to-voltage converter circuit* (also called an *I-V converter* or *transimpedance amplifier*) as shown in Fig. 1.24. The current I_L entering this circuit can be measured as a voltage at the output of the operational amplifier $V_{out} = -I_L R$ by reconnecting the grounded end of L to the inverting input terminal of the *I-V* converter. You can add a small capacitor C across the resistor to form a low-pass filter that reduces high-frequency noise when the measured current is small.

There are some circuits that are so sensitive to the inductor's parasitic resistance that it is difficult to find a suitable inductor or use a sensing resistor to measure the current I_L. In this case, you will need to use the circuit in Fig. 1.25 called a *gyrator* that was invented by Antoniou (1969) to

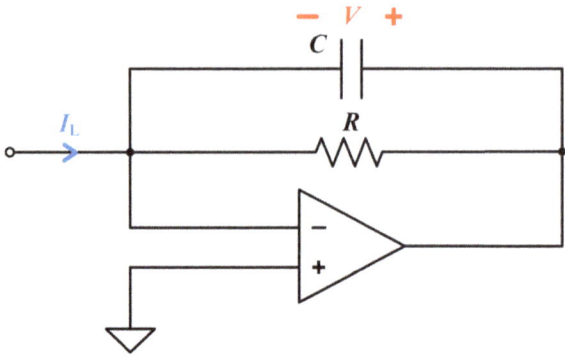

Fig. 1.24   *I-V* converter.

simulate an inductance equal to $L = CR_1R_3R_4/R_2$ with almost no parasitic resistance.

For a 1 H inductor, you can use the component values $C = 1$ µF, $R_1 = R_2 = R_3 = R_4 = 1$ kΩ, which can be verified using an inductance meter when power is supplied to the gyrator. You can then calculate the current I_L by measuring the voltage across R_1 and dividing by its resistance. Note that this gyrator is limited to simulating an inductor with one end grounded or connected to a function generator. While it is possible to construct *floating* gyrators with neither end connected to ground, their construction and analysis is challenging and will not be discussed here.

Once you have a way to acquire V and I_L, you can use the 'XY mode' of the oscilloscope to obtain a phase space plot as shown in Fig. 1.26. Most oscilloscopes are limited to two probes, which makes measuring two components in a chaotic circuit difficult if they are far apart and not both connected to ground. Thus for each circuit, we will tell you when to use tools such as the instrumentation amplifier or a sensing resistor and how to manipulate the measured signals to properly observe the attractor.

If you have obtained oscillations in this circuit, you may notice that the phase space plot is slightly distorted because the operational amplifier saturates when V tries to go beyond the power supply voltage. You can fix this problem by increasing R to about 1.5 kΩ but at the cost of obtaining a more circular limit cycle compared to the numerical plot in Fig. 1.19.

An optional modification is to add a *clipper circuit* as shown in Fig. 1.28 between the comparator and R. This circuit uses *diodes* (explored further in Chapter 2) which act as an open circuit until V_{out} exceeds the sum of

Introduction

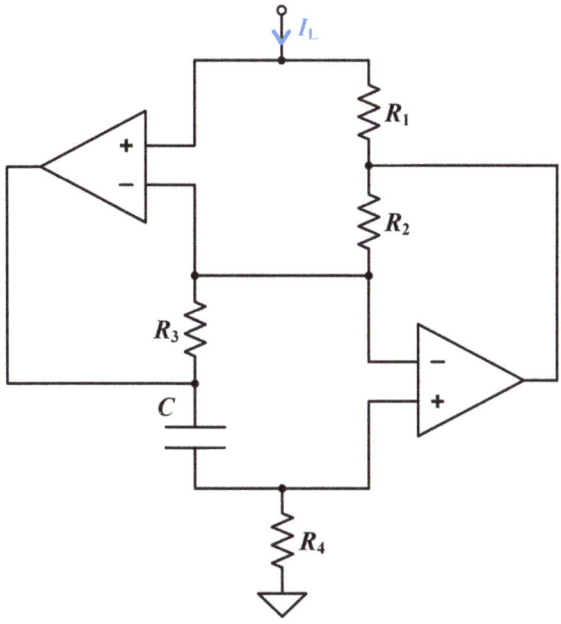

Fig. 1.25 Gyrator circuit.

its voltage drop V_0 and a reference voltage V_{ref} whereupon it conducts, thus limiting the signal between $-V_0 - V_{ref}$ and $+V_0 + V_{ref}$. You can set $R_1 = 10$ kΩ and adjust V_{ref} using the potentiometer between V_{cc} and ground in the circuit of Fig. 1.29 (between V_{ee} and ground for $-V_{ref}$) until the comparator voltage switches symmetrically between ± 1 V. We use this circuit several times throughout the book to set required DC voltages that are different from the power supply voltage.

It is often useful to simulate a circuit before constructing it to confirm the design and ensure that the circuit will operate as expected. We recommend the freeware circuit simulator *LTSpice*® (available from http://www.analog.com/). A SPICE file for each circuit in this book is available to download for free at http://wesleythio.com.

Since differences will arise between theoretical and experimental parameters and some component values are rescaled for convenience, the *Numerical*, *Simulated*, and *Experimental* values that we found from numerical integration of the differential equations, simulated using LTSpice, and used in the constructed circuit, respectively, are shown in a table for each of the circuits.

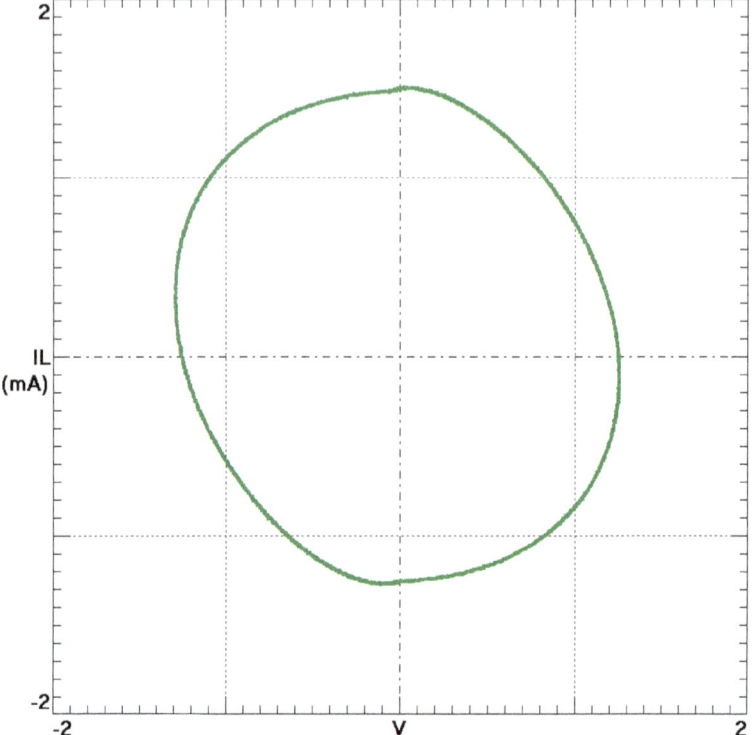

Fig. 1.26 Oscilloscope phase space plot of simple oscillator from Fig. 1.18.

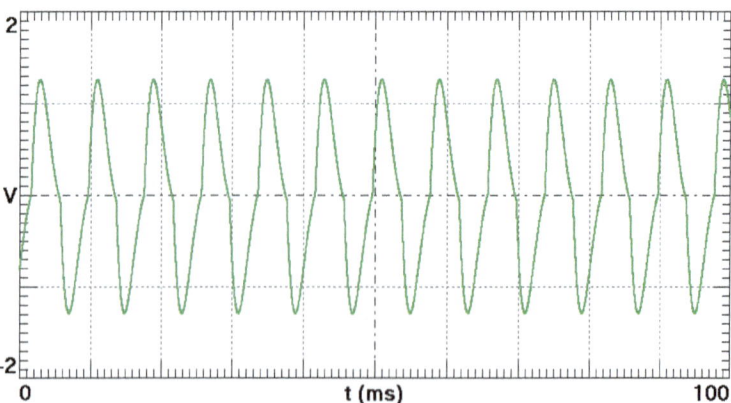

Fig. 1.27 Oscilloscope time plot of simple oscillator from Fig. 1.18.

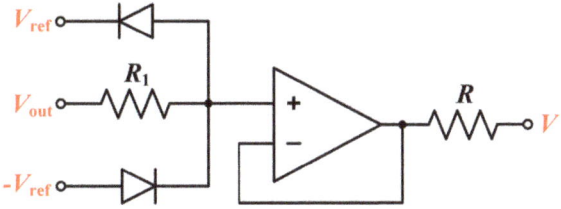

Fig. 1.28   Clipper circuit for simple oscillator.

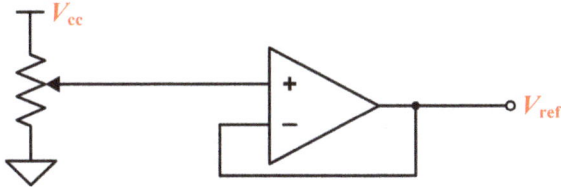

Fig. 1.29   Operational amplifier buffer with voltage divider to set V_{ref}.

Table 1.1   Component values for Fig. 1.18.

Component	Numerical	Simulated	Experimental
R	1 kΩ	1 kΩ	1 kΩ (adjust)
C	1 μF	1 μF	1 μF
L	1 H	1 H	1 H
V_{sat}	1 V	1 V	1 V
Op amp	-	Ideal	LM741

Table 1.1 shows these values for the simple oscillator in Fig. 1.18. We made the oscilloscope plots using an Analog Discovery 2, which is a 14-bit digital oscilloscope that was typically used to collect 1×10^5 samples at a 100 kHz sample rate. However, we varied this for some of the circuits that oscillate at lower or higher frequencies or that have rapid variation in the measured quantities. For small-amplitude signals with high-frequency components, some digital smoothing of the data was necessary to remove noise and digitizer round-off errors.

Chaotic circuits are more unusual than most circuits in that they can behave quite unpredictably and can be challenging even for an experienced circuit builder. For example, it is not unusual to build a circuit exactly to specifications but not obtain chaotic oscillations or any oscillation at all because of the properties of the chaotic system that the circuit implements or due to small nonideal behaviors in the components. We recommend that you replace the resistors with potentiometers and individually adjust their resistances until an oscillation is observed. In each component table, we give a recommendation for the parameter that you might first try adjusting to obtain chaos.

You can also choose to adjust a component so that it becomes a bifurcation parameter in order to study the route to chaos and other dynamical properties of the system. For example, you may notice that your circuit oscillates at a larger value of R than the theoretical value of $R > 0.5$ kΩ from Section 1.9 because the parasitic resistance of the inductor alters the Hopf bifurcation value. We found the onset of oscillations in our circuit to occur at $R \approx 0.7$ kΩ with a 40 Ω parasitic resistance in the inductor, which agrees well with numerical calculations.

If your circuit stops oscillating, try gently tapping it or restart the power supply to make the oscillations return from a stable equilibrium. Something similar can happen with other chaotic circuits when the signal becomes too large and causes the amplifiers to saturate at the supply voltage. This saturation can also be caused by unintentionally jarring the circuit which can push the signal out of the basin of attraction. Restarting the power supply usually fixes this problem, but some circuits require adjusting additional parameters to bring back the oscillations.

We also mentioned earlier that a multistable system may have initial conditions that lead to chaos and others that do not. This can prevent a chaotic circuit from working if its initial conditions (the initial voltages on the capacitors) are not pre-adjusted so that they start near an equilibrium point. Our circuits start near the origin and should work upon startup or by adjusting the parameters, but be aware that other circuits might not have their the origin in the basin of attraction and will require some additional circuitry to set the starting voltage of the capacitors.

1.12 Circuit Elegance

Previously, Sprott (2010) in *Elegant Chaos: Algebraically Simple Chaotic Flows* listed chaotic systems of equations with the fewest number of simple terms with parameters adjusted to have simple numerical values. By analogy, a chaotic circuit can be considered to be elegant if it has the fewest number of simple components and if those components are described by simple algebraic relations and have simple numerical values. As discussed in Section 1.5, every autonomous chaotic circuit requires at least *three* reactive components (inductors or capacitors), an active component, and a nonlinearity. Non-autonomous chaotic circuits require at least *two* reactive components and a passive nonlinearity.

While every chaotic circuit must follow these requirements, the versatility of electronics often allows multiple conditions to be fulfilled in a single device. Thus we describe the physics of these devices to show how such properties arise and formulate a simple model to describe their behavior. Such a model is not meant to cover all the possible behaviors of the device but rather to capture the essential nonlinearity responsible for generating chaos.

There can also be a tradeoff between the number of components used in the circuit and the number of terms in the circuit equations. Some circuits can be made more elegant by adding terms to their governing equations. Elegant equations can also lead to inelegant circuits, and, conversely, elegant circuits may require relatively inelegant equations to describe their operation. This is especially true for the circuits that consider parasitic capacitances in their equations.

We consider an electrical component to be elementary if it is a semiconductor device or cannot be further reduced to discrete components such as resistors or capacitors, the exceptions being for operational amplifiers and analog multipliers which are usually integrated circuits built using many discrete components. Later in the book you will encounter *emulators* that substitute for devices that are no longer or not yet commercially available. These emulators are constructed using discrete components, but we count them as a single component since they are meant to emulate a single device that was once common or should eventually become available.

Many of these standards of elegance are subjective and could be chosen differently. For instance, we consider any resistance in series with an inductor to be parasitic regardless of its magnitude since such an inductor can be fabricated using a material of suitable conductance. In addition, we do not consider voltage and current sources as components in the circuits since a source of electrical energy is required in all the autonomous circuits and is thus readily available.

Finally, the circuits we have included in this book cover only a small subset of the numerous innovative chaotic circuits that have been proposed, constructed, and studied. Generally, we have described only the simplest reported example of a circuit from a family of similar circuits, and it will be left as a challenge for you to find even more elegant examples.

Chapter 2

Conventional Diode Circuits

The diode is the simplest and most fundamental of the semiconductor devices. This chapter will describe ideal diodes and conventional silicon-based diodes. It will also describe some simple circuits using such diodes as the nonlinear component to generate chaos.

2.1 Diode Characteristics

With all electronic devices, we can consider their ideal form in which they obey precise mathematical relations as well as their actual implementation which inevitably deviates from the ideal and is only approximately described by usually more complicated mathematical formulas. So it is with the diode.

2.1.1 *Ideal diode*

The *diode* is a passive two-terminal nonlinear component whose symbol resembles an arrow. Its two terminals are called the *anode* (A) and *cathode* (K), respectively, as shown schematically in Fig. 2.1. The voltage across the diode is V_D, and the current through the diode is I_D.

The diode is capable of regulating current flow by only allowing it to pass in the direction of the arrow, that is from anode to cathode. If current

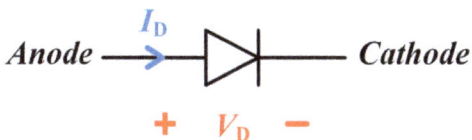

Fig. 2.1 Diode schematic.

passes through the diode in this direction as shown in Fig. 2.2, the diode is said to be *forward-biased*, and it acts like a short circuit that allows current to flow. However, if the diode is reversed as shown in Fig. 2.3, it is *reverse-biased*, and it acts like an open circuit that blocks the current flow. This operation is similar to a water valve that prevents backflow.

If an alternating current (AC) source (V_S) is used instead of a direct current (DC) source as shown in Fig. 2.4, one half of the cycle will be blocked in the reverse-bias condition. This process is called *rectification* and is frequently used in power electronics to convert AC to DC. The rectification process for the circuit in Fig. 2.4 is illustrated in Fig. 2.5. In a practical power supply, the rectified voltage in Fig. 2.5 would be heavily filtered to produce a steady DC voltage with very little time variation (*ripple*).

The ideal diode behaves like a perfect conductor when current is flowing in the direction of the arrow and like a perfect insulator when the voltage across it is negative, and it switches between the two states instantly and completely.

2.1.2 PN junction diode at equilibrium

Real diodes are usually constructed using *PN junctions*, whose properties lead to the rectifying behavior of real diodes. We provide the following brief description of PN junctions since they form the basis for modeling the behavior of many of the devices in this book.

Silicon is one of the materials most commonly used in making PN junctions. Silicon is a *semiconductor*, having a conductivity between that of an insulator and a conductor. This conductivity can be purposely altered by introducing impurities that either increase the concentration of free electrons (e^-) leading to what is called *n-type* material, or increase the concentration of positively charged empty free electron spaces called *holes* (h^+), which is called *p-type* material. These electrons and holes are also called the *majority charge carriers* since there are many of them in their respective n-type and p-type regions, whereas electrons and holes in the opposite regions are called *minority charge carriers* because there are so few of them. If an electron combines with a hole through a process of *recombination*, their charge is neutralized. In fact, a photon can be emitted during this process, forming the basis of the *light emitting diode* (LED).

A PN junction is created by placing a p-type material in contact with an n-type material as shown in Fig. 2.6. As soon as these two materials touch, the large hole concentration in the p-type material diffuses into the

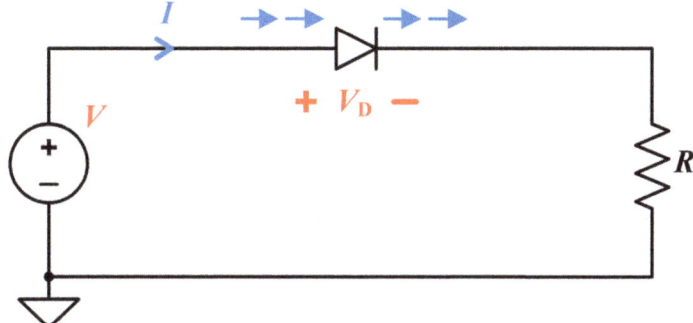

Fig. 2.2 Forward-biased diode with a DC source (current flows).

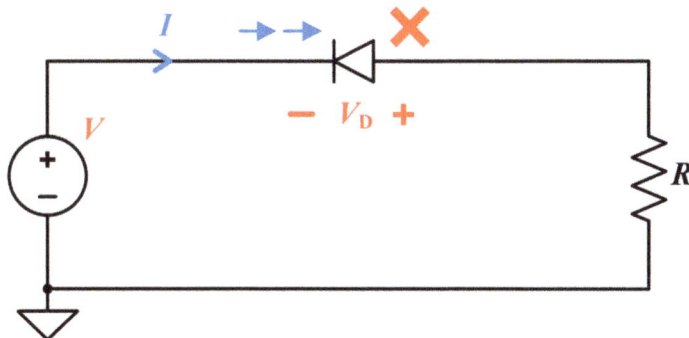

Fig. 2.3 Reverse-biased diode with a DC source (no current flows).

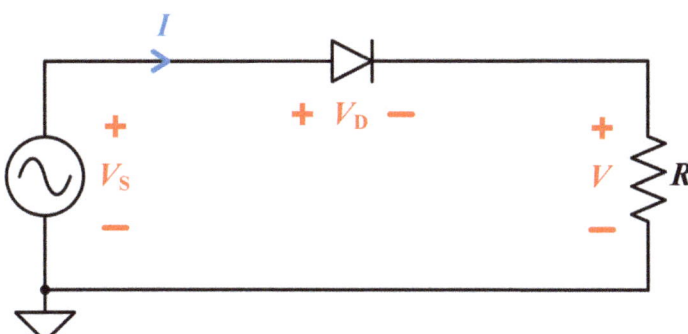

Fig. 2.4 Diode with an AC source (current flows during half of the cycle).

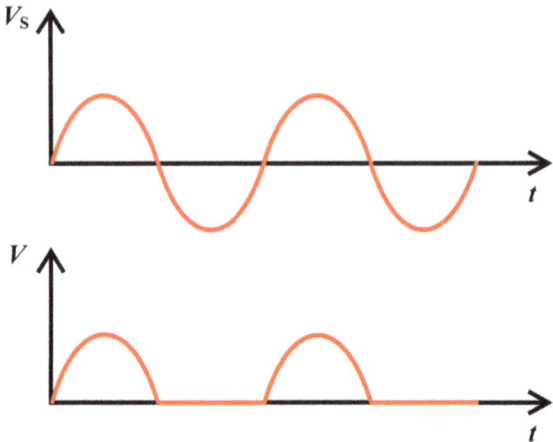

Fig. 2.5   Voltage at V_S and V in Fig. 2.4.

n-type material where there are few holes, while the electrons in the n-type material diffuse into the p-type material. You can think of this as the smoke in one room diffusing into the clean air of an adjacent room when a connecting door is opened. The movement of these majority charge carriers causes a *diffusion current* (I_D) that points in the direction of the positive carriers by convention. Because p-type and n-type materials are normally electrically neutral, the recombination of these charge carriers creates a space in the PN junction called the *depletion region* that has an electric potential called a *barrier voltage* (V_0). The height of this barrier prevents charge carriers with insufficient energy from passing through.

This barrier voltage produces an electric field that causes an opposite movement of pre-existing holes in the n-type region toward the p-type region, as well as pre-existing electrons toward the n-type region. This is called the *drift current* (I_S), and it is extremely small as it involves the minority charge carriers in those regions. Thus in equilibrium, the diffusion and drift current cancel so that no current flows, leading to $I = I_D - I_S = 0$, or $I_D = I_S$.

2.1.3   *PN junction with an applied voltage*

If the PN junction is forward-biased as shown in Fig. 2.7, the applied voltage V subtracts from the barrier voltage by $V_0 - V$ and reduces the barrier charge and width. With this reduction, more electrons can diffuse from the

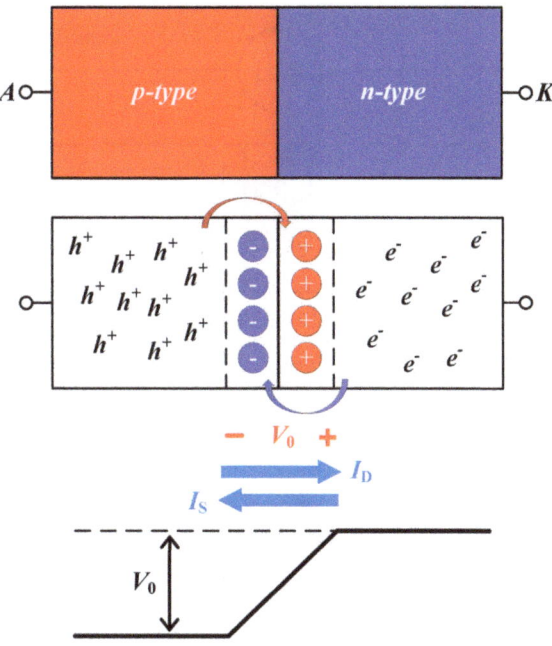

Fig. 2.6 PN junction.

n-type to the p-type region, and more holes can diffuse from the p-region to the n-region leading to an increase in I_D with $I_D > I_S$. The total current is then $I = I_D - I_S$ where the majority charge carriers are mainly responsible for the current.

If the PN junction is reverse-biased as shown in Fig. 2.8, the applied voltage adds to the barrier voltage $V_0 + V$. The increase in barrier voltage effectively makes $I_D \approx 0$ and $I \approx -I_S$ where only the minority charge carriers are flowing. However, it is possible for charge carriers to pass through this barrier by *tunneling*, which we will discuss in later chapters.

2.1.4 *I-V characteristic of the PN junction*

The behavior of the PN junction when forward or reverse biased leads to the following equation called an *I-V characteristic* which describes how the voltage and the current are related

$$I = I_S e^{V/nV_t} - I_S, \tag{2.1}$$

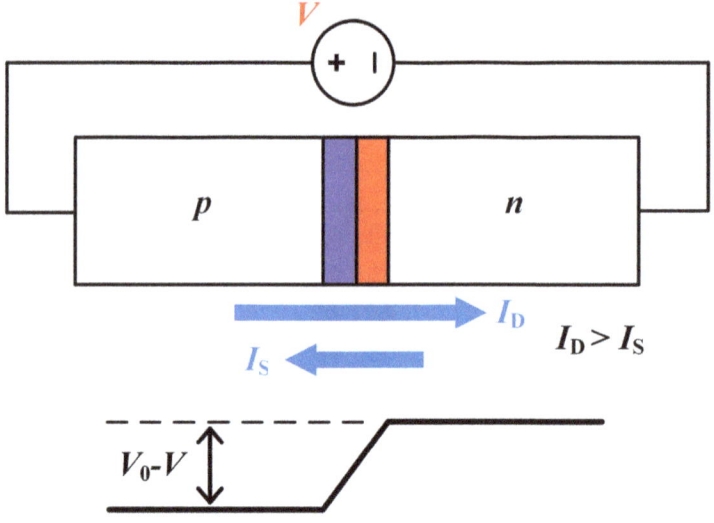

Fig. 2.7 Forward-biased PN junction.

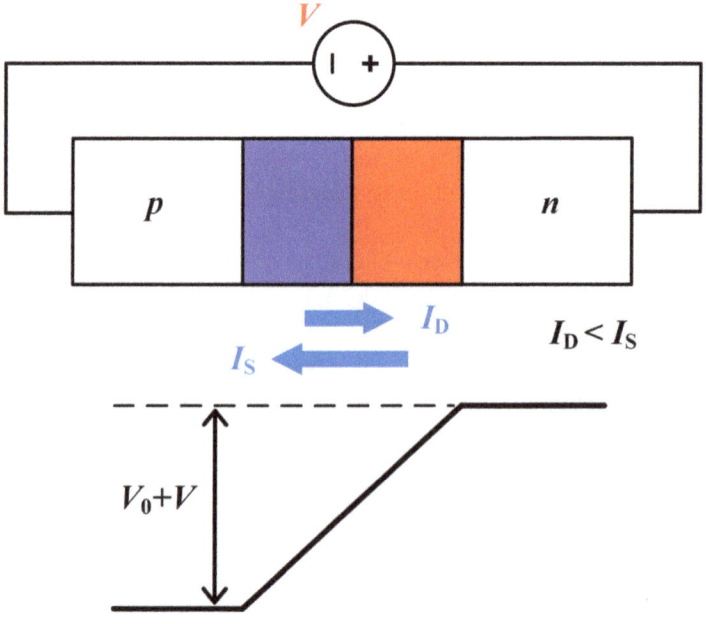

Fig. 2.8 Reverse-biased PN junction.

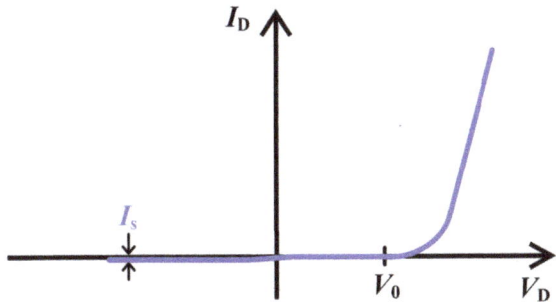

Fig. 2.9 I_D-V_D curve of a diode.

where V_t is the thermal voltage equal to about 25 mV at room temperature, and n is a thermal constant that you can usually take as $n = 1$ for silicon diodes. The e^{V/V_t} factor arises from an important behavior in device physics where the applied voltage V exponentially increases the majority charge carriers diffusing into the p-type and n-type regions when forward-biased. This current becomes much higher than the drift current I_S which is typically 10^{-18} to 10^{-12} A. For negative V, the device is reverse-biased, and so the current is a small value equal to $-I_S$ since only the minority charge carriers are flowing. A graph of Eq. (2.1) is shown in Fig. 2.9, which is called an *I-V* curve or I_D-V_D curve in reference to the diode schematic in Fig. 2.1. You can find a derivation of this formula in most device texts such as Streetman and Banerjee (2016).

Since I_S is much smaller than any current of interest in a typical circuit, it is usually convenient and sufficient to neglect I_S and use a piecewise-linear approximation to Eq. (2.1) given by

$$I = \max(0, V - V_0)/R_D = \begin{cases} 0 & V \leq V_0 \\ (V - V_0)/R_D & V > V_0, \end{cases} \quad (2.2)$$

where R_D is the (usually small) *forward resistance* of the diode and V_0 is its *forward voltage drop*. For many purposes the forward resistance R_D is much smaller than any other resistance in the circuit and can be set to any small value (but not zero!). Setting it too small can cause difficulty with your numerical integrator.

The forward voltage drop V_0 is the value of V for which $I_S e^{V/nV_t}$ is of order unity or $V_0 = -nV_t \log_e I_S \approx 0.6$ volts for a typical silicon diode. In the interest of elegance, we will often set $V_0 = 0$ when the forward voltage drop is no consequence or $V_0 = 1$ when V_0 is the determining voltage to which all others in the circuit are proportional.

For a circuit in which the diode is the only nonlinear component and there are no voltage or current sources, the scale for all the voltages and currents in the circuit is set by the value of V_0. An *ideal diode* is one in which V_0 and R_D are both zero. Equation (2.2) was widely used to analyze circuits before personal computers and modern circuit simulation software became commonplace, and it is still useful in mathematical modeling of circuits. It has the additional advantage that the circuit equations are linear in the regimes $V \leq V_0$ and $V > V_0$, and thus they can be solved analytically with a boundary condition at $V = V_0$.

For some purposes, it is useful to have an expression for the diode voltage V as a function of its current I. Such an equation follows from inverting Eq. (2.1) to obtain

$$V = nV_t \log(1 + I/I_S), \tag{2.3}$$

which you can approximate by a piecewise-linear equation,

$$V = \min(I, I_0)R_D = \begin{cases} IR_D & I < I_0 \\ I_0 R_D & I \geq I_0, \end{cases} \tag{2.4}$$

where $I_0 \approx V_0/R_D$ and R_D is the (usually large) *reverse resistance* of the diode.

2.1.5 Capacitance of the PN junction

The PN junction has a significant parasitic capacitance that can be exploited in a chaotic circuit to act as both a nonlinearity and a reactive component. This capacitance occurs when the applied voltage across a PN junction changes the amount of charge in the depletion region, forming a capacitor C_D whose effective plates are at the edge of the region. Because the capacitance is inversely proportional to the distance between the plates, it will change with bias voltage.

When the diode is reverse-biased, the increasing depletion region width decreases the capacitance. The nonlinear capacitance created by the reverse bias is called the *junction capacitance*, and it is given approximately by

$$C_j = C_{j0}/\sqrt{1 - V/V_t} \tag{2.5}$$

for $V \leq 0$, where V_t is about 25 mV and C_{j0} is the capacitance of the junction with no bias ($V = 0$) and is the order of 100 pF for a typical silicon diode.

However, the more important nonlinearity comes from the forward-bias capacitance called the *diffusion capacitance* given approximately by

$$C_s = C_{s0} e^{V/V_t} \tag{2.6}$$

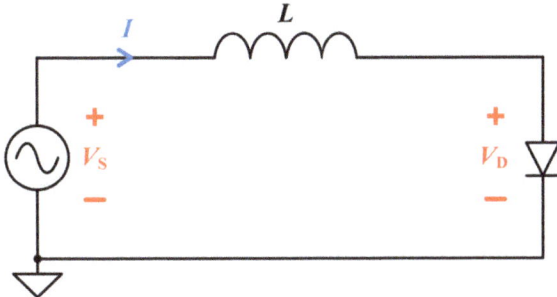

Fig. 2.10 Forced diode resonator described by Eq. (2.7).

Table 2.1 Component values for Fig. 2.10.

Component	Numerical	Simulated	Experimental
R_L	1 kΩ	0 Ω	50 Ω
C_0	1 μF	-	-
L	1 H	0.1 H	0.1 H
V_t	1 V	-	-
V_0	10 V	10 V	10 V
ω	1 rad/ms	40π rad/ms	40π rad/ms (adjust)
Diode	Eqs. (2.5), (2.6)	1N4001	1N4001

for $V > 0$. Forward bias decreases the width of the depletion layer and thus increases the capacitance exponentially to values that can exceed 1 μF. This large capacitance is usually of no consequence since it is in parallel with the low forward resistance of the diode, but it is critical for the operation of some circuits. To avoid a discontinuity in the capacitance at $V = 0$, we will generally assume $C_{j0} = C_{s0} = C_0$.

2.2 Forced Diode Resonator

One circuit that exploits the parasitic capacitance of the PN junction is the forced diode resonator in Fig. 2.10. It is one of the simplest non-autonomous chaotic circuits and the simplest in this book. Along with Chua's circuit, it is also one of the earliest chaotic circuits and was first studied by Linsay (1981) and later by Testa et al. (1982). It has been used as part of a frequency selective tuning circuit in a radio receiver and as a frequency divider at microwave frequencies.

This circuit is particularly remarkable for exhibiting complex behavior with only a diode and inductor, suggesting the ubiquity of chaos in our everyday electronics. In addition, its simplicity makes it an excellent introductory teaching tool for chaos. The behavior of this circuit is also unique in that it uses the capacitance of the diode and not its rectifying behavior as in most of the other chaotic circuits in this chapter.

The equations that describe the circuit in Fig. 2.10 are

$$\dot{V}_D = (I - I_D)/C_D$$
$$\dot{I} = (V_S - V_D - IR_L)/L, \quad (2.7)$$

where C_D is the diode capacitance given by $C_D = C_{j0}/\sqrt{1 - V/V_t}$ from Eq. (2.5) for $V_D \leq 0$ and by $C_D = C_{s0}e^{V/V_t}$ from Eq. (2.6) for $V_D > 0$, $V_S = V_0 \sin \omega t$, and R_L is the parasitic resistance in the circuit (not shown) and is primarily in the inductor and the voltage source. If the frequency is sufficiently high, the resistive current in the diode I_D can be neglected compared with the capacitive current.

Elegant circuit values that give chaos are $R_L = 1$ kΩ, $C_0 = C_{j0} = C_{s0} = 1$ μF, $L = 1$ H, $V_t = 1$ V, $V_0 = 10$ V, and $\omega = 1$ radian/millisecond (≈ 159 Hz). The dimensionless equations are

$$\dot{x} = \gamma y$$
$$\dot{y} = a \sin \tau - x - y, \quad (2.8)$$

where $\gamma = \sqrt{1-x}$ for $x \leq 0$ and $\gamma = e^{-x}$ for $x > 0$, and $\tau = \omega t$ is the dimensionless time.

The parameter $a = 400$ ($= 10/0.025$) gives the attractor in Fig. 2.11 and a time series as shown in Fig. 2.12. The Lyapunov exponents are $(0.0568, 0, -1.0568)$, the Kaplan–Yorke dimension is $D_{ky} = 2.0537$, and the attractor is globally attracting with a Class 1a basin of attraction. However, initial conditions with x large and positive produce an exponentially long-duration periodic transient because of the large diode capacitance. In reality, the forward resistive current in the diode I_D that we neglected would suppress such a transient. Because of the sinusoidal forcing, this system has no equilibrium point. However, in the absence of forcing, the equilibrium at the origin is a stable focus with eigenvalues $-0.5 \pm 0.8660i$.

The experimental phase space plot of the diode voltage and inductor current is shown in Fig. 2.13. To view the inductor current I on an oscilloscope, you can put a small series resistance such as 50 Ω between the diode and ground and then determine the current through the inductor from the voltage across this resistor using Ohm's law. You can measure the diode voltage V shown in Fig. 2.14 by subtracting the cathode voltage from the anode voltage of the diode.

The earliest versions of this circuit used a varactor diode, designed to have a large variation in junction capacitance when reverse biased and can be used to replace the bulky parallel plate capacitor in the tuning stage of a radio receiver. This capacitance was initially believed to be the main source of chaos and can be used to accurately model this circuit.

However, Hunt (1982) and later Rollins and Hunt (1982) identified the important nonlinearity as coming from the *reverse recovery time* of the diode. This effect occurs when the diode is initially forward-biased as shown in Fig. 2.7, increasing the concentration of extra minority charge carriers in both the n-type and p-type regions. If the voltage is immediately reversed (Fig. 2.8), it takes time for these extra minority charge carriers to recombine with electrons or holes and disappear. Thus the diode conducts for a short time even when reverse-biased. The behavior is analogous to the backflow in a water pipe.

Consequently, it is best to use a diode with a long reverse recovery time, preferably in the microsecond range such as the older 1N400X series. Fast-switching Schottky diodes would make a poor choice for this circuit since their metal-semiconductor junction has a negligible recovery time. The recovery time can also vary among diodes, even for the same diode type, which could explain the differences between simulation and experiment. We investigated other fast signal diodes such as the 1N4148, but did not find chaos. However, you can use fast diodes with the other circuits in this book that rely only on the rectification property of the diode.

The most important parameter to adjust for this circuit is ω, corresponding to the frequency of the function generator. You should obtain chaos provided V_S is sufficiently large, but it may require different parameters than ours due to the variation in the reverse recovery time among diodes.

For the inductor, choose one that has a high Q with low parasitic resistance and small eddy current and hysteresis losses. The DC resistance of the 0.1 H inductor we used in this circuit was about 5.3 Ω and can be neglected. Other circuit values and component types are shown in Table 2.1. Without counting voltage sources, this circuit has only two components, an inductor and a diode, and thus it is about as elegant as it gets.

2.3 Vilnius Oscillator

The *Vilnius oscillator*, named after the city in Lithuania of its origin, is a variant of the harmonic oscillator that describes the periodic motion of

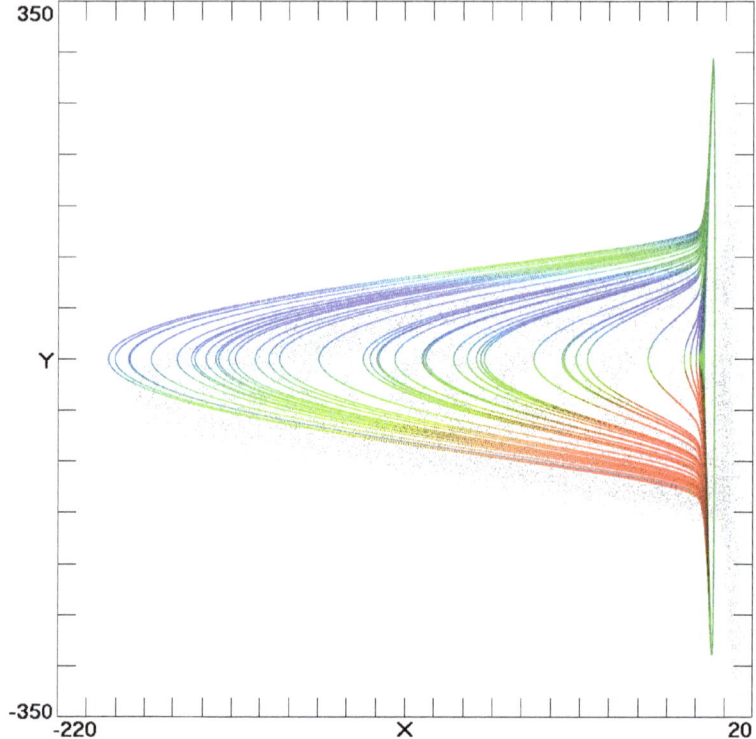

Fig. 2.11 Numerical solution of Eq. (2.8) with $a = 400$.

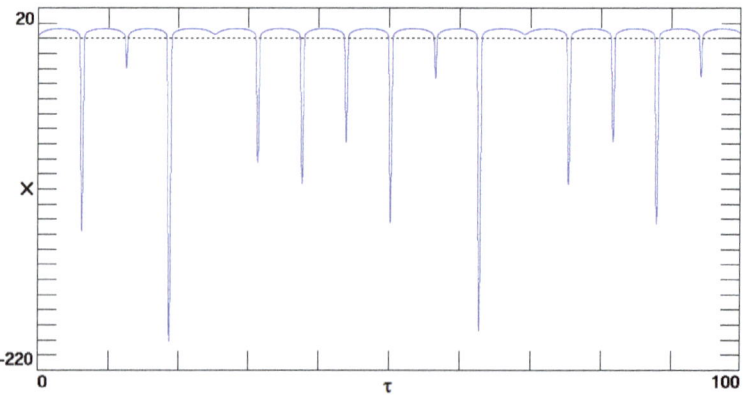

Fig. 2.12 Numerical waveform for $x(\tau)$ from Eq. (2.8) with $a = 400$.

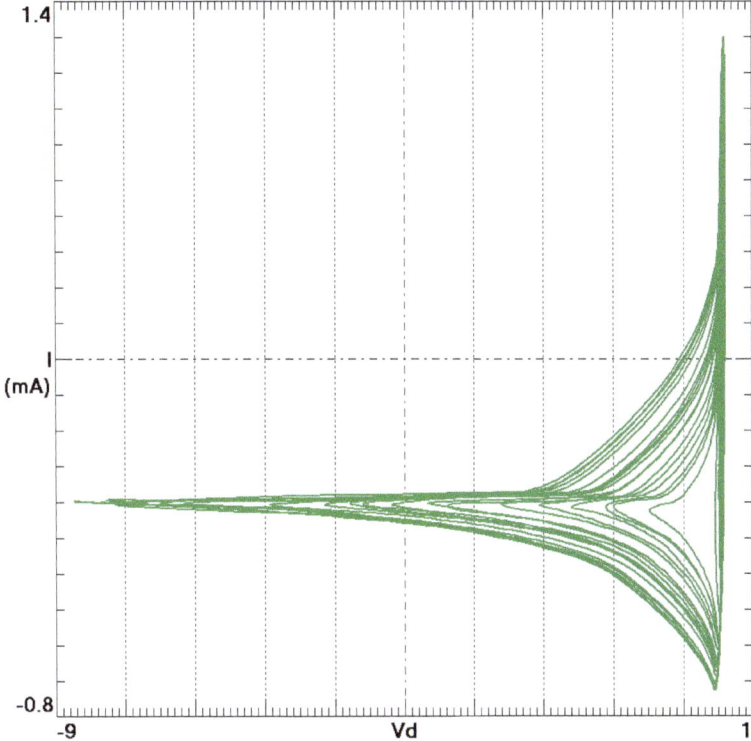

Fig. 2.13 Oscilloscope phase space plot of the forced diode resonator from Fig. 2.10.

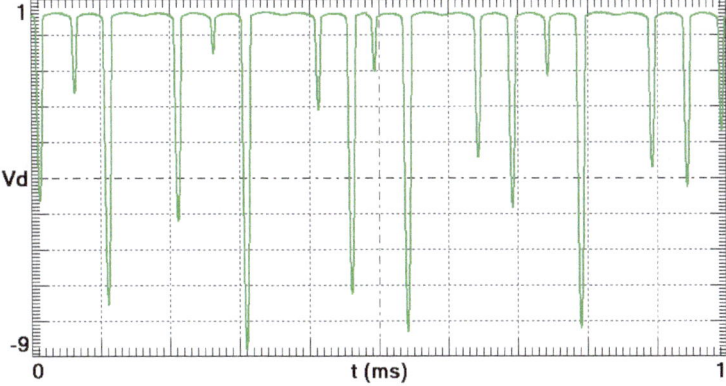

Fig. 2.14 Oscilloscope time plot of the forced diode resonator from Fig. 2.10.

systems such as a gently swinging pendulum. The electrical analog of the harmonic oscillator is the LC circuit, which consists of an inductor and a capacitor with a resonant frequency of $\omega_0 = 1/\sqrt{LC}$. These circuits were used to develop the first operating radio, and they are the basis of communication circuits used for transmitting and receiving signals in cell phones and televisions.

Just as a pendulum will eventually cease to swing due to air resistance, electrical oscillations will decay due to the inevitable resistance as shown previously in Fig. 1.2 and must be sustained through amplification. This is done in the circuit from Tamasevicius (2004) shown in Fig. 2.15 using an operational amplifier that provides a gain of $K = 1 + R_1/R_2$ (also called a *non-inverting operational amplifier*) that will cause the amplitude to increase with time rather than decay. A grounded resistor R is used to convert the inductor current I into a voltage for amplification.

A diode can be added in series with the LC circuit, and this was done in practice for *amplitude modulated* (AM) radio in demodulating the incoming signal filtered by an LC circuit. However, the circuit cannot oscillate since the rectifying behavior of the diode prevents current from flowing in both directions, which is needed in a resonant circuit.

To solve this problem, a technique that is repeated for several of the circuits in this chapter is to use a current source to provide a constant current flow through the diode. This current source can be added to the output of the diode so that the current across the load is $I_D - I_0$. This shifts the I-V curve in Fig. 2.9 down, so that the current at $V = 0$ (called the *quiescent point* or *Q-point*) is no longer zero but is I_0. This method is called *biasing* the diode.

Biasing the diode allows current in the output to flow in both directions but keeps the nonlinearity needed for chaos by rectifying the signal if the current amplitude exceeds I_0. Thus I_0 must be adjusted in such a way that sufficient current flows in the reverse direction while being distorted from rectification. The addition of both the diode and the current source also prevents the amplitude from growing and approaching the voltage of the power supply, which would cause the operational amplifier to saturate.

Finally, a third dynamic element is needed to make the circuit oscillate chaotically. Since the junction capacitance of the diode is usually insufficient to act as a reactive element, a capacitor C_2 is added in parallel with the diode.

Conventional Diode Circuits

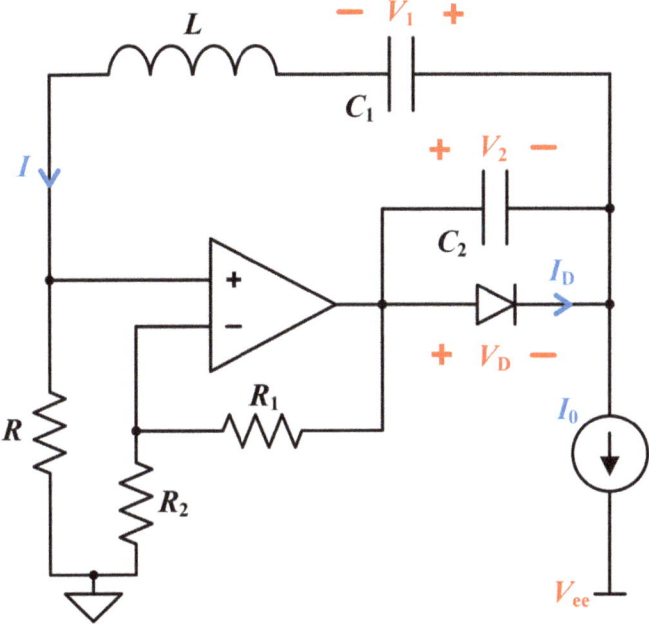

Fig. 2.15 Vilnius oscillator described by Eq. (2.9).

Table 2.2 Component values for Fig. 2.15.

Component	Numerical	Simulated	Experimental
R	1 kΩ	0.5 kΩ	0.5 kΩ (adjust)
R_1	1 kΩ	1 kΩ	1 kΩ
R_2	1 kΩ	1 kΩ	1 kΩ
R_D	0.5 kΩ	-	-
C_1	1 μF	1 μF	1 μF
C_2	0.1 μF	0.1 μF	0.1 μF
L	1 H	1 H	1 H
V_0	1 V	-	-
I_0	1 mA	1 mA	1 mA
Op amp	-	Ideal	LM741
Diode	Eq. (2.2)	1N4148	1N4148

The equations that describe the circuit in Fig. 2.15 are

$$\dot{V}_1 = I/C_1$$
$$\dot{V}_2 = (I_0 + I - I_D)/C_2 \qquad (2.9)$$
$$\dot{I} = [(K-1)RI - V_1 - V_2]/L,$$

where the gain of the noninverting operational amplifier is $K = 1 + R_1/R_2$ and the diode current is $I_D = \max(0, V_2 - V_0)/R_D$ from Eq. (2.2).

Elegant circuit values that give chaos are $R = R_1 = R_2 = 1$ kΩ, $R_D = 0.5$ kΩ, $C_1 = 1$ μF, $C_2 = 0.1$ μF, $L = 1$ H, $V_0 = 1$ V, and $I_0 = 1$ mA. The dimensionless equations are

$$\dot{x} = z$$
$$\dot{y} = a[b + z - c\max(0, y - 1)] \qquad (2.10)$$
$$\dot{z} = z - x - y.$$

The parameters $a = 10, b = 1$, and $c = 2$ give the attractor shown in Fig. 2.17 and a time series as shown in Fig. 2.18. The Lyapunov exponents are $(0.1460, 0, -14.3059)$, the Kaplan–Yorke dimension is $D_{ky} = 2.0102$, and the attractor has a Class 3 basin with $P \approx 9/r^{2.2}$. The equilibrium at $(-1.5, 1.5, 0)$ is an unstable saddle focus with eigenvalues $(0.2569 \pm 0.9793i, -19.5137)$.

You can build the current source used to bias the diode with a LM334 three-terminal adjustable current source with the wiring shown in Fig. 2.16 and calculate the value of the current I_0 from the resistance R_{set} using the formula given in the data sheet. However, we prefer to use an ammeter and adjust R_{set} until I_0 reaches the desired value. The source V_{ee} can come from the same negative power supply used for the operational amplifier.

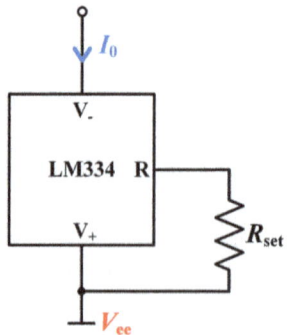

Fig. 2.16   LM334 current source.

The polarity of the diode is particularly important in all the circuits using a biased diode. The diode current direction must be opposite to the direction of the current source so that the diode is biased by $I_D - I_0$ allowing current to flow in the reverse bias direction. If both the diode and current source point in the same direction (e.g., into a node), a reverse bias current will not be able to flow, and no oscillation will occur.

The experimental phase space plot of I versus V_1 is shown in Fig. 2.19, and the time plot of $V_1(t)$ is shown in Fig. 2.20. We used the instrumentation amplifier circuit from Fig. 1.23 to obtain V_1 by measuring the difference in potential across capacitor C_1. You can obtain the current I by measuring the voltage at the non-inverting input of the operational amplifier and dividing it by R. There will be several other circuits where some manipulation of the voltage or current variables is needed to view the desired attractor.

Component values are shown in Table 2.2 where R requires some adjustment to obtain chaos. This circuit has three resistors, two capacitors, an inductor, an operational amplifier, and a diode, for a total of eight components. We do not consider current sources as additional components since they are a dual of the voltage source.

2.4 Banlue–Rattikarn Circuit

The circuit designed by Srisuchinwong and Treetanakorn (2014) shown in Fig. 2.21 is particularly interesting because it uses an unconventional method for producing oscillations. Whereas most circuits use an operational amplifier with a gain of $K > 1$ to sustain the oscillation (such as the previous circuit), this circuit uses an operational amplifier with a gain of $K = 1$ that does not increase the amplitude of the input signal. This type of amplifier circuit is called a *buffer*.

The oscillations are instead initiated from I_0, which charges C_2 and causes the LC network consisting of L, C_1, and C_2 to begin oscillating at $\omega_0 = 1/\sqrt{LC}$ where $C = C_1 C_2 /(C_1 + C_2)$. A feedback path is constructed by taking the LC circuit's output and feeding back a fraction of that output to the LC circuit through a capacitive voltage divider. This type of feedback path is a characteristic of the Colpitts oscillator previously shown in Fig. 1.4. Without this feedback, the circuit will eventually reach a (non-oscillating) steady state due to the DC source.

The feedback is constructed using a buffer that prevents the feedback path from drawing current from the LC circuit and thereby preventing

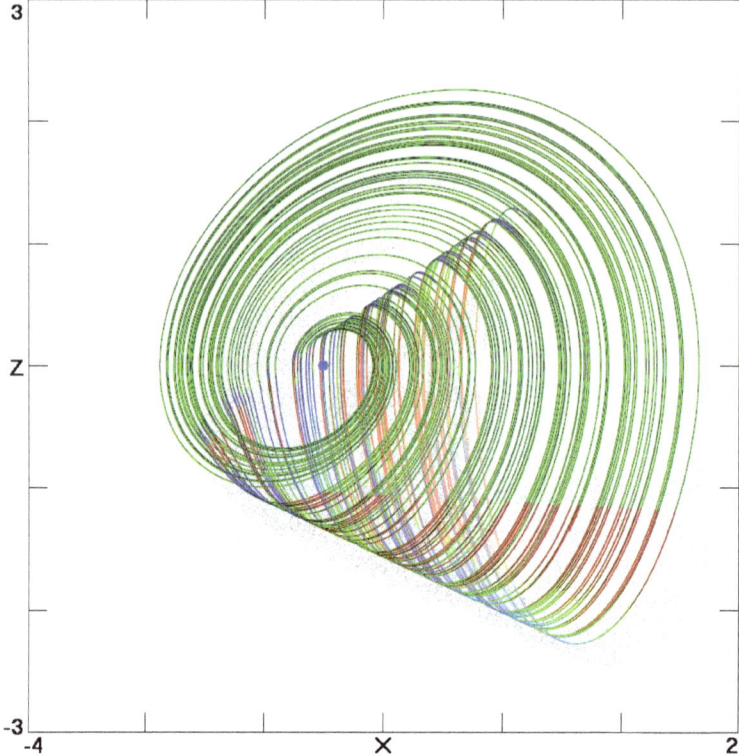

Fig. 2.17   Numerical solution of Eq. (2.10) with $a = 10$, $b = 1$, and $c = 2$.

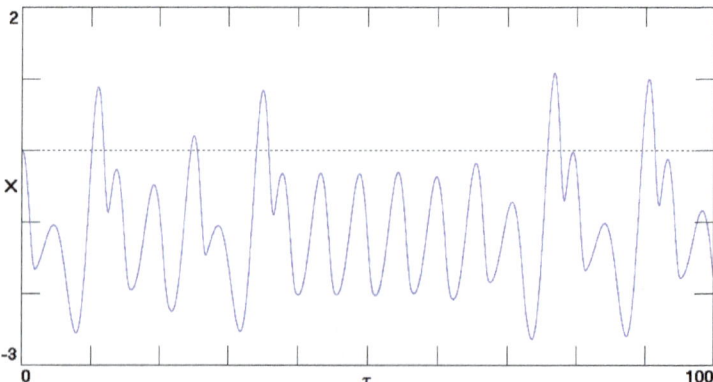

Fig. 2.18   Numerical waveform for $x(\tau)$ from Eq. (2.10) with $a = 10$, $b = 1$, and $c = 2$.

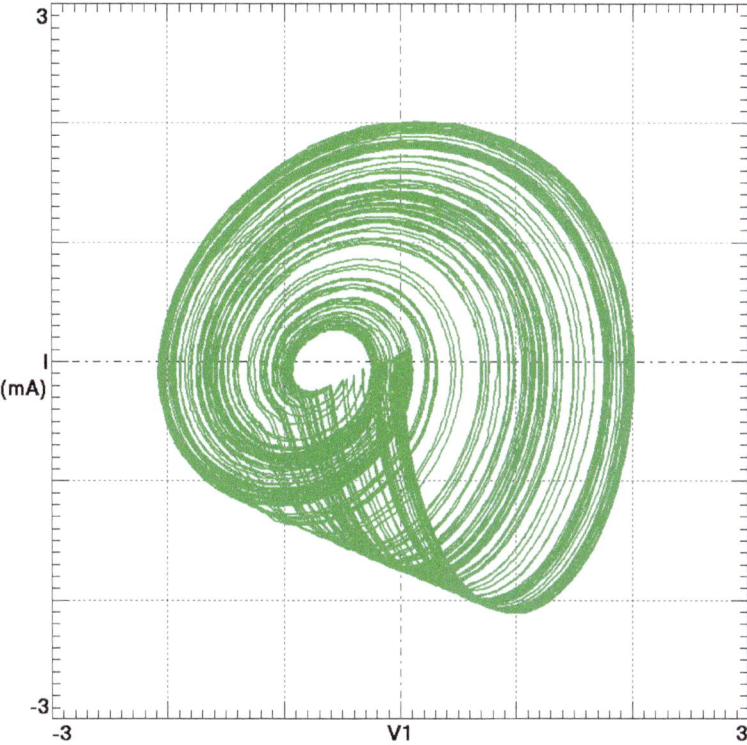

Fig. 2.19 Oscilloscope phase space plot of the Vilnius oscillator from Fig. 2.15.

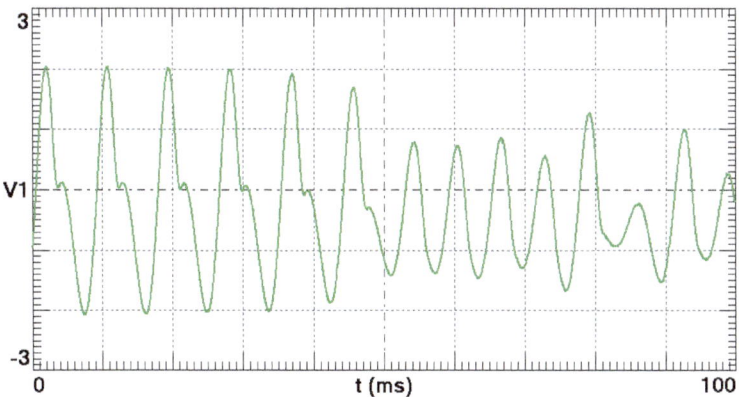

Fig. 2.20 Oscilloscope time plot of the Vilnius oscillator from Fig. 2.15.

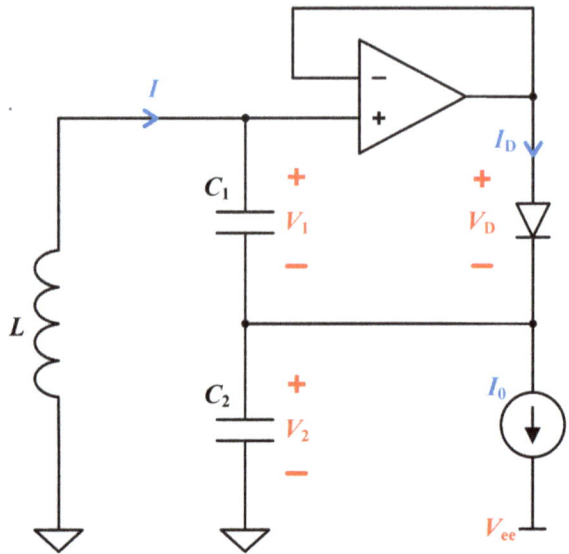

Fig. 2.21 Banlue-Rattikarn circuit described by Eq. (2.11).

Table 2.3 Component values for Fig. 2.21.

Component	Numerical	Simulated	Experimental
R_L	1 kΩ	1 kΩ	1 kΩ
R_D	0.03 kΩ	-	-
C_1	1 μF	1 μF	1 μF
C_2	1 μF	1 μF	1 μF
L	1 H	1 H	1 H
V_0	1 V	-	-
I_0	1 mA	0.5 mA	0.5 mA (adjust)
Op amp	-	Ideal	LM741
Diode	Eq. (2.2)	1N4148	1N4148

additional losses. Since the feedback voltage is 90° out of phase with V_2, it will work as positive feedback and add to the voltage, causing the oscillations to grow in amplitude until it becomes limited by the voltage supply powering the buffer (saturation). The addition of a load (in this case a diode) limits the amount of positive feedback and prevents saturation. The use of a biased diode in this circuit initiates the resonant oscillations and limits the positive feedback in the oscillator.

The equations that describe the circuit in Fig. 2.21 are
$$\dot{V}_1 = I/C_1$$
$$\dot{V}_2 = (I + I_D - I_0)/C_2 \qquad (2.11)$$
$$\dot{I} = -(V_2 + V_1 + IR_L)/L,$$
where $I_D = \max(0, V_1 - V_0)/R_D$ from Eq. (2.2) and R_L is the parasitic resistance in the inductor.

Elegant circuit values that give chaos are $R_L = 1$ kΩ, $R_D = 0.03$ kΩ, $C_1 = C_2 = 1$ μF, $L = 1$ H, $V_0 = 1$ V, and $I_0 = 1$ mA. The dimensionless equations are
$$\dot{x} = z$$
$$\dot{y} = z + a\max(0, x - 1) - b \qquad (2.12)$$
$$\dot{z} = -(y + x + z).$$

The parameters $a = 33$ and $b = 1$ give the attractor shown in Fig. 2.22 and a time series as shown in Fig. 2.23. The Lyapunov exponents are $(0.0697, 0, -1.0697)$, the Kaplan–Yorke dimension is $D_{ky} = 2.0652$, and the attractor has a global Class 1a basin of attraction. The equilibrium at $(1.0303, -1.0303, 0)$ is an unstable saddle focus with eigenvalues $(1.1737 \pm 2.9121i, -3.3475)$.

The current source used to supply the circuit with power and bias the diode can be built using the LM334 circuit in Fig. 2.16. Similar to the previous circuit, the diode current direction must be opposite to the direction of the current source which starts the oscillations so that the diode is biased correctly and reverse bias currents can flow. If the diode is reversed, the direction of the current source must also be reversed. This current source is also the main parameter to adjust in Table 2.3 where you can change R_{set} until you see chaos. In addition, you may need to add a 1 kΩ resistor in series with the 1 H inductor since its internal DC resistance is likely insufficient for R_L (ours was 40 Ω).

The oscilloscope phase space plot of V_2 versus V_1 is shown in Fig. 2.24, and the time plot of $V_1(t)$ is shown in Fig. 2.25. We used our digital oscilloscope to measure the voltage at the top of C_1 in Fig. 2.21 as well as C_2 and took the difference between the signals to obtain V_1. The circuit has two capacitors, an inductor with a parasitic resistance, an operational amplifier, and a diode, giving a total of five components.

2.5 Banlue–Buncha Diode Circuit

The next circuit relies on the *electronic filter*, which is an important circuit that is used to pass electrical signals at certain frequencies or frequency

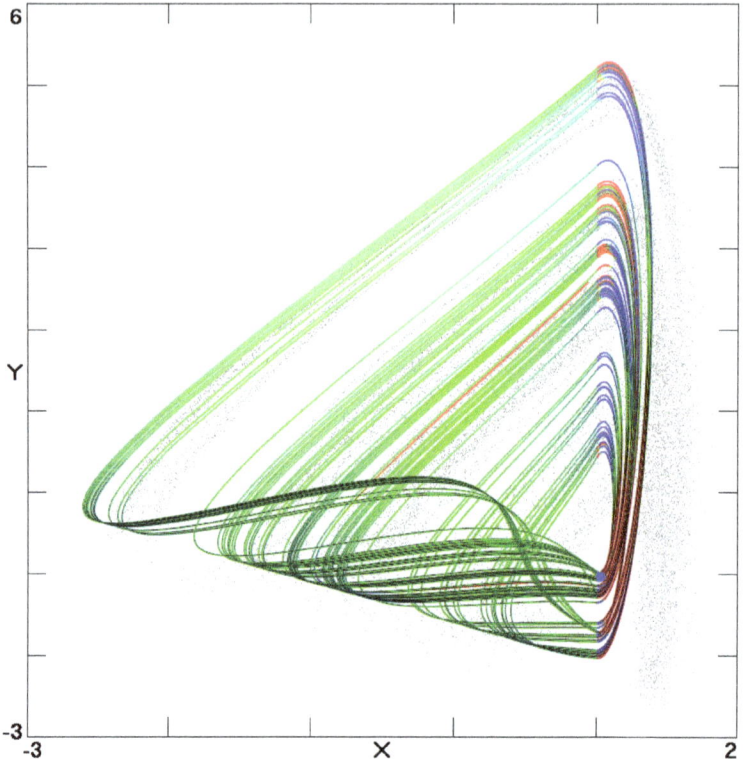

Fig. 2.22 Numerical solution of Eq. (2.12) with $a = 33$ and $b = 1$.

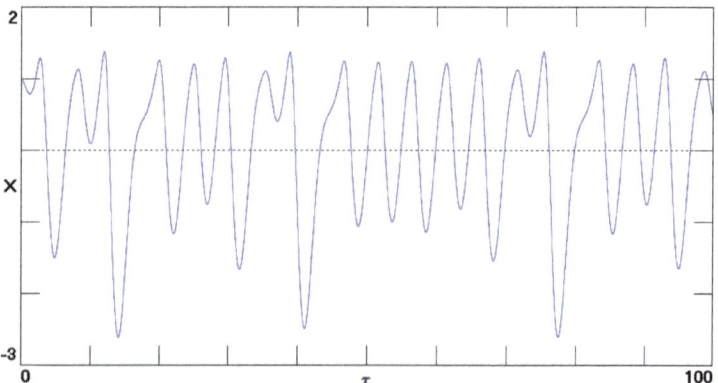

Fig. 2.23 Numerical waveform for $x(\tau)$ from Eq. (2.12) with $a = 33$ and $b = 1$.

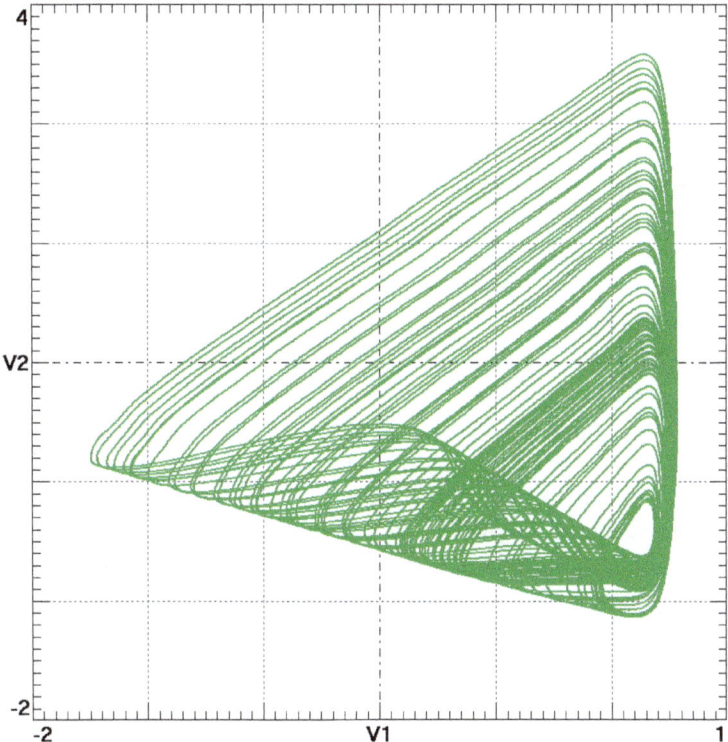

Fig. 2.24 Oscilloscope phase space plot of the Banlue-Rattikarn circuit from Fig. 2.21.

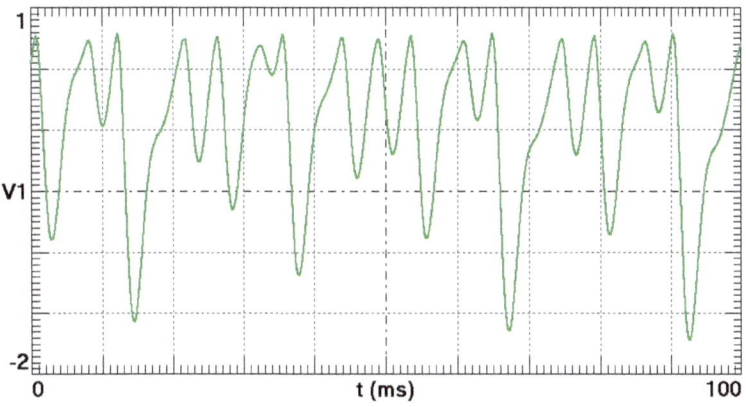

Fig. 2.25 Oscilloscope time plot of the Banlue-Rattikarn circuit from Fig. 2.21.

ranges while preventing the passage of others. Its most common application is in radio frequency communications circuits as a way to select the desired frequency while suppressing others, and it is also used in power supplies to remove noise and power line ripple. There are several types of filters, which can be categorized as either *active* or *passive*. An active filter is a device usually using a network of resistors and capacitors built around an operational amplifier, while passive filters use only inductors, capacitors, and resistors. Active filters generally perform better than passive ones, but require a power supply and are limited by the specifications of their operational amplifier.

The circuit in Fig. 2.26 contains one active and one passive filter connected together to produce oscillations. The passive filter consists of L and C_1, where the input is taken across the series combination of inductor and capacitor while the output is taken across the capacitor. If an oscillating voltage is applied to the input of this filter, the filter will pass frequencies below its resonant frequency $\omega_0 = 1/\sqrt{LC_1}$ while rejecting higher frequencies at the output. Thus it is called a *resonant filter*. In addition, if the input voltage to this filter oscillates at the resonant frequency ω_0, it will be amplified and phase shifted by $-90°$ at the output of the capacitor. The total impedance at resonance is then R_L from the series resistance in the inductor.

In this particular circuit, the input and output of the resonant filter are switched so that the input is taken across the capacitor while the output is across the series combination of inductor and capacitor. The behavior is largely the same, except that the phase shift is instead $+90°$ at the output of the inductor.

The active filter is made up of the operational amplifier, the capacitor C_2, and the series resistance of the inductor R_L. This circuit is another low-pass filter where any oscillating input applied to R_L will pass below a frequency cutoff of $\omega_0 = 1/R_L C_2$. Additionally, the operational amplifier will integrate the input voltage signal, invert it, and amplify it by a factor of $1/R_L C_2$ at the output of the operational amplifier. Thus this circuit is called an *operational amplifier integrator*.

A sinusoidal oscillator can be created with these two filters in the circuit, where the frequency is determined by the low-pass resonant filter. The resonant filter will oscillate at a frequency of $\omega_0 = 1/\sqrt{LC_1}$ with a phase shift of $+90°$ at the output, which is then fed into the active filter that integrates (a $+90°$ phase shift) and inverts the signal (a $+180°$ phase shift). The output of the active filter is then $+360°$, which is in phase with the

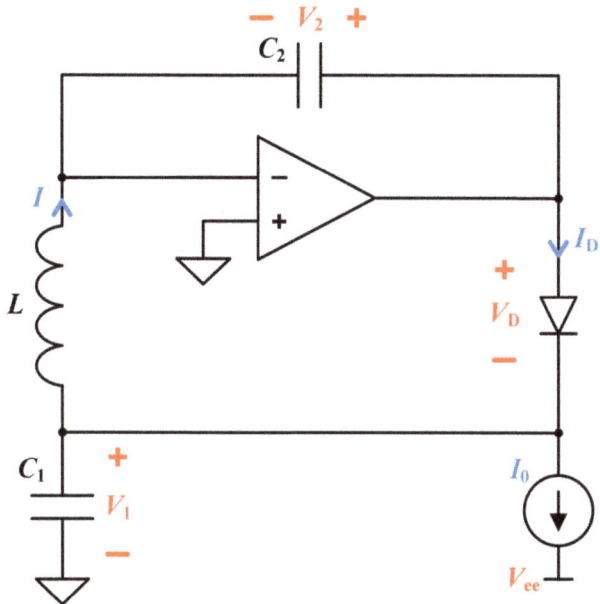

Fig. 2.26 Banlue–Buncha diode circuit described by Eq. (2.13).

Table 2.4 Component values for Fig. 2.26.

Component	Numerical	Simulated	Experimental
R_L	0.5 kΩ	0.5 kΩ	0.5 kΩ
R_D	0.5 kΩ	-	-
C_1	1 µF	1 µF	1 µF
C_2	0.05 µF	0.05 µF	0.05 µF
L	1 H	1 H	1 H
V_0	0.6 V	-	-
I_0	0.05 mA	0.05 mA	0.05 mA (adjust)
Op amp	-	Ideal	LM741
Diode	Eq. (2.2)	1N4148	1N4148

input of the resonant filter and will act as positive feedback. Since the oscillation frequency is largely determined by the resonant filter, the cutoff frequency of the active filter must be set above $\omega_0 = 1/\sqrt{LC_1}$ to prevent filtering out the resonant oscillations.

In addition, the active filter must amplify the signal so that the oscillations can be sustained in the presence of energy dissipation in R_L. As a result, the oscillations will grow in amplitude until limited by saturation of the operational amplifier, which can be prevented by adding a resistive load, or in this case a biased diode.

A similar circuit was first introduced by Piper and Sprott (2010) who used a comparator as the nonlinearity rather than a biased diode as Srisuchinwong and Munmuangsaen (2012) used. Srisuchinwong and Munmuangsaen (2012) further explored other rearrangements of the circuit by interchanging the position of the inductor and diode. You can remove the current source if desired by placing the diode in parallel with the inductor as San-Um et al. (2014) showed, but with the addition of a resistor at the inverting input of the operational amplifier integrator.

The circuit in Fig. 2.26 is described by the equations

$$\dot{V}_1 = (I_D - I_0 - I)/C_1$$
$$\dot{V}_2 = -I/C_2 \qquad (2.13)$$
$$\dot{I} = (V_1 - IR_L)/L,$$

where $I_D = \max(0, V_2 - V_1 - V_0)/R_D$ from Eq. (2.2).

Elegant circuit values that give chaos are $R_L = R_D = 0.5$ kΩ, $C_1 = 1$ μF, $C_2 = 0.05$ μF, $L = 1$ H, $V_0 = 0.6$ V, and $I_0 = 0.05$ mA. The dimensionless equations are

$$\dot{x} = a\max(0, y - x - 0.6) - b - z$$
$$\dot{y} = -cz \qquad (2.14)$$
$$\dot{z} = x - dz.$$

The parameters $a = 2, b = 0.05, c = 20$, and $d = 0.5$ give the attractor shown in Fig. 2.27 and a time series as shown in Fig. 2.28. The Lyapunov exponents are $(0.0501, 0, -0.7519)$, the Kaplan–Yorke dimension is $D_{ky} = 2.0666$, and the attractor is a global attractor with a Class 1a basin of attraction. The equilibrium point at $(0, 0.625, 0)$ is an unstable saddle focus with eigenvalues $(0.8734 \pm 2.9421i, -4.2469)$.

Unlike many diode circuits, the operation of this circuit depends critically on the fact that the diode is not ideal. In particular, the behavior depends on the existence of a forward voltage drop in the diode. Furthermore, the forward resistance of the diode plays a role in the analysis and cannot be taken arbitrarily small. That makes the analysis somewhat difficult because the numerical integrator must be capable of correctly resolving the sharp bend in the diode I-V curve.

In this case, we used a fourth-order Runge-Kutta integrator with an adaptive step size and error control patterned after the one in Press et al.

(2007). It is well worth your while to implement such a numerical integrator since it can be used for the analysis of all the circuits in this book, and it will protect you from errors that are all too common in the published literature where the authors used integrators with a small but fixed step size.

The oscilloscope plot of V_2 versus V_1 is shown in Fig. 2.29, and the time plot of $V_1(t)$ is shown in Fig. 2.30. You can observe V_2 by directly probing the output of the operational amplifier since the other terminal of C_2 is connected to the inverting input of the operational amplifier, which is a *virtual ground* (a point whose voltage is maintained near zero by the negative feedback without drawing any current).

Component values are given in Table 2.4 where the main parameter to adjust for chaos is the current source I_0 from Fig. 2.16 which can be changed through R_{set}. The inductor we used in this circuit has an internal DC resistance of 39.7 Ω, and so we added a 500 Ω series resistor to increase R_L. This circuit has two capacitors, an inductor with parasitic resistance, an operational amplifier, and one diode for a total of five components.

2.6 Chaotic Wien Bridge Oscillator

Although integrated circuit technology has excelled at miniaturizing resistors and capacitors for microelectronics, it is still difficult to reduce the size of inductors without introducing parasitics or degrading their quality factor. This limitation has motivated the development of inductorless waveform generating circuits.

One such circuit is the *Wien bridge oscillator* shown in Fig. 2.31 which can generate sine waves by tuning a network of resistors and capacitors (called *branches*). This oscillator is formed by two RC networks, a parallel RC network formed by R_1 and C_1 and a series RC network formed by R_2 and C_2. The parallel RC network acts as a passive low-pass filter, blocking any frequencies above $1/R_1C_1$ while the series RC network acts as a passive high-pass filter that blocks any frequencies that are below $1/R_2C_2$. Thus if $R = R_1 = R_2$ and $C = C_1 = C_2$, the RC network will filter out all frequencies other than the resonant frequency of $\omega_0 = 1/RC$. This type of filter is also called a *band-pass filter*. Since the amplitude of the oscillation at this frequency is attenuated, a non-inverting operational amplifier is required to amplify the resonant oscillation before feeding it back to the RC network.

There are various ways to convert the Wien bridge oscillator into a chaotic circuit. Perhaps the most obvious is to replace an existing function

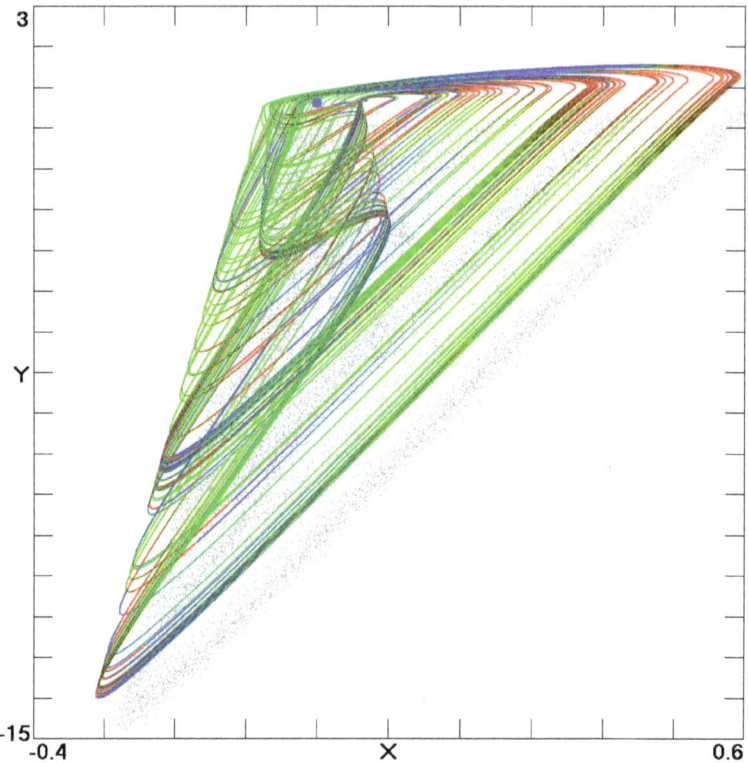

Fig. 2.27 Numerical solution of Eq. (2.14) with $a = 2$, $b = 0.05$, $c = 20$, and $d = 0.5$.

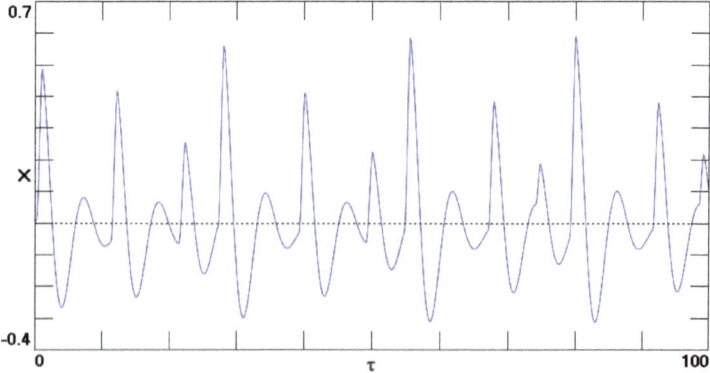

Fig. 2.28 Numerical waveform for $x(\tau)$ from Eq. (2.14) with $a = 2$, $b = 0.05$, $c = 20$, and $d = 0.5$.

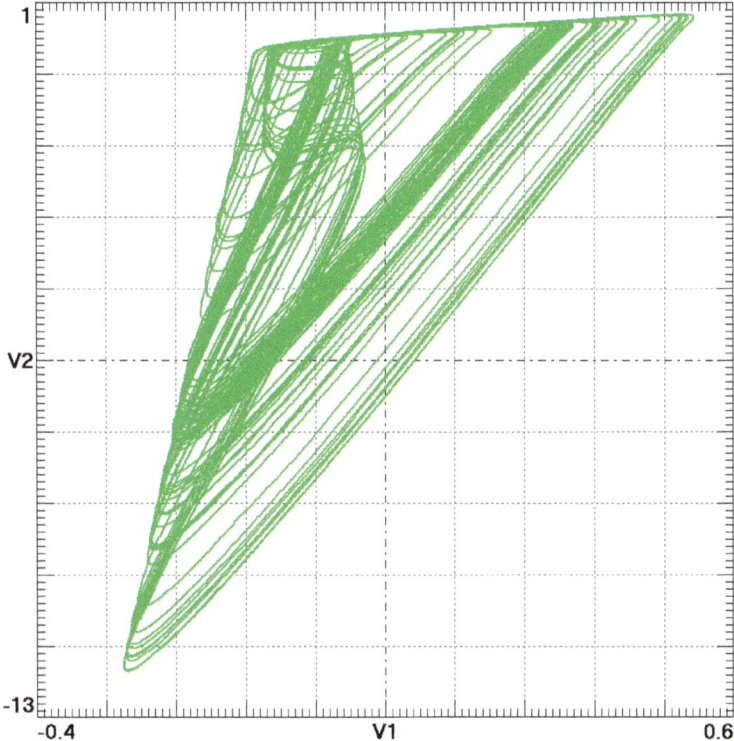

Fig. 2.29 Oscilloscope phase space plot of the Banlue–Buncha diode circuit from Fig. 2.26.

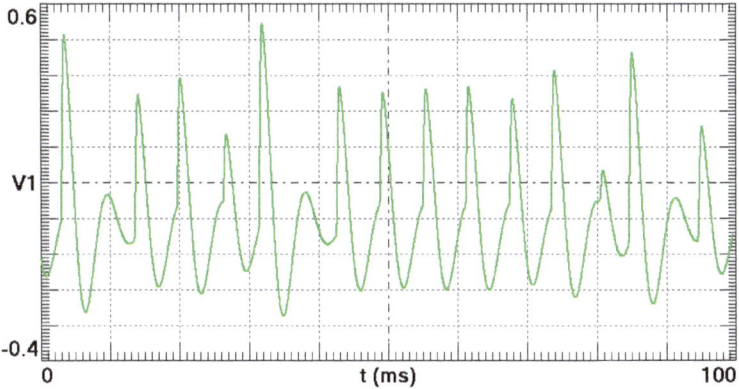

Fig. 2.30 Oscilloscope time plot of the Banlue–Buncha diode circuit from Fig. 2.26.

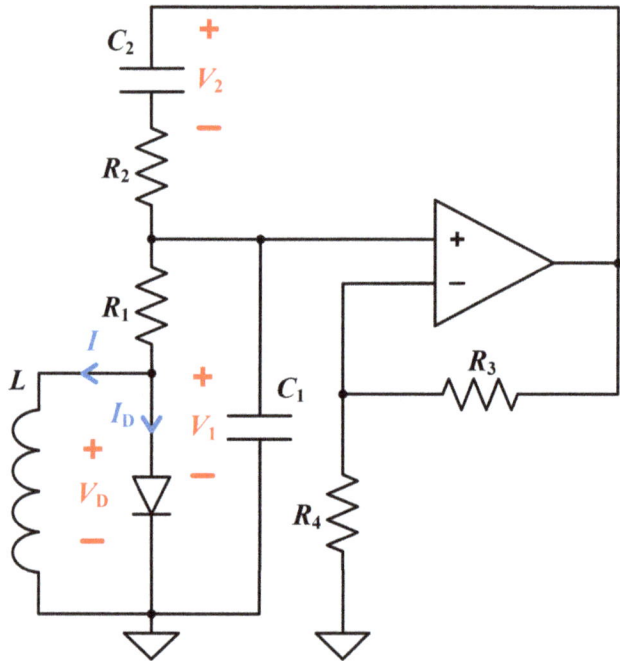

Fig. 2.31 Chaotic Wien bridge oscillator described by Eq. (2.15).

Table 2.5 Component values for Fig. 2.31.

Component	Numerical	Simulated	Experimental
R_1	1 kΩ	1 kΩ	1 kΩ (±1%)
R_2	1 kΩ	1 kΩ	1 kΩ (±1%)
R_3	1 kΩ	1 kΩ	1 kΩ (±1%)
R_4	1 kΩ	1 kΩ	1 kΩ (±1%)
R_D	0.01 kΩ	-	-
C_1	0.4 μF	0.4 μF	0.4 μF
C_2	1 μF	1 μF	1 μF
L	1 H	1.43 H	1.43 H
V_0	1 V	-	-
Op amp	-	Ideal	LM741
Diode	Eq. (2.2)	1N4148	1N4148

generator in a nonautonomous circuit (such as the forced diode resonator) with a Wien bridge oscillator to convert it into an autonomous circuit. It can also be used to replace a passive resonant element, which was done by Morgul (1995) who used a Wien bridge oscillator to replace the inductor in Chua's circuit.

Probably the most elegant implementation of the chaotic Wien bridge oscillator was proposed by Namajunas and Tamasevicius (1995) who added a *junction field effect transistor* (JFET) in the feedback path of the Wien bridge oscillator. This device will be explored in more detail in Chapter 3, but it behaves much like the biased diode used in earlier circuits. Their circuit inspired the chaotic Wien bridge oscillators described by Elwakil and Soliman (1997a), where they examined the different types of Wien bridge oscillators described by Raadhakrishnan et al. (1992) and converted them into chaotic circuits by adding a junction field effect transistor.

A few diode-based chaotic Wien bridge oscillators were also proposed during this time, the first by Namajunas and Tamasevicius (1996) who combined the Wien bridge with another operational amplifier circuit containing the diode. Tamasevicius et al. (1996b) slightly simplified this circuit by modifying the operational amplifier circuit with additional diodes to produce hysteresis.

Elwakil and Kennedy (1998) pursued a different route by inserting an inductor in parallel with the diode to the parallel RC circuit branch as shown in Fig. 2.31. Both this circuit and the JFET-based oscillator later led Elwakil and Kennedy (2000a, 2001) to discover several new chaotic circuits by conjecturing that they could be created through the addition of either nonlinear device to a periodic oscillator. We will explore one of their examples in the next section.

This circuit in Fig. 2.31 is described by the equations

$$\begin{aligned}
\dot{V}_1 &= [(K - 1 - R_2/R_1)V_1 - V_2 + R_2V_D/R_1]/R_2C_1 \\
\dot{V}_2 &= [(K - 1)V_1 - V_2]/R_2C_2 \\
\dot{I} &= V_D/L \\
\dot{V}_D &= [V_1 - V_D - R_1(I + I_D)]/R_1C_D,
\end{aligned} \quad (2.15)$$

where $K = 1 + R_3/R_4$, $I_D = \max(0, V_D - V_0)/R_D$ as given by Eq. (2.2), and C_D is the capacitance of the diode junction (assumed constant).

Elegant circuit values that give chaos are $R_1 = R_2 = R_3 = R_4 = 1$ kΩ, $R_D = 0.01$ kΩ, $C_1 = 0.4$ μF, $C_2 = C_D = 1$ μF, $L = 1$ H, and $V_0 = 1$ V.

The dimensionless equations are

$$\begin{aligned}\dot{x} &= a(u - y) \\ \dot{y} &= x - y \\ \dot{z} &= u \\ \dot{u} &= x - u - z - b\max(0, u - 1).\end{aligned} \quad (2.16)$$

The parameters $a = 2.5$ and $b = 100$ give the attractor shown in Fig. 2.32 and a time series as shown in Fig. 2.33. The Lyapunov exponents are $(0.0433, 0, -0.8369, -28.7516)$, the Kaplan–Yorke dimension is $D_{ky} = 2.0518$, and the attractor is a global attractor with a Class 1a basin of attraction. The origin is an unstable saddle focus with eigenvalues $(0.2971 \pm 0.9409i, -1.2971 \pm 0.9409i)$.

The experimental phase space plot of V_2 versus V_1 is shown in Fig. 2.34, and the time plot of $V_1(t)$ is shown in Fig. 2.35. An easy way to obtain both V_1 and V_2 without an instrumentation amplifier is to probe the output of the operational amplifier (which is KV_1) and the negative terminal of C_2. You can then find V_1 by dividing by $K = 2$ and V_2 by taking the potential difference between the two probes. Alternatively, you can switch the position of C_2 and R_2 and directly take the difference across C_2 so that you have both V_1 and V_2.

Since this circuit is very sensitive to a variation in the parameters deviating from those given in the Table 2.5 (a 0.61% variation in parameters is likely to destroy the chaos) you should use ±1% tolerance resistors for all the 1 kΩ resistors. The exact values of the capacitors are less critical at ±10% tolerance. The circuit should oscillate chaotically as soon as it is powered on without any need to adjust the resistances.

The experimental inductance value deviates from the numerical value, and we found that a value of 1.43 H will produce an attractor that more resembles the numerical plot. However, this exact value is not critical to obtain chaos, and 1.5 H can also be used. We connected several inductors in series to obtain this value with a total DC parasitic resistance of 59.7 Ω. The circuit has four resistors, two capacitors, one inductor, an operational amplifier, and a diode for a total of nine components.

2.7 Elwakil–Kennedy Diode Oscillator

So far, we have used 'voltage' operational amplifiers, which amplify the difference between their terminals by $V_{out} = A(V_+ - V_-)$. This voltage is fed back to the negative input so that the voltages at the input terminals are approximately equal ($V_+ \approx V_-$). Another type of operational amplifier is the *current feedback operational amplifier* or 'current feedback amplifier' (CFA) where the noninverting input has a high impedance, and the inverting input is a low-impedance output terminal. The current flowing out of the inverting input is called the *error current* (I_e) and is amplified by the operational amplifier so that the voltage output is $V_{out} = I_e Z_t$, where Z_t is the *open loop transimpedance gain*.

Current feedback amplifiers can be used to replace voltage operational amplifiers without much difference in operation. In fact, the original chaotic Wien bridge circuit from Elwakil and Kennedy (1998) used a current feedback amplifier in place of a voltage operational amplifier, but the two circuits are equally elegant. However, the elegance can be improved if we use a type of current feedback amplifier called a *second generation current conveyor* invented by Sedra and Smith (1970) that has an additional terminal from which the error current flows. It is called the high impedance terminal (TZ) because you can optionally add a large resistor to it to represent Z_t, whose voltage V_{out} is measured by the current conveyor using an internal buffer. This high impedance terminal can be configured in a variety of ways to design circuits that might not otherwise be possible with a voltage operational amplifier. For example, Svoboda (1989) described a Wien bridge oscillator using this current conveyor that is easier to fabricate into an integrated circuit because it uses grounded capacitors [Bhushan and Newcomb (1967)].

This circuit can be converted into a chaotic circuit by introducing either a junction field effect transistor or an inductor-diode pair in one of the branches of the circuit as described by Elwakil and Kennedy (2000a,b), or through a biased diode as Munmuangsaen and Srisuchinwong (2013) showed. A slightly more elegant circuit shown in Fig. 2.36 can be created by replacing the junction field effect transistor with a biased diode in the circuit as described by Elwakil and Kennedy (2000b) but with the drawback of making the circuit less practical due to the presence of an inductor.

The equations that describe the circuit in Fig. 2.36 are

$$\begin{aligned} \dot{V}_1 &= (I_D - I)/C_1 \\ \dot{V}_2 &= (I_D - I_0)/C_2 \\ \dot{I} &= (V_1 - IR_L)/L, \end{aligned} \quad (2.17)$$

where $I_D = \max(0, V_1 - V_2)/R_D$ from Eq. (2.2). Note that no current flows into the non-inverting input, whereas the current flowing out of TZ is mirrored to the diode current I_D flowing out of the inverting input. The

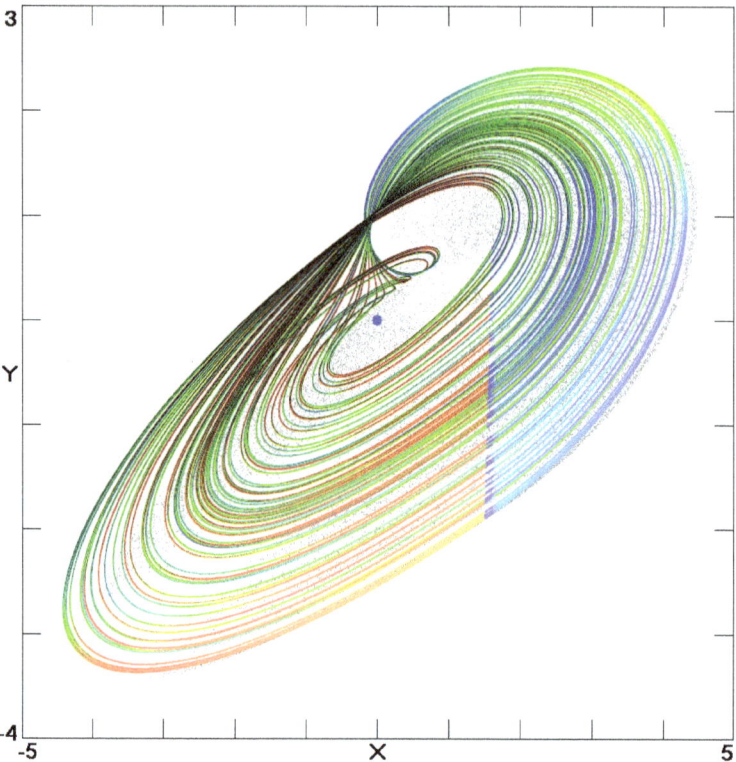

Fig. 2.32 Numerical solution of Eq. (2.16) with $a = 2.5$ and $b = 100$.

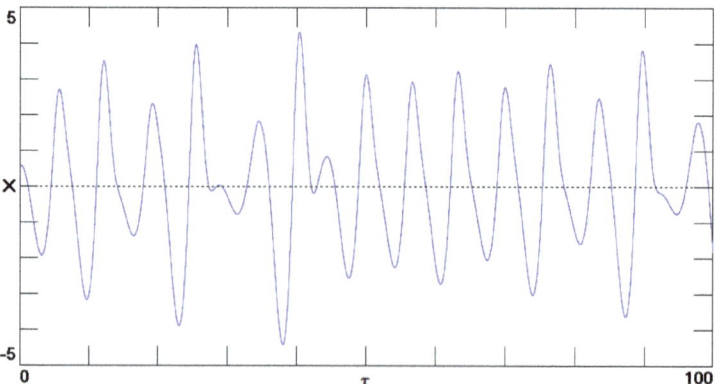

Fig. 2.33 Numerical waveform for $x(\tau)$ from Eq. (2.16) with $a = 2.5$ and $b = 100$.

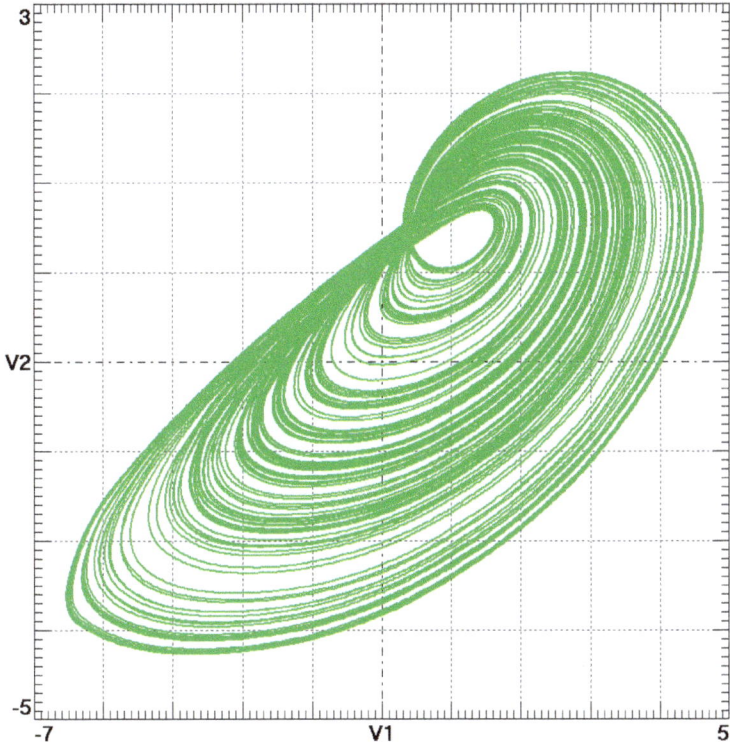

Fig. 2.34 Oscilloscope phase space plot of the chaotic Wien bridge oscillator from Fig. 2.31.

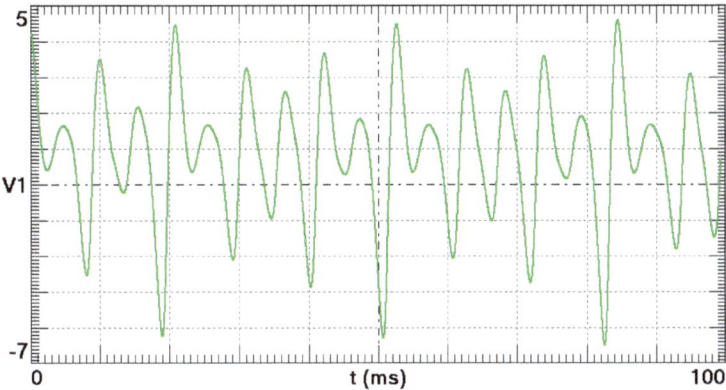

Fig. 2.35 Oscilloscope time plot of the chaotic Wien bridge oscillator from Fig. 2.31.

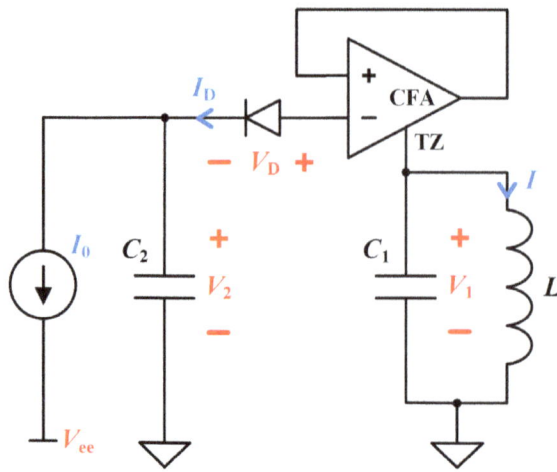

Fig. 2.36 Elwakil–Kennedy diode oscillator described by Eq. (2.17).

Table 2.6 Component values for Fig. 2.36.

Component	Numerical	Simulated	Experimental
R_L	0.5 kΩ	0.5 kΩ	0.5 kΩ
R_D	0.1 kΩ	-	-
C_1	1 µF	1 µF	1 µF
C_2	1 µF	1 µF	1 µF
L	1 H	1 H	1 H
I_0	1 mA	0.1 mA	0.2 mA (adjust)
Op amp	-	AD844	AD844
Diode	Eq. (2.2)	1N4148	1N4148

voltage output is equal to the voltage applied to TZ, which is fed back to the non-inverting input so that the voltage at the inverting input is V_1.

Elegant circuit values that give chaos are $R_D = 0.1$ kΩ, $R_L = 0.5$ kΩ, $C_1 = C_2 = 1$ µF, $L = 1$ H, and $I_0 = 1$ mA. The dimensionless equations are

$$\begin{aligned} \dot{x} &= a \max(0, x - y) - z \\ \dot{y} &= a \max(0, x - y) - 1 \\ \dot{z} &= x - bz. \end{aligned} \quad (2.18)$$

Note that I_0 is an amplitude parameter [Li and Sprott (2013)] that only affects the size of the attractor provided it is sufficiently large that the forward voltage drop V_0 in the diode can be ignored.

The parameters $a = 10$ and $b = 0.5$ give the attractor shown in Fig. 2.37 and a time series as shown in Fig. 2.38. The Lyapunov exponents are $(0.0687, 0, -0.5687)$, the Kaplan–Yorke dimension is $D_{ky} = 2.1208$, and the attractor has a global Class 1a basin of attraction. This system has an unstable saddle focus at $(0.05, 0.04, 0.1)$ with eigenvalues $(0.8337 \pm 1.9796i, -2.1674)$.

The experimental phase space plot of V_2 versus V_1 is shown in Fig. 2.39, and the time plot of $V_1(t)$ is shown in Fig. 2.40. The main component to adjust for chaos in Table 2.6 is the current source I_0 built using the circuit in Fig. 2.16. You can adjust R_{set} and observe how it controls the size of the attractor since it acts as an amplitude parameter. This circuit has two capacitors, one inductor, a current feedback amplifier, and one diode for a total of five components.

2.8 Saito Family Diode Circuit

In addition to amplifying or filtering a signal, operational amplifiers can be used to design unusual components that would not otherwise exist as physical devices. One of these is the *active resistor*, which has a negative resistance of $-R$, causing it to add power to the circuit rather than dissipate it. Since active resistors are linear, a chaotic circuit still requires an additional nonlinear device.

Shinriki et al. (1981) reported one of the first chaotic circuits that used an active resistor with diodes, although they did not recognize it as chaos, instead calling it a 'multimode' or 'random' oscillation that Freire et al. (1984) eventually showed was chaotic. Both of these studies included the case where the operational amplifier used to implement the active resistor would saturate (see Chapter 6) and become nonlinear. Inaba et al. (1987) later found that this saturation was unnecessary and that chaos can occur for a linear active resistor, leaving the diode as the nonlinearity responsible for chaos.

Saito (1989) simplified this circuit to contain an active resistor, two capacitors, an inductor, and one diode, and showed these components could be rearranged in many ways leading to many types (or a 'family') of chaotic circuits with one diode. This family of circuits is large and includes circuits described by Nishio et al. (1990a) and by Tamasevicius et al. (1996a).

One of the circuits from this family is shown in Fig. 2.41 as Inaba and Mori (1989) described. This circuit is one of the simplest autonomous chaotic oscillators in this book since it has the minimum number of

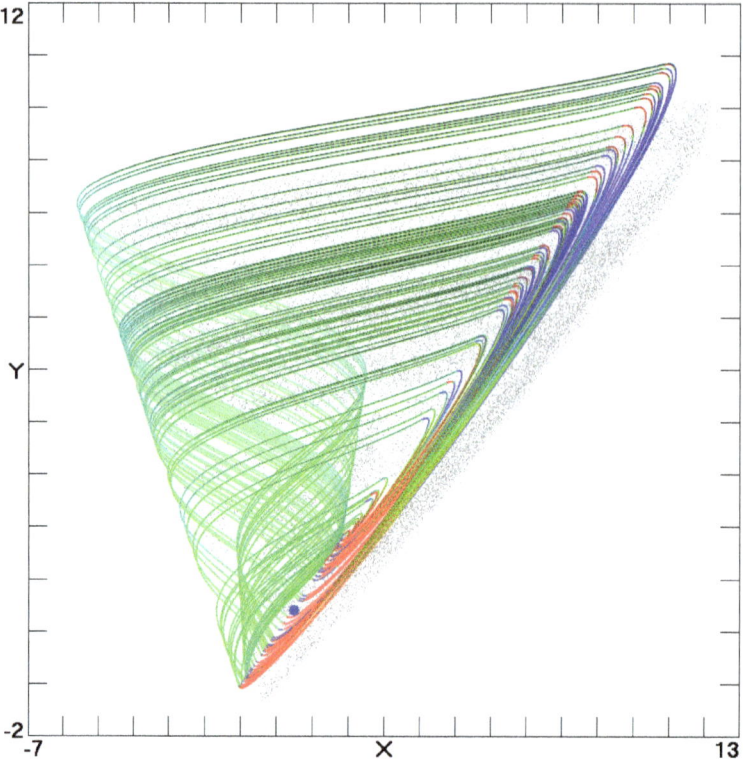

Fig. 2.37 Numerical solution of Eq. (2.18) with $a = 10$ and $b = 0.5$.

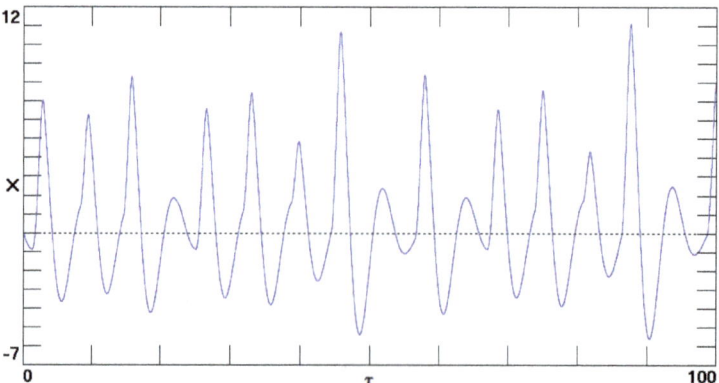

Fig. 2.38 Numerical waveform for $x(\tau)$ from Eq. (2.18) with $a = 10$ and $b = 0.5$.

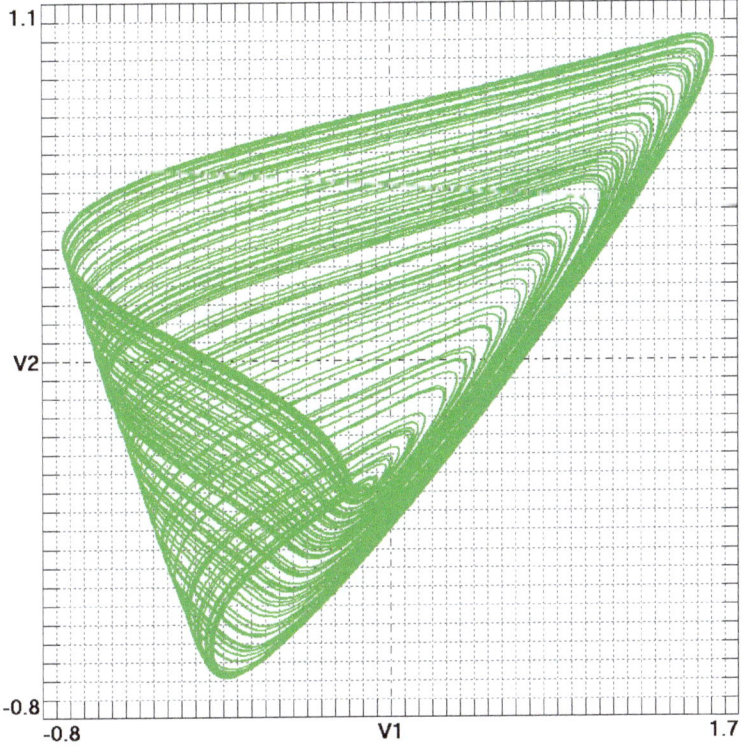

Fig. 2.39 Oscilloscope phase space plot of the Elwakil–Kennedy diode oscillator from Fig. 2.36.

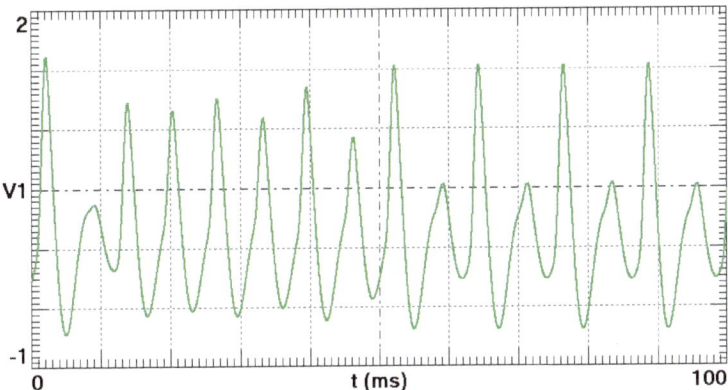

Fig. 2.40 Oscilloscope time plot of the Elwakil–Kennedy diode oscillator from Fig. 2.36.

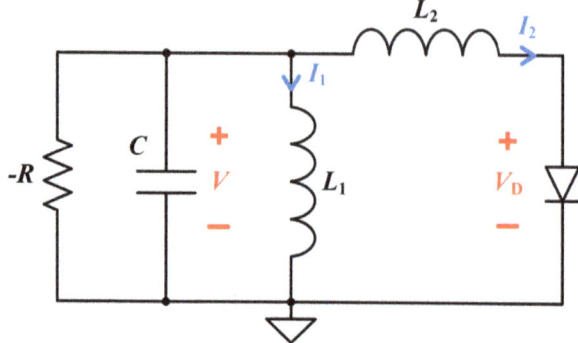

Fig. 2.41 Saito family diode circuit described by Eq. (2.19).

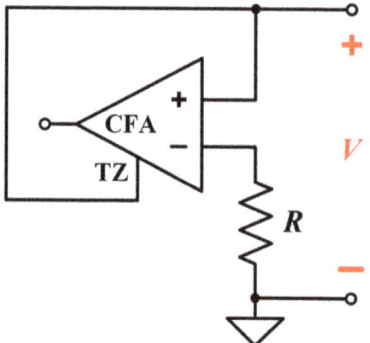

Fig. 2.42 Active resistor circuit.

Table 2.7 Component values for Fig. 2.41.

Component	Numerical	Simulated	Experimental
R	1 kΩ	1 kΩ	1 kΩ (adjust)
R_D	2 kΩ	-	-
C	1 μF	1 μF	1 μF
L_1	1 H	1 H	1 H
L_2	0.1 H	0.1 H	0.1 H
V_0	1 V	-	-
Op amp	-	-	AD844
Diode	Eq. (2.4)	1N4148	1N4148

components needed for chaos and will be revisited several times with other nonlinear devices.

Unlike the previous circuits, the diode is current-controlled rather than voltage-controlled. This leads to the circuit equations,

$$\dot{V} = (V/R - I_1 - I_2)/C$$
$$\dot{I}_1 = V/L_1 \qquad (2.19)$$
$$\dot{I}_2 = (V - V_D)/L_2,$$

where the current-controlled diode equation from Eq. (2.4) is

$$V_D = \min(I_2, I_0)R_D = \begin{cases} I_2 R_D & I_2 < I_0 \\ I_0 R_D & I_2 \geq I_0, \end{cases} \qquad (2.20)$$

where $I_0 \approx V_0/R_D$ and R_D is the (large) reverse resistance of the diode.

Elegant circuit values that give chaos are $R = 1$ kΩ, $R_D = 2$ kΩ, $C = 1$ μF, $L_1 = 1$ H, $L_2 = 0.1$ H, and $V_0 = 1$ V. The dimensionless equations are

$$\dot{x} = x - y - z$$
$$\dot{y} = x \qquad (2.21)$$
$$\dot{z} = ax - b\min(z, 1).$$

The parameters $a = 10$ and $b = 20$ give the attractor shown in Fig. 2.43 and the time series in Fig. 2.44. The Lyapunov exponents are $(0.1460, 0, -14.3051)$, the Kaplan–Yorke dimension is $D_{ky} = 2.0102$, and the attractor has a Class 3 basin with $P \approx 6.4/r^{2.1}$ and a smooth basin boundary. The origin is an unstable saddle focus with eigenvalues $(0.2569 \pm 0.9793i, -19.5137)$.

There are several ways to implement the active resistance, the simplest being with a AD844 current feedback amplifier and an ordinary resistor as shown in Fig. 2.42. In this circuit, the error current flowing out of the high impedance node (TZ) is equal to $-V/R$ and is fed back to the circuit so that the current flows opposite to the applied voltage V. This resistor R is the main component in Table 2.7 that you should adjust to obtain chaos.

The experimental phase space plot of I_1 versus V is shown in Fig. 2.45, and the time plot of $V(t)$ is shown in Fig. 2.46. I_1 was viewed by adding a 50 Ω resistor in series between L_1 and ground. In this circuit, there is one resistor, one capacitor, two inductors, one current feedback amplifier for the active resistor, and one diode for a total of six components.

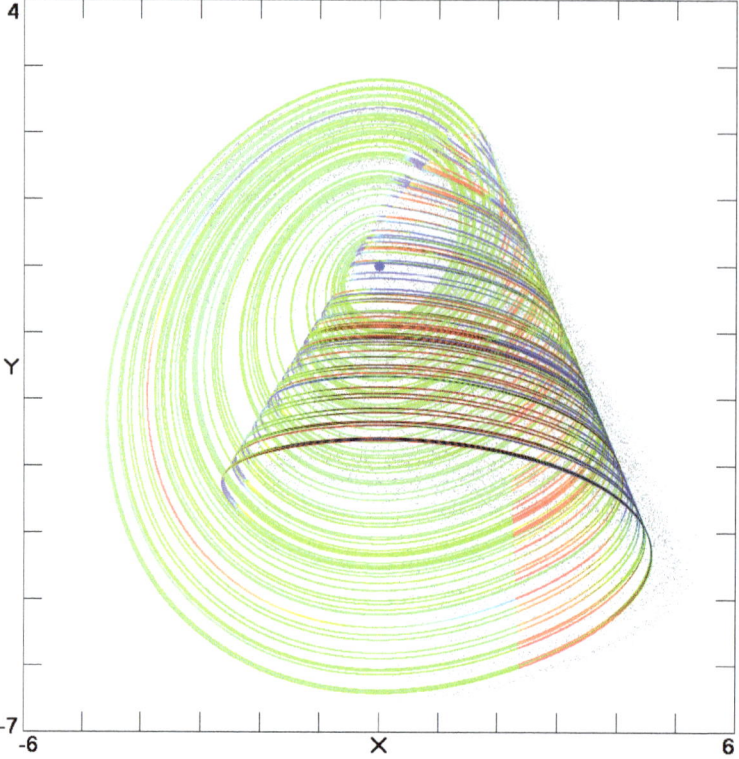

Fig. 2.43 Numerical solution of Eq. (2.21) with $a = 10$ and $b = 20$.

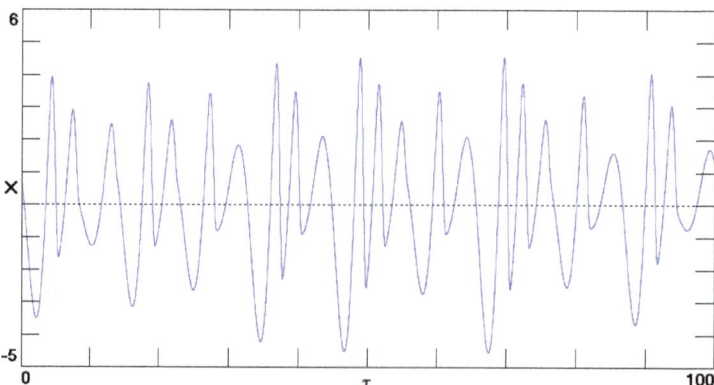

Fig. 2.44 Numerical waveform for $x(\tau)$ from Eq. (2.21) with $a = 10$ and $b = 20$.

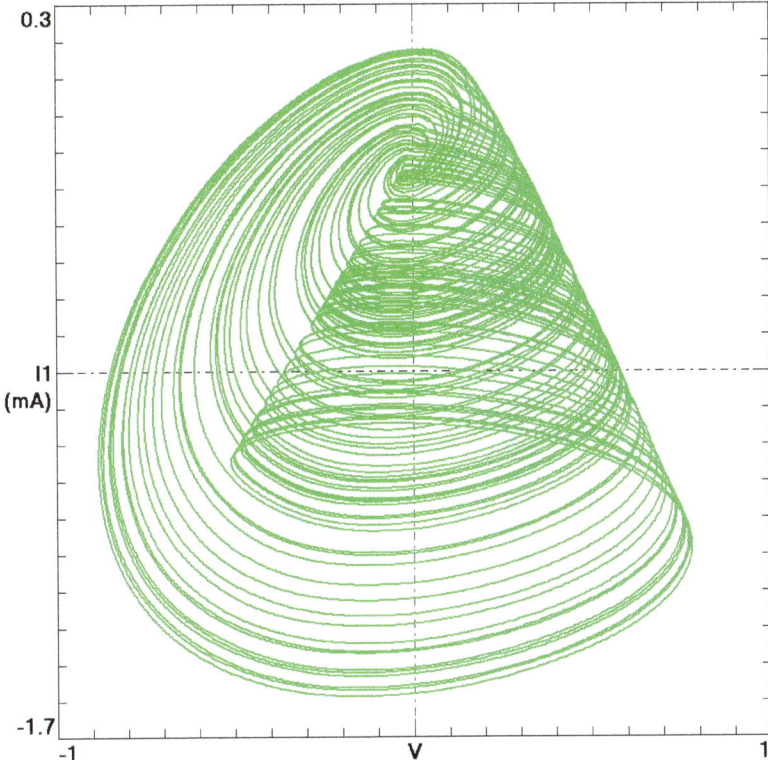

Fig. 2.45 Oscilloscope phase space plot of the Saito family diode circuit from Fig. 2.41.

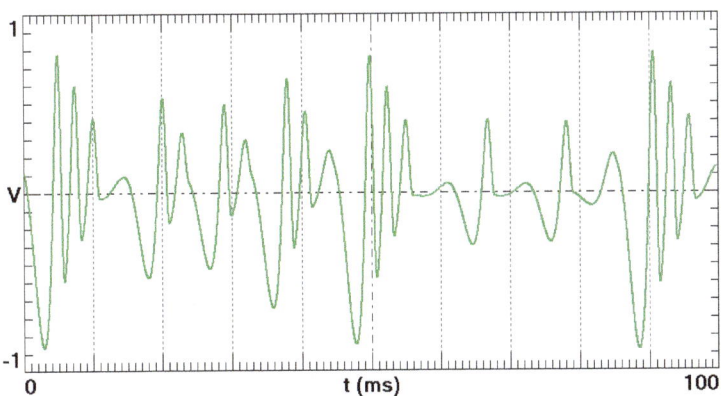

Fig. 2.46 Oscilloscope time plot of the Saito family diode circuit from Fig. 2.41.

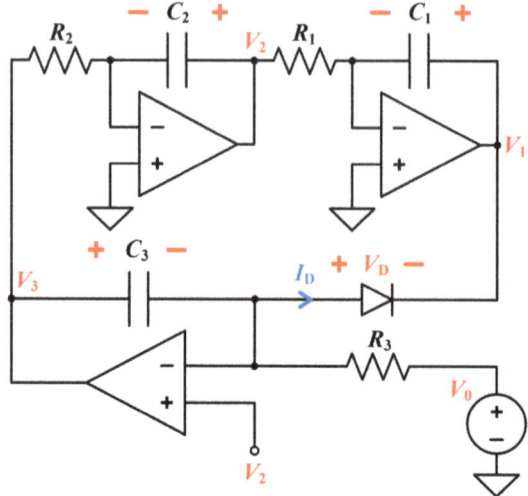

Fig. 2.47 Diode jerk circuit described by Eq. (2.22).

Table 2.8 Component values for Fig. 2.47.

Component	Numerical	Simulated	Experimental
R_1	0.2 kΩ	0.1 kΩ	0.1 kΩ (±1%)
R_2	4 kΩ	2 kΩ	2 kΩ (±1%)
R_3	1 kΩ	1 kΩ	1 kΩ (±1%)
R_D	1 kΩ	-	-
C_1	1 μF	1 μF	1 μF
C_2	1 μF	1 μF	1 μF
C_3	1 μF	1 μF	1 μF
V_0	1 V	0.04 V	0.04 V
Op amp	-	Ideal	LM348
Diode	Eq. (2.2)	1N4148	1N4148

2.9 Diode Jerk Circuit

We conclude this chapter with an example of a circuit that was inspired by a systematic search for simple equations with chaotic solutions [Sprott (1994, 1997b)]. Although this circuit is not as elegant as most of the previous examples, the governing equations are especially simple, and the circuit does not require inductors. What the circuit lacks in elegance is compensated by the elegance of the mathematical model.

This circuit was motivated by what is probably the simplest differential equation with a piecewise-linear nonlinearity whose solutions are chaotic [Linz and Sprott (1999)]. The original version of the circuit had seventeen components including two diodes [Sprott (2000a)], but it was subsequently simplified to the single diode circuit in Fig. 2.47 [Sprott (2000b)].

The equations that govern the circuit are

$$\dot{V}_1 = -V_2/R_1C_1$$
$$\dot{V}_2 = -V_3/R_2C_2 \qquad (2.22)$$
$$\dot{V}_3 = -V_3/R_2C_2 + [I_D + (V_2 - V_0)/R_3]/C_3,$$

where $I_D = \max(0, V_2 - V_1)/R_D$ from Eq. (2.2). Note that V_3 is the output voltage of the operational amplifier rather than the voltage across C_3, which is $V_3 - V_2$.

Elegant circuit values that give chaos are $R_1 = 0.2$ kΩ, $R_2 = 4$ kΩ, $R_3 = 1$ kΩ, $R_D = 1$ kΩ, $C_1 = C_2 = C_3 = 1$ μF, and $V_0 = 1$ V. The dimensionless equations are

$$\dot{x} = -ay$$
$$\dot{y} = -bz \qquad (2.23)$$
$$\dot{z} = -bz + \max(0, y - x) + y - 1,$$

which can be written more compactly in terms of the single variable x and its time derivatives as

$$\dddot{x} = -b[\ddot{x} - \dot{x} - \min(0, \dot{x} + ax) - a]. \qquad (2.24)$$

This is an example of a *jerk equation* since it involves the third derivative $\dddot{x}$ as a function of $\ddot{x}$, $\dot{x}$, and x. In fact, any explicit autonomous three-dimensional system of ordinary differential equations with a single nonlinearity, including many of the circuit models in this book, can be written in jerk form through an appropriate transformation of variables [Eichhorn et al. (1998)], but this is a particularly simple example.

The parameters $a = 5$ and $b = 0.25$ give the attractor shown in Fig. 2.48 and the time series in Fig. 2.49. The Lyapunov exponents are $(0.0332, 0, -0.2832)$, the Kaplan–Yorke dimension is $D_{ky} = 2.1172$, and the attractor has a Class 3 basin with $P \approx 3.6/r^{0.36}$. There is an unstable saddle focus at $(-1, 0, 0)$ with eigenvalues $(0.3750 \pm 1.0533i, -1)$.

This circuit is somewhat delicate (a 10% change in parameters is likely to destroy the chaos). Depending on the chosen diode, it may be necessary to add some series resistance if the forward resistance of the diode is too small. However, the forward voltage drop of the diode is not problematic, only causing a small shift of the attractor in the x (or V_1) direction.

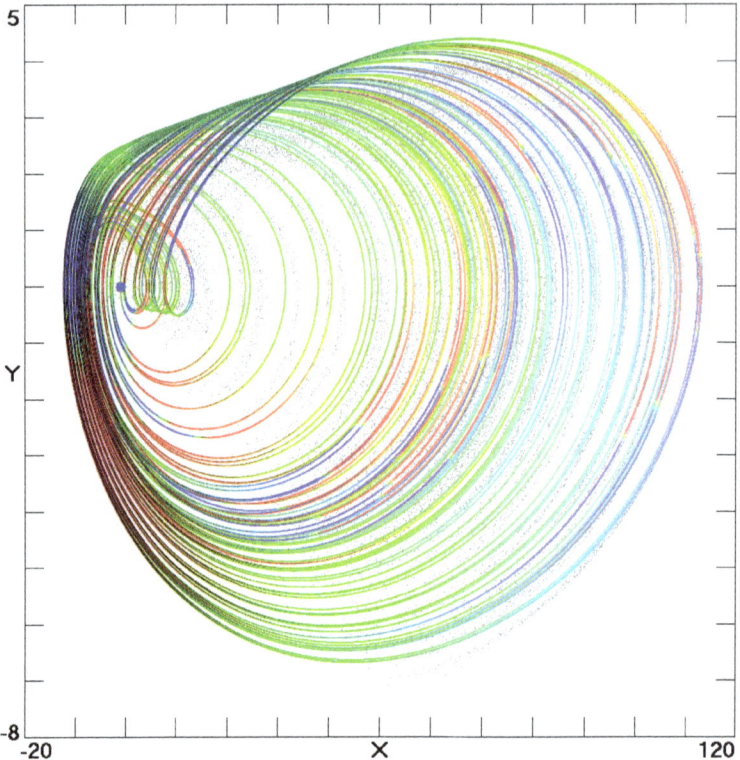

Fig. 2.48 Numerical solution of Eq. (2.23) with $a = 5$ and $b = 0.25$.

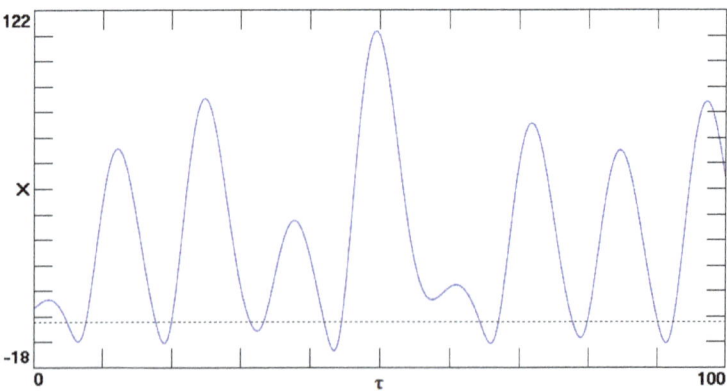

Fig. 2.49 Numerical waveform for $x(\tau)$ from Eq. (2.23) with $a = 5$ and $b = 0.25$.

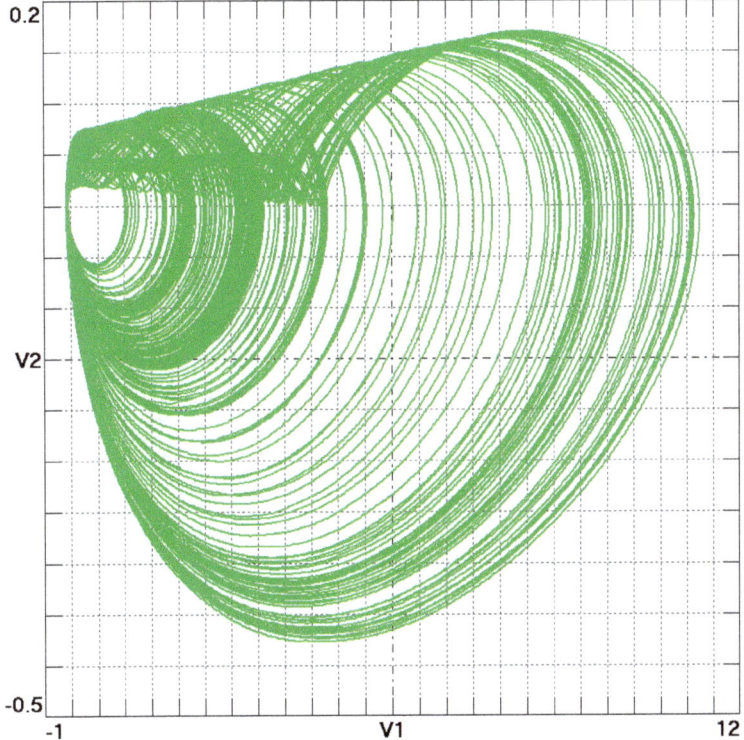

Fig. 2.50 Oscilloscope phase space plot of the diode jerk circuit from Fig. 2.47.

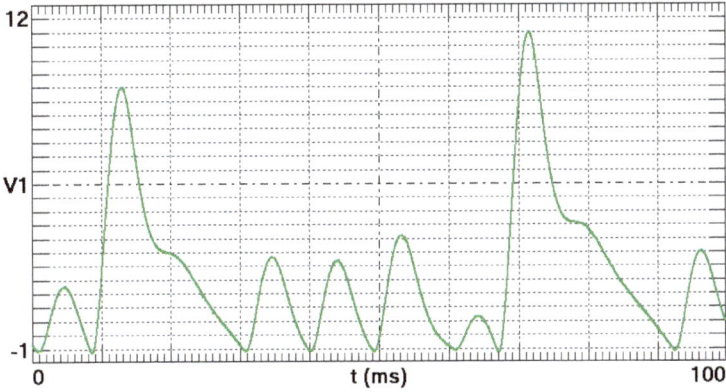

Fig. 2.51 Oscilloscope time plot of the diode jerk circuit from Fig. 2.47.

The voltages everywhere in the circuit are entirely determined by and are proportional to V_0. You will need to reduce V_0 to the order of 0.1 V to avoid saturating the operational amplifiers. Note that you cannot simply increase the series resistance R_3 with a large value of V_0 to maintain a given current since the inverting input of the lower operational amplifier is not a virtual ground, unlike most of the other operational amplifier circuits in this book.

You can obtain V_0 by using a voltage divider from the positive power supply with an operational amplifier buffer whose output is V_0 with a low output impedance as shown in Fig. 1.29. The voltage V_0 is the only parameter that you will need to vary through the voltage divider, and you can measure its value using a voltmeter while adjusting the circuit.

If you increase V_0 to the point where the amplitude of V_3 approaches the source voltage of +15 V, the amplifier will saturate, and adjusting V_0 will not bring back the chaos. If this occurs, turn V_0 to zero, restart the power supply, and slowly increase V_0 to make the circuit chaotic again.

The experimental phase space plot of V_2 versus V_1 is shown in Fig. 2.50, and the time plot for $V_1(t)$ is shown in Fig. 2.51. Component values are given in Table 2.8, where all resistors have a $\pm 1\%$ tolerance. You may want to use an LM348 containing four operational amplifiers to make it easier to construct the circuit. This circuit has three resistors, three capacitors, three operational amplifiers, and one diode, for a total of ten components.

Other jerk circuits using a single diode have been proposed and studied [Sprott (2000a, 2011)], but they have more than ten components, and so they are less elegant than the case described here and thus will not be further discussed, although we will show several other jerk circuits using nonlinear devices other than diodes in later chapters.

Chapter 3

Transistor Circuits

The transistor is the oldest three-terminal semiconductor device, and its development revolutionized the electronics industry. This chapter describes the device and shows how it can operate as an amplifier or as a switch to produce chaotic signals.

3.1 Transistor Characteristics

Hailed as one of the most important inventions of the 20th century, the transistor is largely responsible for transforming technology into what we know today. It is a fundamental building block of integrated circuits, which help perform the computing tasks needed to run the devices we use to work, live, and play. Many types of transistors exist for different applications, two of the most common of which we now describe.

3.1.1 *Bipolar junction transistor (BJT)*

The earliest and best known type of transistor is the *bipolar junction transistor* (BJT) whose schematic is shown in Fig. 3.1. There are two types of bipolar junction transistors, one being the *NPN transistor* which consists of an n-type *emitter* (E), a p-type *base* (B), and an n-type *collector* (C) as shown in Fig. 3.2. The other type of bipolar junction transistor is the *PNP transistor*, which contains a p-type emitter, n-type base, and a p-type collector. Each transistor has two junctions, the *emitter-base junction* (EBJ) and the *collector-base junction* (CBJ).

The transistor has several modes of operation that depend on the polarity of the voltage bias across the EBJ and CBJ as summarized in Table 3.1. The most important operation occurs in the *active* mode where the bipolar junction transistor acts as an amplifier for small signals, whereas the *cut-off* and *saturation* modes allow the transistor to be used as an electronic

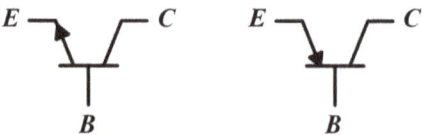

Fig. 3.1 Bipolar junction transistors, NPN and PNP.

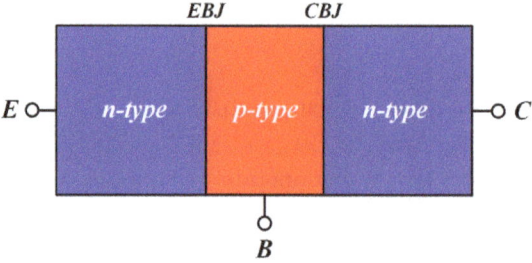

Fig. 3.2 Structure of an NPN transistor.

Table 3.1 Bipolar junction transistor modes of operation.

Mode	EBJ	CBJ
Cutoff	Reverse	Reverse
Active	Forward	Reverse
Saturation	Forward	Forward
Reverse-Active	Reverse	Forward

switch. In the next section, we will develop a model for the transistor that encompasses these modes to use in the chaotic circuits of this chapter.

3.1.2 BJT I-V characteristic

In this section, we will develop an idealized model of the bipolar junction transistor for use in our chaotic circuits by examining what the bipolar junction transistor does in each mode of operation. In Fig. 3.3, the transistor is biased in its active mode through V_{BE} and V_{CE}. Since the EBJ acts as a forward-biased diode, the electrons flow from the emitter to the base to recombine with the holes in that region. Thus the current through the base follows the model of an idealized diode described by Eq. (2.2).

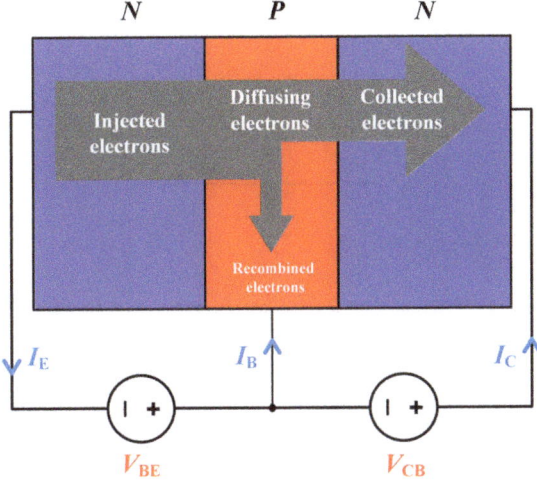

Fig. 3.3 Current flow through an NPN transistor in its active mode.

This model also incorporates the cutoff mode since $I_B = 0$ when the EBJ is reverse biased and is given by

$$I_B = \max(0, V_{BE} - V_0)/R = \begin{cases} 0 & V_{BE} \leq V_0 \\ (V_{BE} - V_0)/R & V_{BE} > V_0. \end{cases} \quad (3.1)$$

When current flows through the EBJ, most of the electrons entering the base from the emitter do not recombine because a transistor is purposely fabricated to have few holes in the base. As a result, the majority of these electrons are instead swept into the collector region due to the positive bias from V_{CB} and causes the base current to become only a small fraction of the larger collector current. This relation can be expressed by a dimensionless parameter β, which varies among transistors, but typically has a value of order 100. The collector current in the active region can then be simply modeled by

$$I_C = \beta I_B. \quad (3.2)$$

Both the active and cutoff modes can be modeled by the equivalent circuit in Fig. 3.4. Here the EBJ is modeled using an idealized diode that follows Eq. (3.1), and the CBJ is modeled using a current-controlled current source that follows Eq. (3.2).

If V_{CB} is reversed, the CBJ becomes forward biased and causes the transistor to enter saturation mode. Holes are injected by V_{CB} into the

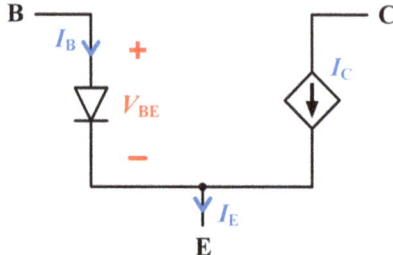

Fig. 3.4 Equivalent circuit model for the transistor in its active mode.

base and recombine with the electrons entering from the emitter, which decreases the collector current, leading to $I_C < \beta I_B$. Saturation mode can be used as the 'ON' (conducting) state in an electronic switch because both the EBJ and CBJ are forward biased, allowing current to flow from collector to emitter almost as if it were a short circuit. The 'OFF' (nonconducting) state then occurs in cutoff mode.

You can observe the cutoff, active, and saturation regions by sweeping V_{CE} for a given V_{BE} and measuring I_C as shown in Fig. 3.5. At low values of V_{CE}, the transistor is in saturation mode because the collector voltage becomes less than the base voltage and forward biases the CBJ. Increasing V_{CE} eventually reverse biases the CBJ and causes the transistor to enter the active mode where the collector current is controlled by the base current dependent on V_{BE}. The collector current can be modeled to include the saturation mode using

$$I_C = \min(\beta I_B, V_{CE}/R). \tag{3.3}$$

It is important to note that this model is idealized and neglects many characteristics of actual transistors. For example, the active region in Fig. 3.5 would have a slope due to the *Early effect* [Early (1952)] where the effective width of the base varies with the applied base-to-collector voltage, but this additional accuracy is generally unnecessary. Furthermore, parasitic capacitances are usually neglected, although they are required for some circuits.

Cutoff, active, and saturation are the most commonly used modes of the bipolar junction transistor, and they are used to perform amplification and switching in practical applications. Less commonly used is the *reverse-active* mode where the EBJ is reverse biased and the CBJ is forward biased. Although the collector current is still controlled by the base current in

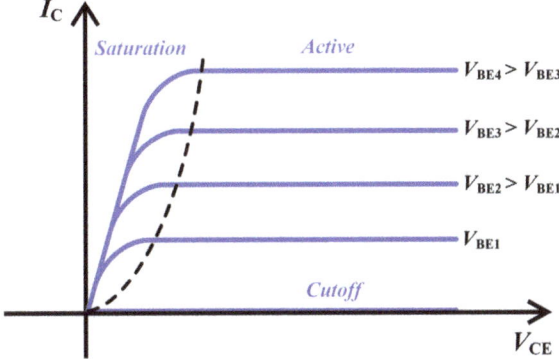

Fig. 3.5 I_C-V_{CE} curve of the transistor for different values of V_{BE}.

this mode, the gain is much smaller because the manufactured size of the emitter limits the number of electrons that can flow through and results in $I_C = \beta_R I_B$ where $\beta_R \ll \beta$. The reverse-active mode has few applications, although some chaotic circuits involve switching between the active and reverse-active mode.

You can use the *Ebers–Moll model* [Ebers and Moll (1954)] to model this transition as shown in Fig. 3.6 and given by

$$\begin{aligned} I_C &= \alpha_F I_F - I_R \\ I_E &= I_F - \alpha_R I_R \\ I_F &= \max(0, V_{BE} - V_0)/R_D \\ I_R &= \max(0, V_{BC} - V_0)/R_D. \end{aligned} \quad (3.4)$$

This model includes all four behaviors of the bipolar junction transistor, but it is less elegant and usually unnecessary.

There are two diodes in this model corresponding to the EBJ and the CBJ. The current that flows through the EBJ diode (I_F) is amplified by $\alpha_F I_F = \beta/(\beta+1)$ in the collector, where α_F represents the fraction of the emitter current that reaches the collector. Since most of the emitter current reaches the collector, α_F is slightly less than 1. On the other hand, any current that flows through the CBJ diode (I_R) is amplified by a factor of $\alpha_R I_R$ where α_R represents the fraction of the collector current that reaches the emitter. Since very little of the collector current reaches the emitter, α_R is only slightly greater than 0.

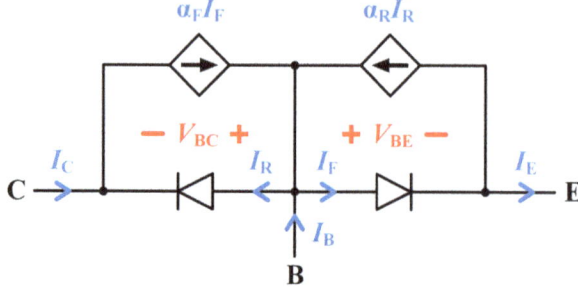

Fig. 3.6   Ebers–Moll model.

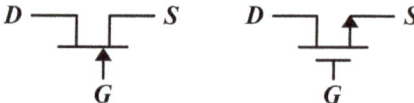

Fig. 3.7   n-channel JFET and MOSFET.

3.1.3   *Field effect transistor*

The bipolar junction transistor is a type of current-controlled device since it uses the base current to control the collector current ($I_C = \beta I_B$). Another important type of transistor is the *field effect transistor* (FET) which has three terminals called the *source* (S), *gate* (G), and *drain* (D). Unlike the bipolar junction transistor, it is the electric field created by the voltage applied to these terminals, rather than the current, that controls the current through the transistor. This makes the field effect transistor a voltage-controlled device in contrast to the current-controlled bipolar junction transistor.

The simplest type of field effect transistor is the *junction field effect transistor* (JFET) shown in Fig. 3.7 with the structure described in Fig. 3.8. The drain and source are connected to the same n-type semiconductor, and a current I_{DS} can freely flow between them through a path called a *channel*.

The gate is connected to a p-type semiconductor than surrounds the channel, and whose purpose is to close the channel when certain conditions are met. One condition is applying a reverse-bias voltage to the gate-source junction which creates a depletion region (see Fig. 2.8) that blocks I_{DS}. Increasing this reverse-bias voltage eventually makes the depletion region

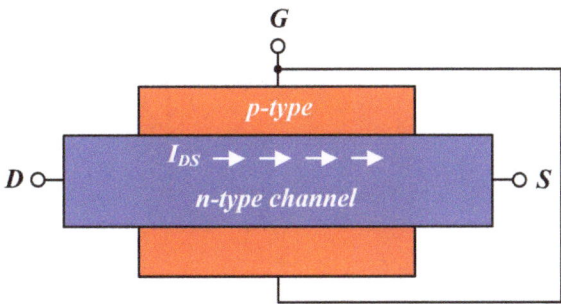

Fig. 3.8 Structure of an n-channel JFET.

so large that no additional current can flow. The value of V_{GS} at which the current stops flowing is called the *pinch-off voltage* (V_P or $V_{GS(off)}$). An analogy to this device is a water hose that normally allows water to pass through between the source and drain until a foot (the gate) is continuously pressed down on it to lower the water flow. The water stops flowing at the *cutoff region* when $V_{GS} \leq V_P$ with $V_P < 0$.

Junction field effect transistors are now mostly obsolete because the reverse bias current (I_S) flowing through the gate leads to a high power consumption. In addition, this current can change with temperature as described in Eq. (2.2). For these reasons, other technologies are preferred such as the *metal oxide semiconductor field effect transistor* (MOSFET) shown in Fig. 3.7.

A brief discussion of the MOSFET is important since chaotic circuits built using these transistors can be easily fabricated into integrated circuits. The n-channel *enhancement-mode* MOSFET initially blocks current from drain to source, but current can be made to flow by applying a positive voltage at the gate to attract electrons that form a conductive channel called an *inversion layer* allowing I_{DS} to flow. There are other MOSFETs called depletion-mode MOSFETs that initially conduct like a JFET, but they are less frequently used.

The main advantage of the MOSFET over the JFET is that no current can flow through the gate since it consists of an insulating layer of oxide. This layer prevents the electrons forming the inversion layer from leaving the device and breaking the conducting path for I_{DS}. However, both field effect transistors can be modeled in a similar way as we will now show.

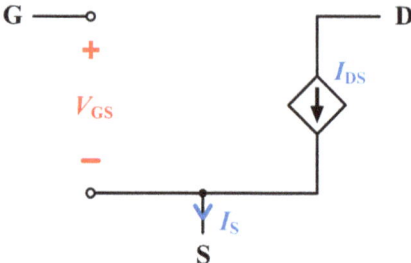

Fig. 3.9 FET model.

3.1.4 FET I-V characteristic

One important difference between the field effect transistor and the bipolar junction transistor is that no current can flow from gate to source in the FET. In the junction field effect transistor, the gate and source are reverse biased, and so little current flows, while in a MOSFET there is an insulating oxide layer in the gate that prevents any current from flowing through it. Thus the gate-source junction can be treated as an open circuit as shown in Fig. 3.9. However, in some JFET-based chaotic circuits, V_{GS} can be forward biased, and it is necessary to incorporate Eq. (2.2) as follows

$$I_{GS} = \max(0, V_{GS} - V_0)/R = \begin{cases} 0 & V_{GS} \leq V_0 \\ (V_{GS} - V_0)/R & V_{GS} > V_0. \end{cases} \qquad (3.5)$$

The I-V characteristics of all field effect transistors are identical except for the values of certain constants and thresholds of operation. The field effect transistor has three modes of operation which can be observed if V_{GS} is held at a fixed voltage and V_{DS} is varied. Cutoff mode occurs for the JFET when V_{GS} is less than V_P since the channel is blocked by the depletion region.

In a MOSFET, no channel is formed if V_{DG} is less than the threshold voltage V_T (a positive value). If V_{GS} reaches this threshold, the field effect transistor enters the *triode* region where increasing V_{DS} increases I_{DS} because of the conducting channel in the JFET or the formed inversion layer in the MOSFET.

Further increase of V_{DS} also increases the reverse-bias voltage across the junction field effect transistor's gate-drain junction (V_{GD}) and forms a depletion region that limits the current through the channel. When V_{GD} drops below the pinch-off voltage ($V_{GD} \leq V_P$), the depletion layer is so

Table 3.2   n-channel FET modes of operation.

Mode	JFET	MOSFET
Cutoff	$V_{GS} \leq V_P$	$V_{GS} \leq V_T$
Saturation	$V_{DS} \geq V_{GS} - V_P$ $V_{GS} > V_P$	$V_{DS} \geq V_{GS} - V_T$ $V_{GS} > V_T$
Triode	$V_{DS} < V_{GS} - V_P$ $V_{GS} > V_P$	$V_{DS} < V_{GS} - V_T$ $V_{GS} > V_T$

large at the drain that current ceases to increase, and the junction field effect transistor enters *saturation* mode. This condition occurs when $V_{DS} \geq V_{GS} - V_P$.

A similar effect occurs in the MOSFET, where $V_{GD} \leq V_T$ (or $V_{DS} \geq V_{GS} - V_T$) causes the inversion layer to become thinner at the drain and prevents further current increase. The cutoff, triode, and saturation behavior for a field effect transistor are qualitatively similar to Fig. 3.5 and are summarized in Table 3.2 and given approximately by

$$I_{DS} = \begin{cases} 0 & \text{Cutoff} \\ K(V_{GS} - V_P)^2 & \text{Saturation} \\ K[2V_{DS}(V_{GS} - V_P) - V_{DS}^2] & \text{Triode,} \end{cases} \quad (3.6)$$

where $V_{DS} = V_{DG} + V_{GS}$ and V_P can be replaced with V_T for a MOSFET.

For an n-channel junction field effect transistor, $V_P < 0$ and $K = I_{DSS}/V_P^2$ with units of mA/V^2, and I_{DSS} is the maximum value of I_{DS} when the gate is connected directly to the source. For an n-channel enhancement-mode MOSFET, $V_T > 0$ and K is a more complicated parameter based on the physical and electrical properties of the MOSFET.

3.2   Chaotic Colpitts Oscillator

The *Colpitts oscillator* was invented and patented (US patent No. 1,624,537) by Edwin H. Colpitts as a simple way to generate sine waves using batteries and vacuum tubes, and it was frequently used in radio frequency communication circuits. Although there have been many changes and improvements in the design over time, the modern version of these oscillators still work on the same principle.

The Colpitts oscillator generates oscillations using an LC circuit with two capacitors C_1 and C_2 and one inductor L. These capacitors control the oscillation frequency by $\omega_0 = 1/\sqrt{LC}$ where $C = C_1 C_2/(C_1 + C_2)$ is the

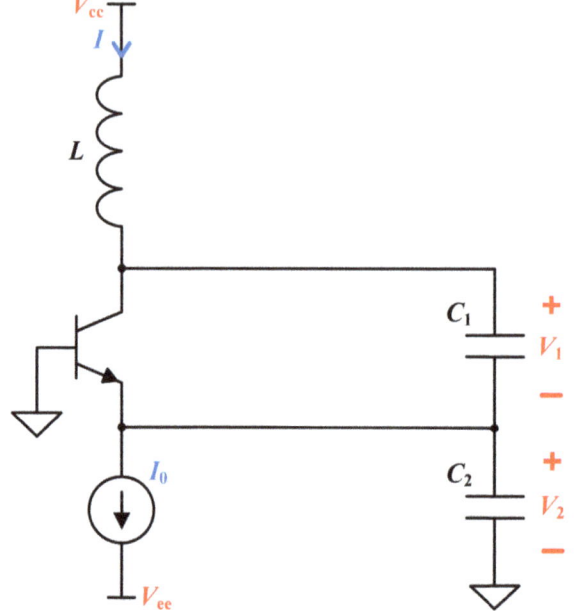

Fig. 3.10 Chaotic Colpitts oscillator.

Table 3.3 Component values for Fig. 3.10.

Component	Numerical	Simulated	Experimental
R_L	1 kΩ	1.1 kΩ	1.1 kΩ (adjust)
R_D	2 kΩ	1 kΩ	1 kΩ
C_1	1 μF	1 μF	1 μF
C_2	1 μF	1 μF	1 μF
L	1 H	1 H	1 H
V_{cc}	5 V	5 V	5 V
V_0	1 V	1 V	1 V
I_0	0.2 mA	0.2 mA	0.2 mA
β	100	-	-
Transistor	Eqs. (3.1), (3.2)	2N3904	2N3904

series combination of C_1 and C_2. A voltage divider is formed by these two capacitors and feeds back a fraction of the output voltage to the input.

You can amplify the signal of an LC circuit by using one of the three types of transistor configurations that allow the transistor to work as an amplifier: the *common emitter, common-base,* and *common-collector* circuits, which are named based on which terminal is grounded ('common' being another term for 'ground'). The particular transistor configuration in Fig 3.10 is a common-base amplifier since the base of the transistor is connected to ground.

In the common-base configuration, the input signal is applied to the emitter of the transistor and the output signal is taken at the collector. The circuit then sustains oscillations by feeding back a fraction of the output to the input through the capacitive divider. There is no phase shift between the input and output in this configuration, and so the feedback is positive.

The transistor must operate in the active mode to provide amplification, and so we use a current source I_0 at the emitter to bias the collector current through $I_C = \alpha I_E$, where α is the *common-base current gain* and is close to 1.0. We choose the value of I_C so that the operating point (or 'Q-point') in Fig. 3.5 is in the active region, making it unnecessary to incorporate the saturation mode into the model.

Although the common-base Colpitts oscillator has had a long history and application, chaos for this circuit was only much later reported by Nguyen (1991) who called it a 'multi oscillation phenomenon.' Only later did Kennedy (1994) correctly characterize the behavior as chaotic.

Several other different transistor-based variants of the Colpitts oscillator have been designed to produce chaos such as the *Clapp oscillator* [Clapp (1948)]. This circuit is similar to the common emitter Colpitts oscillator but with an additional capacitor and has a chaotic version first examined by Srisuchinwong *et al.* (2019). Another example is the Hartley oscillator, previously seen in Fig. 1.3, which uses two inductors for the voltage divider and can also be modified to produce chaos as we will discuss in Section 3.8.

Since the Colpitts oscillator is capable of oscillating at very high frequencies, there have been many investigations of generating high frequency chaotic signals, first explored for the radio frequency range by Wegener and Kennedy (1995). This circuit also inspired one of the few quasi-systematic procedures for designing chaotic oscillators described by Elwakil and Kennedy (1999a), who examined existing sinusoidal oscillators that used two capacitors and made them chaotic by adding an inductor and a junction field effect transistor.

The circuit in Fig. 3.11 is described by the equations

$$\dot{V}_1 = (I - I_C)/C_1$$
$$\dot{V}_2 = (I - I_0 + I_B)/C_2 \quad (3.7)$$
$$\dot{I} = (V_{cc} - IR_L - V_1 - V_2)/L,$$

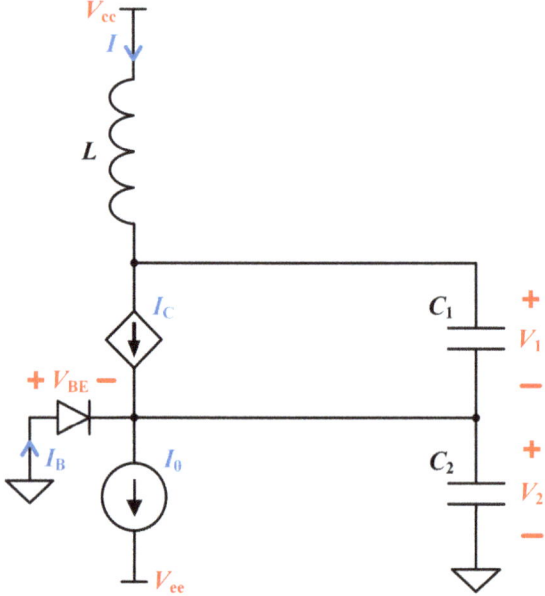

Fig. 3.11 Chaotic Colpitts oscillator with transistor model described by Eq. (3.7).

where the base current from Eq. (3.1) is $I_B = \max(0, -V_2 - V_0)/R_D$ and the collector current from Eq. (3.2) is $I_C = \beta I_B$.

Elegant circuit values that give chaos are $R_L = 1$ kΩ, $R_D = 2$ kΩ, $C_1 = C_2 = 1$ μF, $L = 1$ H, $V_{cc} = 5$ V, $V_0 = 1$ V, $I_0 = 0.2$ mA, and $\beta = 100$. The dimensionless equations are

$$\dot{x} = z - \beta a \max(0, -y - 1)$$
$$\dot{y} = z - b + a \max(0, -y - 1) \qquad (3.8)$$
$$\dot{z} = c - z - x - y.$$

The parameters $a = 0.5, b = 0.2, c = 5$, and $\beta = 100$ give the attractor shown in Fig. 3.12 and a time series as shown in Fig. 3.13. The Lyapunov exponents are $(0.0392, 0, -1.1174)$, the Kaplan–Yorke dimension is $D_{ky} = 2.0351$, and the attractor is a global attractor with a Class 1a basin of attraction. The single equilibrium point at $(5.8056, -1.0040, 0.1980)$ is an unstable saddle focus with eigenvalues $(1.2565 \pm 3.3175i, -4.0129)$.

The experimental phase space plot of V_2 versus V_1 is shown in Fig. 3.14, and the time plot of $V_1(t)$ is shown in Fig. 3.15. Even though no operational amplifier is used in this circuit, you will still need to construct the dual power supply circuit from Fig. 1.21 which will supply a voltage of $V_{cc} = 5$ V and $V_{ee} = -5$ V to operate the current source described in Fig. 2.16.

Component values are given in Table 3.3 where you will also need to add a resistor in series with the inductor since the parasitic resistance is likely less than $R_L = 1$ kΩ. You should be able to observe chaos for $R_L = 1$ kΩ, but there is better agreement with the numerical simulation at $R_L = 1.1$ kΩ. This circuit has two capacitors, an inductor with a parasitic resistance, and a bipolar junction transistor for a total of four components.

3.3 Minati Circuit

Most of the chaotic oscillators described so far have been modifications of existing periodic oscillator circuits such as the Colpitts oscillator or the Wien bridge oscillator. Elwakil and Kennedy (2000a, 2001) postulated that nearly any sinusoidal oscillator can be converted into a chaotic oscillator by adding an inductor or capacitor and a nonlinear component such as a diode.

On the other hand, original chaotic circuits that are not based on existing sinusoidal oscillators are relatively rare. Minati (2013) discovered several new transistor-based chaotic circuits through a random search where the simplest one had five components [Minati (2014)]. Minati et al. (2017) then searched again with additional component restrictions that excluded those in Minati (2013) and Minati (2014) and discovered several new circuits with five or six components which are listed in the supplementary of Minati et al. (2017). The most elegant of these circuits is shown in Fig. 3.16.

Since this circuit was found by random search and is not based on any existing oscillator, further investigation is needed to understand how chaos arises in the circuit and its variants. It may also be possible to further simplify this circuit by using the parasitic capacitance between the base and emitter as was done by Tekam et al. (2019).

There are several other factors that make Minati's circuit unusual. For example, his circuit can generate different attractors with different parameters including one similar to the Rössler (1976) attractor. We explore a few other circuits from Minati et al. (2017) with additional components and unique behavior in the following sections.

The equivalent circuit in Fig. 3.17 is described by the equations

$$\begin{aligned}
\dot{V} &= [(V_S - V)/R - I_1 - I_2]/C \\
\dot{V}_{CE} &= (I_1 - I_C)/C_{CE} \\
\dot{I}_1 &= (V - V_{CE})/L_1 \\
\dot{I}_2 &= (V - V_{BE})/L_2,
\end{aligned} \qquad (3.9)$$

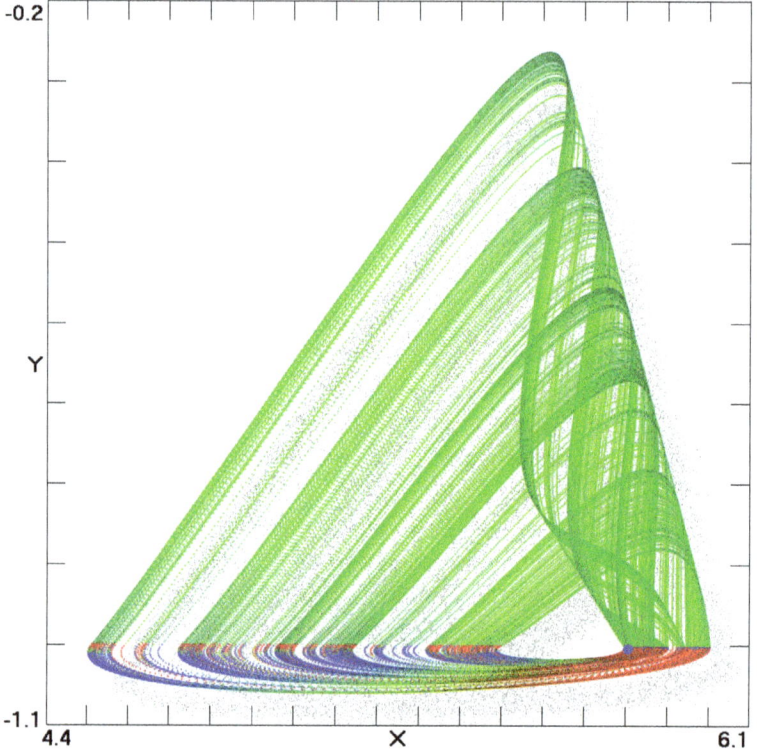

Fig. 3.12 Numerical solution of Eq. (3.8) with $a = 0.5, b = 0.2, c = 5$, and $\beta = 100$.

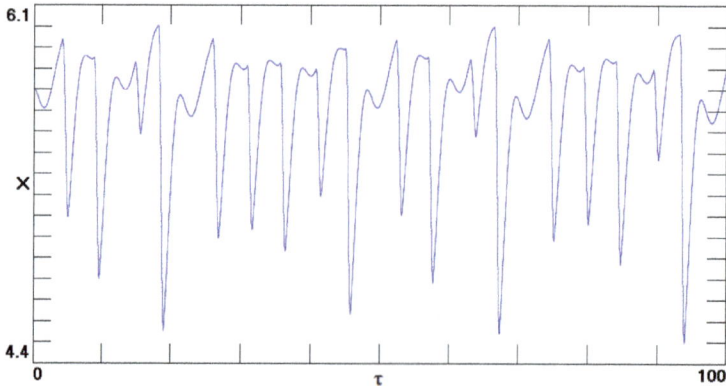

Fig. 3.13 Numerical waveform for $x(\tau)$ from Eq. (3.8) with $a = 0.5, b = 0.2, c = 5$, and $\beta = 100$.

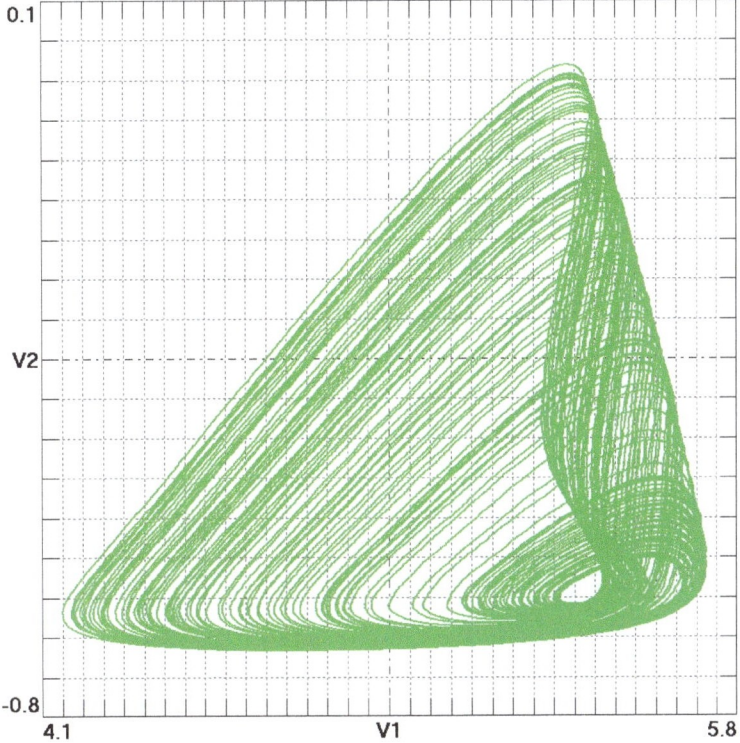

Fig. 3.14 Oscilloscope phase space plot of chaotic Colpitts oscillator from Fig. 3.10.

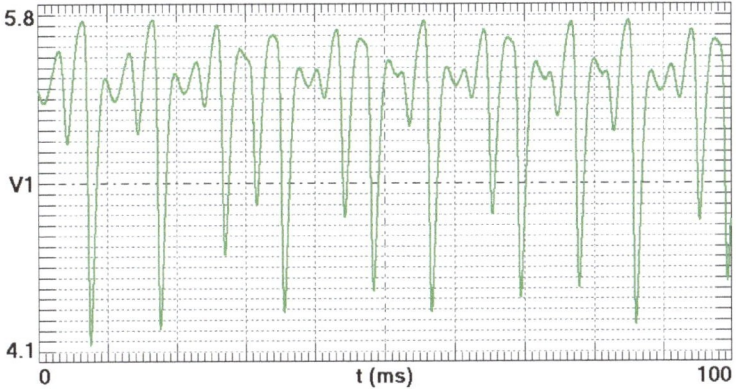

Fig. 3.15 Oscilloscope time plot of chaotic Colpitts oscillator from Fig. 3.10.

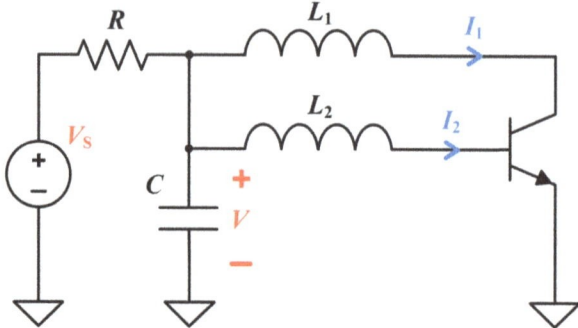

Fig. 3.16 Minati circuit.

Table 3.4 Component values for Fig. 3.16.

Component	Numerical	Simulated	Experimental
R	2.28 kΩ	1 kΩ	2 kΩ (adjust)
R_D	0.2 kΩ	-	-
C	1 μF	1 μF	1 μF
C_{CE}	1 μF	-	-
L_1	1 H	1 H	1 H
L_2	1 H	1 H	1 H
V_S	10 V	10 V	10.2 V (adjust)
V_0	1 V	-	-
β	100	-	-
Transistor	Eq. (3.3)	2N3904	2N3904

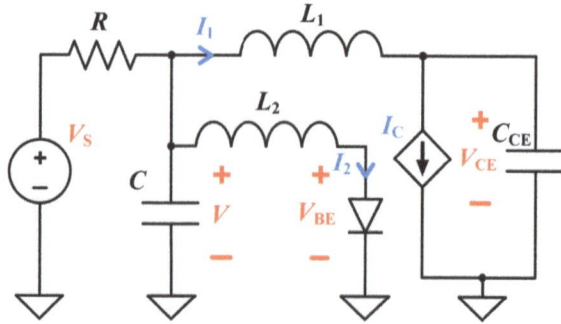

Fig. 3.17 Minati circuit with transistor model described by Eq. (3.9).

where C_{CE} is the parasitic capacitance across the collector and emitter, and the collector current is $I_C = \min(\beta I_2, V_{CE}/R_D)$ from Eq. (3.3). Since the main nonlinearity comes from this collector current, V_{BE} can be modeled as a constant voltage drop given by $V_{BE} = V_0$.

Elegant circuit values that give chaos are $R = 2.28$ kΩ, $R_D = 0.2$ kΩ, $C = C_{CE} = 1$ μF, $L_1 = L_2 = 1$ H, $V_S = 10$ V, $V_0 = 1$ V, and $\beta = 100$. The dimensionless equations are

$$\begin{aligned} \dot{x} &= (a-x)/b - z - u \\ \dot{y} &= x - \min(\beta u, cy) \\ \dot{z} &= x - y \\ \dot{u} &= x - 1. \end{aligned} \qquad (3.10)$$

The parameters $a = 10, b = 2.28, c = 5$, and $\beta = 100$ give the attractor shown in Fig. 3.18 and a time series as shown in Fig. 3.19. The Lyapunov exponents are $(0.0610, 0, -0.4925, -4.4873)$, the Kaplan–Yorke dimension is $D_{ky} = 2.1239$, and the attractor has a global Class 1a basin of attraction. The single equilibrium point at $(1, 1, 3.9083, 0.0391)$ is an unstable saddle focus with eigenvalues $(1.9622 \pm 2.3940i, -2.1815 \pm 2.4045i)$.

A property of this circuit as well as the other circuits found by Minati's method is that R and V_S are more important in obtaining chaos than the other parameters in Table 3.4. A general procedure is to keep V_S at a low value and carefully adjust R with a potentiometer until a limit cycle is observed. Avoid turning this resistance too low (less than about 100 Ω), since the high current could damage the transistor. Once you find a limit cycle, you can adjust V_S to obtain chaos.

The circuit is sensitive to a change in parameters (a 17% change in R is likely to destroy the chaos), and we only found chaos for a specific value of $R = 2$ kΩ and $V_S = 10.2$ V. However, your R could be different since the circuit relies on parasitic effects in the transistor. We also found that adjusting V_S or R beyond these values caused the circuit to saturate and required restarting the power supply.

The experimental phase space plot of I_1 versus V is shown in Fig. 3.20, and the time plot of $V_1(t)$ is shown in Fig. 3.21. Agreement between the numerical, simulated, and experimental results is poor since C_{CE} is a parasitic capacitance that is not constant. Additionally, there could be other experimental parasitics that contribute to different values of C_{CE}, such as the arrangement of the components on the breadboard. It is also likely that parasitics in the transistor other than C_{CE} that are not included in the model are contributing to the poor agreement.

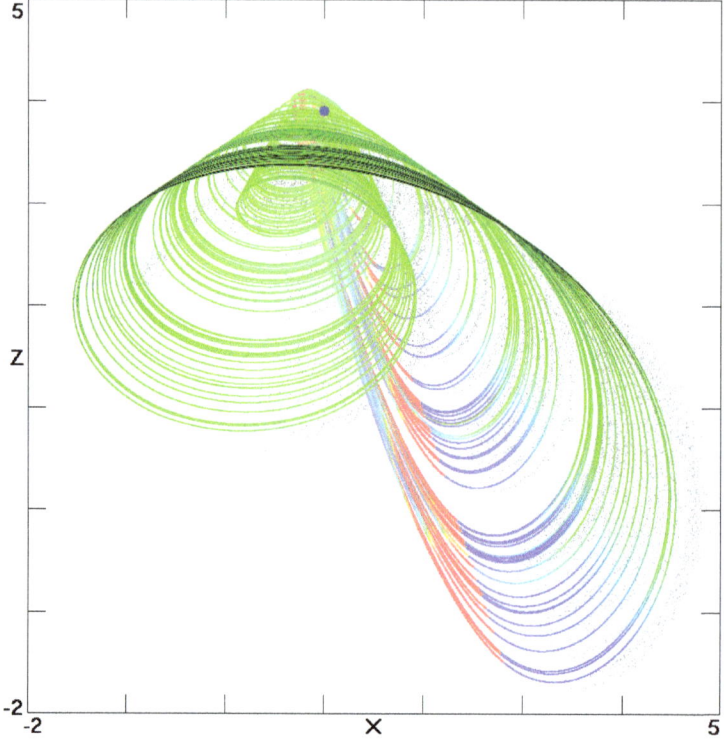

Fig. 3.18 Numerical solution of Eq. (3.10) with $a = 10$, $b = 2.28$, $c = 5$, and $\beta = 100$.

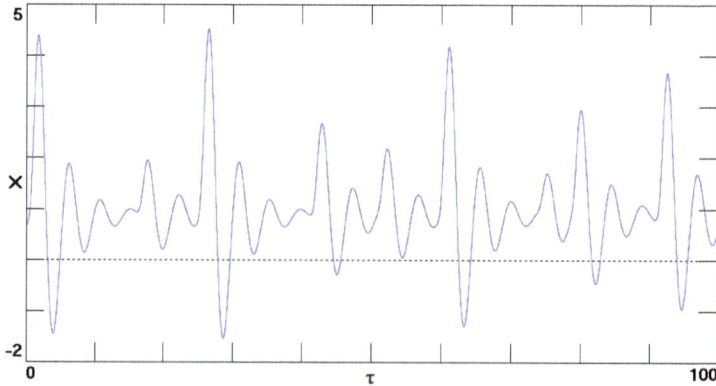

Fig. 3.19 Numerical waveform for $x(\tau)$ from Eq. (3.10) with $a = 10$, $b = 2.28$, $c = 5$, and $\beta = 100$.

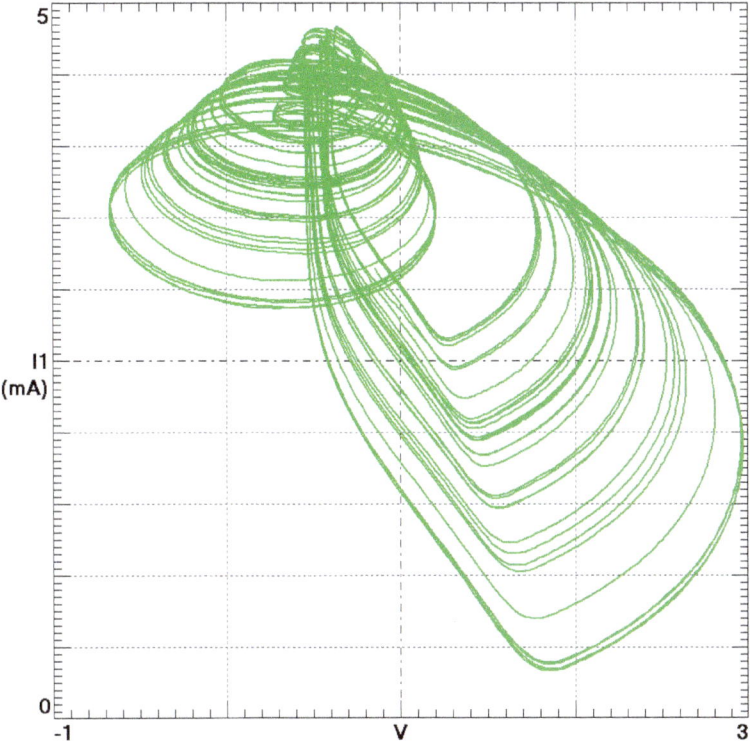

Fig. 3.20 Oscilloscope phase space plot of Minati circuit from Fig. 3.16.

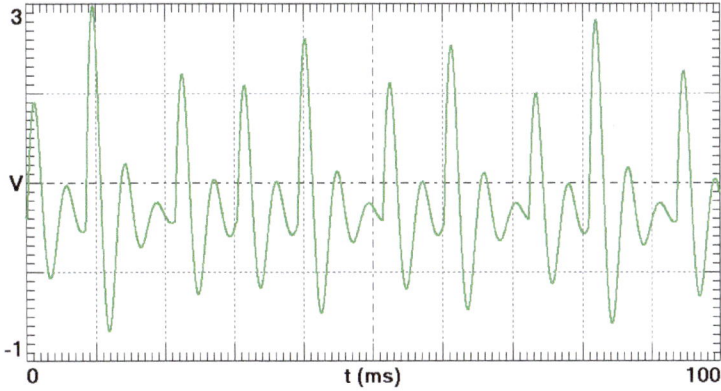

Fig. 3.21 Oscilloscope time plot of Minati circuit from Fig. 3.16.

Since C_{CE} is a small value, it is difficult to observe V_{CE}. Instead, we chose to measure I_1 by adding a 50 Ω resistor in series with L_1 with one end at V. You can then calculate the current from $I_1 = (V - V_A)/50$, where V_A is the voltage at the other node of this resistor. This current is very noisy because of its small amplitude, so we smoothed it in the phase space plot. The circuit has one resistor, one capacitor, two inductors, and a bipolar junction transistor for a total of five components.

3.4 Minati–Frasca Double-scroll Circuit

Minati et al. (2017) found several other simple transistor circuits with some unique behaviors that are worth mentioning. One of these circuits is the most elegant transistor circuit that can produce a *multiscroll* attractor and is shown in Fig. 3.22 and in reduced form in Fig. 3.23. Because this circuit was discovered through a random search, further investigation is needed to understand how the chaos arises. Note that the top right transistor is reversed from collector to emitter compared to the original circuit from Minati et al. (2017).

There have been only a few transistor circuits that can produce multiscrolls, one example being the transistor version of Chua's circuit from Matsumoto et al. (1986) that can produce two scrolls. Another transistor chaotic circuit by Keuninckx et al. (2014) is based on a phase-shift oscillator that consists of an RC network with two transistors. However, the circuit in Fig. 3.22 uses fewer components than both of these circuits and is not based on any existing oscillator.

The equivalent circuit in Fig. 3.23 is described by the equations

$$\begin{aligned}
\dot{V} &= I_1/C \\
\dot{V}_{CE1} &= (I_2 - I_{C1} - I_{B2})/C_{CE1} \\
\dot{V}_{CE2} &= [(V_S - V_{CE2})/R - I_1 - I_2 - I_{C2} - I_{B1}]/C_{CE2} \\
\dot{I}_1 &= (V_{CE2} - V)/L_1 \\
\dot{I}_2 &= (V_{CE2} - V_{CE1})/L_2,
\end{aligned} \quad (3.11)$$

where

$$\begin{aligned}
I_{C1} &= \min(\beta I_{B1}, V_{CE1}/R_D), \\
I_{C2} &= \min(\beta I_{B2}, V_{CE2}/R_D), \\
I_{B1} &= \max(0, V_{CE2} - V_0)/R_D, \text{ and} \\
I_{B2} &= \max(0, V_{CE1} - V_0)/R_D,
\end{aligned}$$

Transistor Circuits

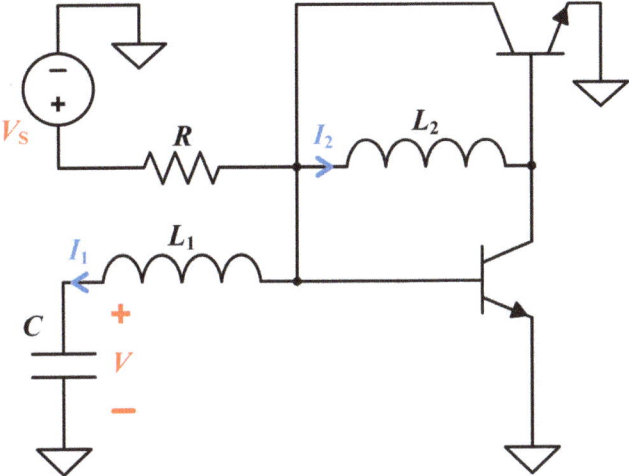

Fig. 3.22 Minati–Frasca double-scroll circuit.

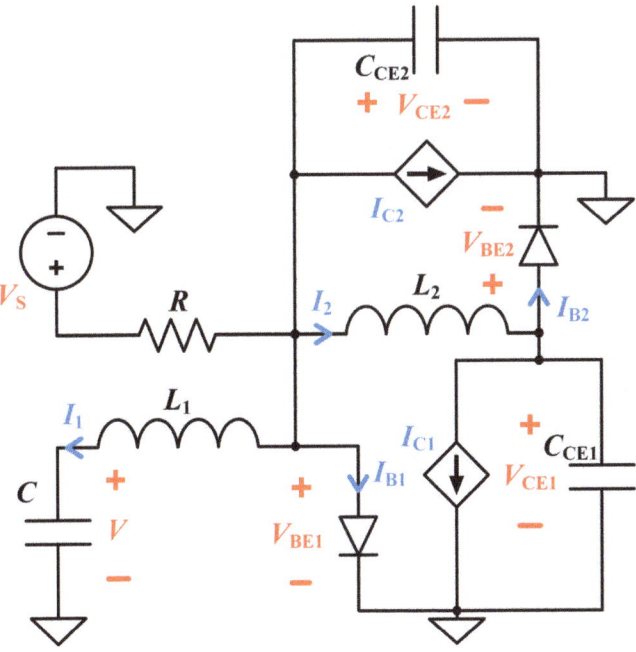

Fig. 3.23 Minati–Frasca double-scroll circuit with transistor model described by Eq. (3.11).

Table 3.5 Component values for Fig. 3.22.

Component	Numerical	Simulated	Experimental
R	1 kΩ	679 Ω	775 Ω (adjust)
R_D	0.02 kΩ	-	-
C	1 μF	0.1 μF	0.1 μF
C_{CE1}	0.01 μF	-	-
C_{CE2}	0.01 μF	-	-
L_1	1 H	0.1 H	0.1 H
L_2	1 H	0.1 H	0.1 H
V_S	5.2 V	5 V	5 V
V_0	1 V	-	-
β	100	-	-
Transistor	Eqs. (3.1), (3.3)	2N3904	2N3904

where I_{B1} and I_{B2} are from Eq. (3.1) and I_{C1} and I_{C2} are from Eq. (3.3). Chaos will occur even if saturation is omitted from the model, but the resulting attractor will not be a double scroll.

Elegant circuit values that give chaos are $R = 1$ kΩ, $R_D = 0.02$ kΩ, $C = 1$ μF, $C_{CE1} = C_{CE2} = 0.01$ μF, $L_1 = L_2 = 1$ H, $V_S = 5.2$ V, $V_0 = 1$ V, and $\beta = 100$. The dimensionless equations are

$$\dot{x} = u$$
$$\dot{y} = a[v - \min(\beta c \max(0, z - 1), cy) - c\max(0, y - 1)]$$
$$\dot{z} = a[b - z - u - v - \min(\beta c \max(0, y - 1), cz) - c\max(0, z - 1)] \quad (3.12)$$
$$\dot{u} = z - x$$
$$\dot{v} = z - y.$$

The capacitors C_{CE1} and C_{CE2} are the small parasitic capacitances of the transistors that control the rate at which the dynamics switch from one scroll to the other, and thus their value is not critical. It should be possible to construct a three-dimensional discontinuous but less realistic model with $C_{CE1} = C_{CE2} = 0$ that switches instantaneously.

The parameters $a = 100, b = 5.2, c = 50$, and $\beta = 100$ give the attractor shown in Fig. 3.24 and a time series as shown in Fig. 3.25. The time-lagged x for a (small) lag τ_0 is estimated from $w(\tau) = x(\tau - \tau_0) \approx x(\tau) - \tau_0 \dot{x}(\tau) + \tau_0^2 \ddot{x}(\tau)/2 = x(\tau)/2 + z(\tau)/2 - u(\tau)$ for $\tau_0 = 1$, which avoids having to save past values of $x(\tau)$.

The chaos exists in a narrow range of the parameters surrounded by regions with a long-period limit cycle. The Lyapunov exponents are

(0.0223, 0, −0.0427, −241.7200, −276.8807), the Kaplan–Yorke dimension is $D_{ky} = 2.5217$, and the attractor has a global Class 1a basin of attraction. The single equilibrium point near (1, 1, 1, 0, 2.1) is an unstable saddle focus with eigenvalues (494950, 0.0004, $\pm i$, −505050).

Agreement between the numerical, simulated, and experimental results is poor since the circuit relies on several parasitic properties of the components including C_{CE1} and C_{CE2}. We had to add a series parasitic resistance and parallel parasitic capacitance in the inductors of the simulated circuit to obtain chaos, which were arbitrarily chosen as $R_{L1} = R_{L2} = 5\ \Omega$ and $C_{L1} = C_{L2} = 10$ pF.

We also found the simulated circuit to be more sensitive to high parasitic resistances in the inductor, which made it difficult to use large inductors for lower frequencies. Thus the parameters of the experimental circuit shown in Table 3.5 were chosen to operate at approximately ten times the frequency of the model.

Despite these differences, you are likely to have an easier time building this circuit since the physical components have parasitics that are helpful for obtaining chaos. The chaos in the circuit will arise for different values of R, one of which we found to be around 775 Ω. Since the circuit is very sensitive to this value of R, it may be easier for you to obtain chaos by using a 1 kΩ potentiometer. The value of R you require is likely to be different due to varying parasitic effects in the circuits.

Other resistance values that we found to give chaos in our circuit were around 1.6 kΩ and 3.4 kΩ, and it is probably easier for you to obtain chaos around those values. Avoid turning the resistance to a low value (less than about 100 Ω), since the resulting high current could damage the transistors.

It is easiest to observe the double scroll by plotting a delayed version of V_1 versus V_1 as shown in Fig. 3.26, which we did in both the simulation and experiment by delaying V_1 by 0.15 ms. You could alternatively construct an additional external circuit to obtain a delayed V_1 or directly manipulate the collected digital data. The time plot of $V_1(t)$ is shown in Fig. 3.27. This circuit has a resistor, a capacitor, two inductors, and two bipolar junction transistors for a total of six components.

3.5 Minati–Frasca Spiking Circuit

Neurons are a basic building block of the brain that send signals using electrical impulses called *action potentials*. These action potentials occur only when the neuron reaches a threshold that results in a spike of a nearly fixed

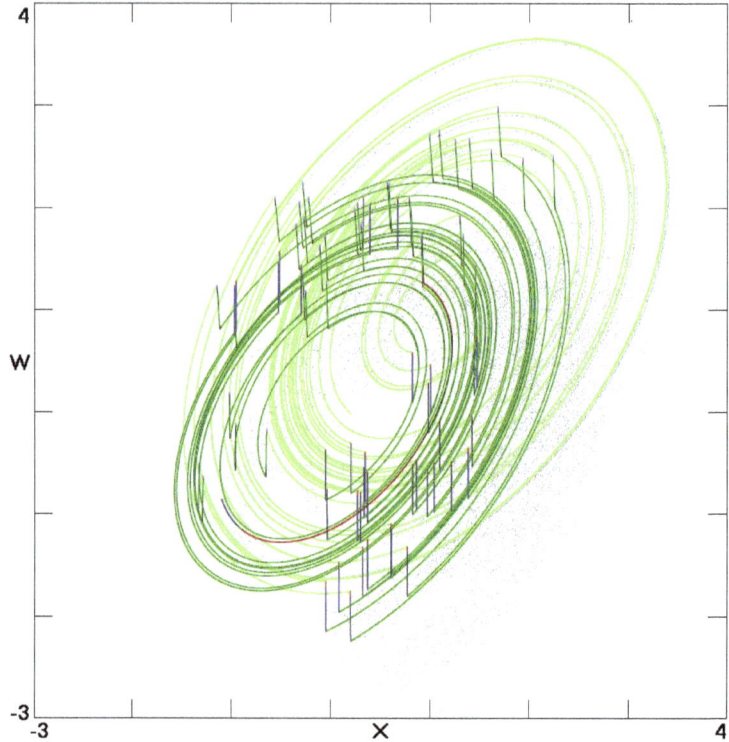

Fig. 3.24 Numerical solution of Eq. (3.12) with $a = 100$, $b = 5.2$, $c = 50$, and $\beta = 100$ (time lag = 1.0).

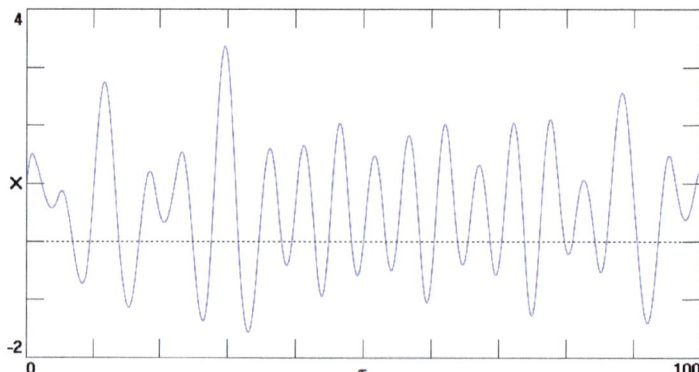

Fig. 3.25 Numerical waveform for $x(\tau)$ from Eq. (3.12) with $a = 100$, $b = 5.2$, $c = 50$, and $\beta = 100$.

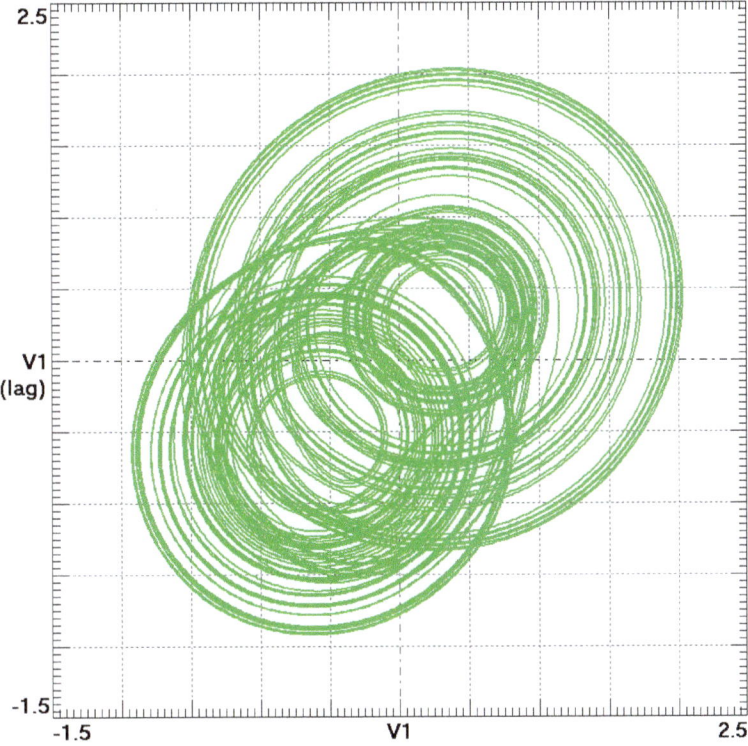

Fig. 3.26 Oscilloscope phase space plot of Minati–Frasca double-scroll circuit from Fig. 3.22 (time lag = 0.15 ms).

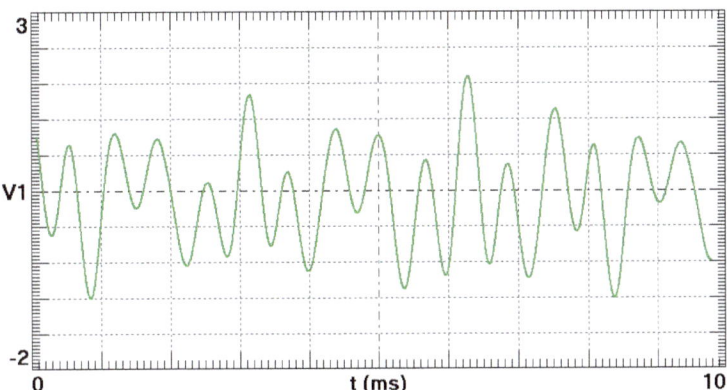

Fig. 3.27 Oscilloscope time plot of Minati–Frasca double-scroll circuit from Fig. 3.22.

size. The important dynamics that occur in the brain come from the intervals between these spikes since individual spikes are similar to one another. One of the dynamics that many neurons exhibit is *bursting oscillations*, where a group of action potentials fire in rapid succession followed by a resting period of no action potentials. Bursting is a fundamental property of neural signaling, and most neurons can have bursting behavior. Some advantages of using bursts is that they are often more reliable than single spikes and have more information content [Izhikevich (2003)].

Bursting has been studied for decades, and there have been several models such as the Hindmarsh and Rose (1984) model to understand this behavior. Complex circuits have also been developed to generate bursting firing patterns [Wijekoon and Dudek (2008)], but those circuits use many transistors, which is not ideal for our purposes since such circuits are meant to be used in large networks containing thousands of neurons.

The search by Minati et al. (2017) yielded the simple chaotic bursting circuit shown in Fig. 3.28 and in equivalent circuit form in Fig. 3.29 where the bursting behavior can be observed by probing V_{CE}. Note that the lower-right transistor is reversed from collector to emitter compared to the original circuit. However, this circuit does not oscillate close to *criticality*, which is an operating regime at the interface between order and disorder in which several brain functions could be optimized, an important feature in bursting oscillations. The elegance and potential applications of this circuit make it ripe for further studies.

The equations that describe the circuit in Fig. 3.29 are

$$\dot{V} = (I_1 + I_{C2} - I_3)/C$$
$$\dot{V}_{BE1} = (I_3 - I_{C2} - I_{B1})/C_{BE1}$$
$$\dot{V}_{BE2} = (I_1 - I_2 - I_{B2})/C_{BE2}$$
$$\dot{V}_{CE} = [(V_S - V_{CE})/R - I_{C1} - I_1]/C_{CE} \quad (3.13)$$
$$\dot{I}_1 = (V_{CE} - V - V_{BE2})/L_1$$
$$\dot{I}_2 = V_{BE2}/L_2$$
$$\dot{I}_3 = (V - V_{BE1})/L_3,$$

where

$$I_{B1} = \max(0, V_{BE1} - V_0)/R_D,$$
$$I_{B2} = \max(0, V_{BE2} - V_0)/R_D,$$
$$I_{C1} = \min(\beta I_{B1}, V_{CE}/R_D), \text{ and}$$
$$I_{C2} = \min(\beta I_{B2}, (V_{BE1} - V)/R_D),$$

where I_{B1} and I_{B2} are from Eq. (3.1), and I_{C1} and I_{C2} are from Eq. (3.3).

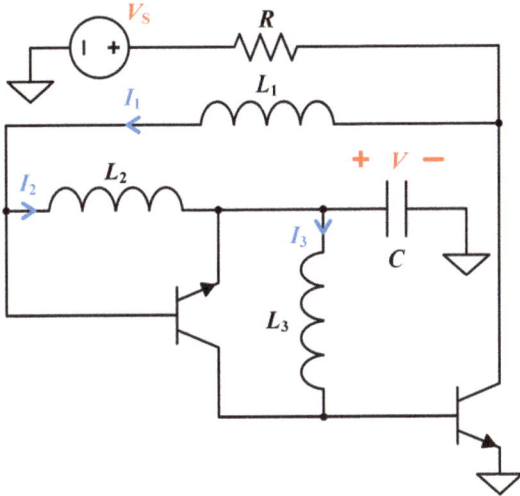

Fig. 3.28 Minati–Frasca spiking circuit.

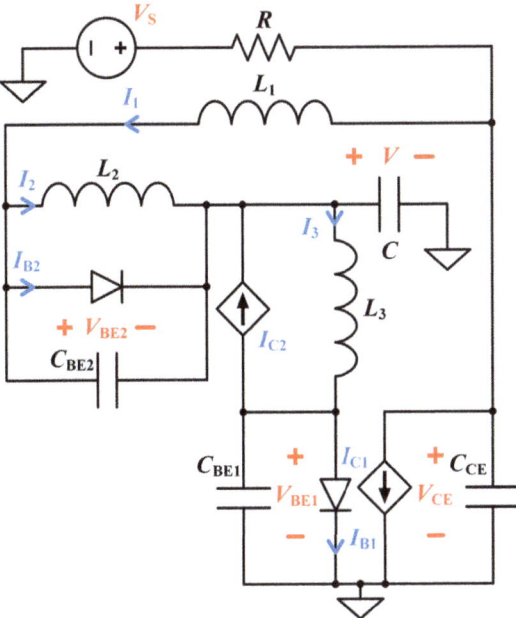

Fig. 3.29 Minati–Frasca spiking circuit with transistor model described by Eq. (3.13).

Table 3.6 Component values for Fig. 3.28.

Component	Numerical	Simulated	Experimental
R	4 kΩ	3 kΩ	3 kΩ (adjust)
R_D	0.1 kΩ	-	-
C	1 μF	10 μF	10 μF
C_{BE1}	0.1 μF	-	-
C_{BE2}	0.1 μF	-	-
C_{CE}	0.1 μF	-	-
L_1	1 H	0.1 H	0.1 H
L_2	1 H	0.1 H	0.1 H
L_3	1 H	1 H	1 H
V_S	20 V	20 V	20 V
V_0	1 V	-	-
β	100	-	-
Transistor	Eqs. (3.1), (3.3)	2N3904	2N3904

The capacitors C_{BE1}, C_{BE2}, and C_{CE} are the small parasitic capacitances of the transistors that control the narrow width of the bursting spikes, and thus their values are not critical. They are taken as 0.1 μF, but any small values should suffice except that they slow the computation if they are taken too small, and they introduce additional structure in the numerical plots that are absent in the experiment.

Elegant circuit values that give chaos are $R = 4$ kΩ, $R_D = 0.1$ kΩ, $C = 1$ μF, $C_{BE1} = C_{BE2} = C_{CE} = 0.1$ μF, $L_1 = L_2 = L_3 = 1$ H, $V_S = 20$ V, $V_0 = 1$ V, and $\beta = 100$. The dimensionless equations are

$$\begin{aligned}
\dot{x} &= v + \iota_4 - s \\
\dot{y} &= c(s - \iota_4 - \iota_1) \\
\dot{z} &= c(v - w - \iota_2) \\
\dot{u} &= c[a(d - u) - \iota_3 - v] \\
\dot{v} &= u - x - z \\
\dot{w} &= z \\
\dot{s} &= x - y,
\end{aligned} \qquad (3.14)$$

where

$$\begin{aligned}
\iota_1 &= b \max(0, y - 1), \\
\iota_2 &= b \max(0, z - 1), \\
\iota_3 &= \min(\beta \iota_1, bu), \text{ and} \\
\iota_4 &= \min[\beta \iota_2, b(y - z)].
\end{aligned}$$

The parameters $a = 0.25, b = c = 10, d = 20$, and $\beta = 100$ give the attractor shown in Fig. 3.30 and a time series as shown in Fig. 3.31. The Lyapunov exponents are (0.2652, 0, −0.6137, −2.0313, −19.470, −81.284, −167.31), the Kaplan–Yorke dimension is $D_{ky} = 2.4321$, and the attractor has a global Class 1a basin of attraction. The single equilibrium at (10.0470, 10.0470, 0, 10.0470, 0.4702, 0.4702, 0.4702) is an unstable saddle focus with eigenvalues (5.6162 ± 12.9378i, 0.0032 ± 3.1353i, −0.2000, −17.0536, −151.4851).

We opted to show I_2 versus V in the phase space plot in Fig. 3.32 because you can obtain both variables using a 50 Ω resistor in series with L_2 with one end at V to measure I_2. The time plot of $V_{CE}(t)$ in Fig. 3.33 shows a mix of bursting and regular spiking behavior, but you might find other parameters than the ones in Table 3.6 where the bursting is more pronounced.

As with the previous circuits by Minati et al. (2017) that we explored, the agreement between the numerical and experimental plots is poor since the circuit relies on several parasitic properties of the transistors. Our model exhibits the spiking and bursting behavior of this circuit, but the phase space plots are rather different and may require a more complicated but less elegant model for better agreement.

We experimentally observed the spiking and bursting behavior in Fig. 3.33 to occur around $R = 3$ kΩ. You may require a different value due to differing parasitics, but take care when adjusting this resistor since values below about 100 Ω produce a large current that can damage the transistors. The circuit has one resistor, one capacitor, three inductors, and two bipolar junction transistors for a total of seven components.

3.6 Chaotic BJT Switch

Transistors are commonly used as switches to control the power to other devices. One example is the circuit in Fig. 3.34 where a bipolar junction transistor controls the current through an inductor which could be a motor or part of a relay controlling a high-voltage circuit. The bipolar junction transistor itself is controlled by a square wave signal V_S where if the amplitude is zero, current is blocked from flowing through L since the transistor is turned OFF in cutoff mode. If the square wave voltage V_S is then sufficiently high to put the bipolar junction transistor into saturation mode, the bipolar junction transistor turns ON and allows the inductor current I to flow from collector to emitter.

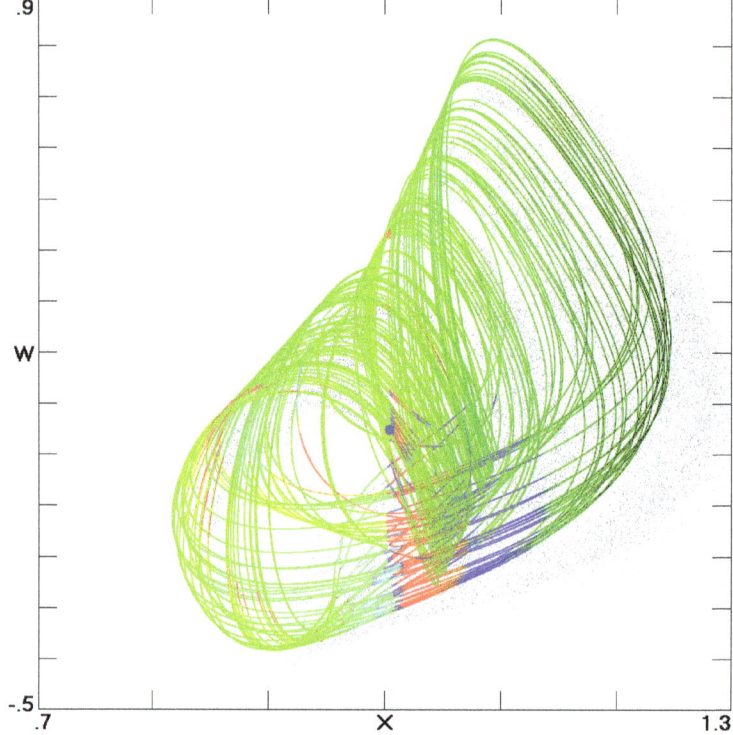

Fig. 3.30 Numerical solution of Eq. (3.14) with $a = 0.25$, $b = c = 10$, $d = 20$, and $\beta = 100$.

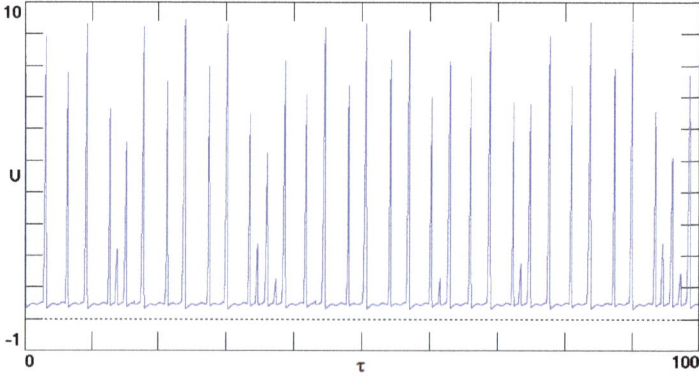

Fig. 3.31 Numerical waveform for $u(\tau)$ from Eq. (3.14) with $a = 0.25$, $b = c = 10$, $d = 20$, and $\beta = 100$.

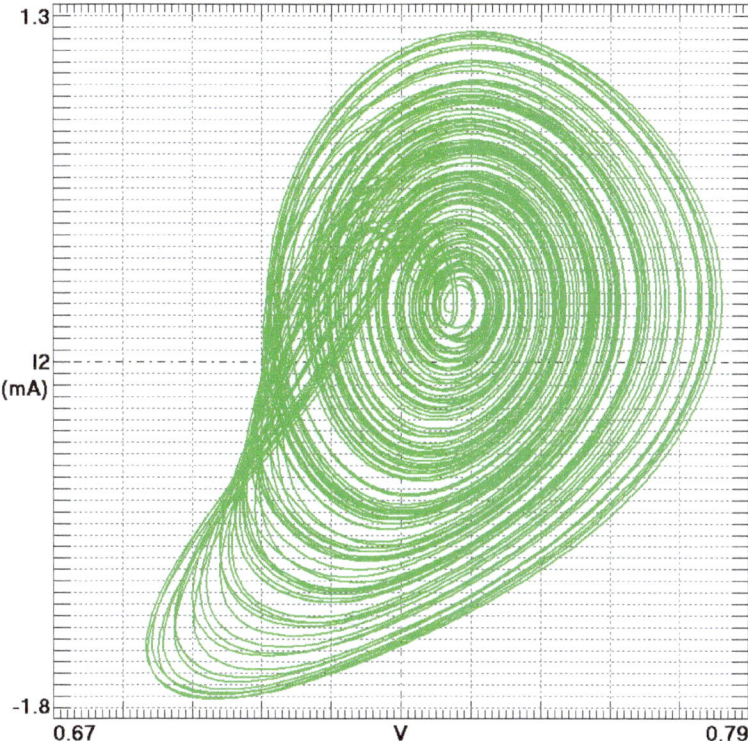

Fig. 3.32   Oscilloscope phase space plot of Minati–Frasca spiking circuit from Fig. 3.28.

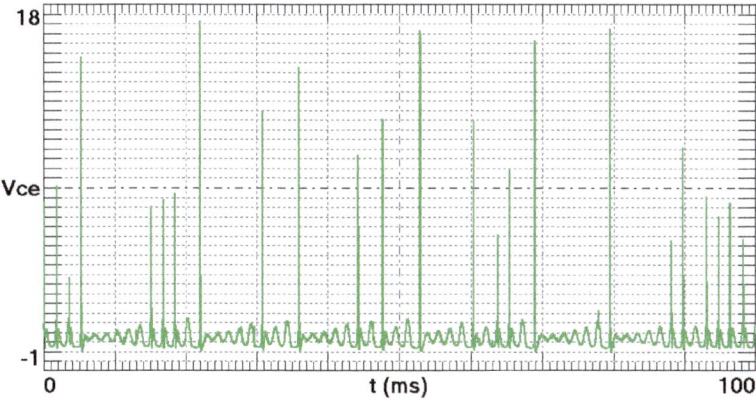

Fig. 3.33   Oscilloscope time plot of Minati–Frasca spiking circuit from Fig. 3.28.

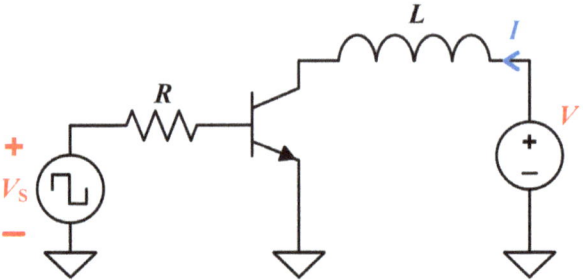

Fig. 3.34 Chaotic BJT switch.

Table 3.7 Component values for Fig. 3.34.

Component	Numerical	Simulated	Experimental
R	0.1 kΩ	5 kΩ	3 kΩ (adjust)
R_D	1 kΩ	-	-
C_{BE}	4 µF	-	-
C_0	0.5 µF	-	-
L	1 H	0.1 H	0.1 H
V	10 V	20 V	20 V
V_0	1 V	-	-
V_{S0}	10 V	5 V	2 V (adjust)
V_t	1 V	-	-
ω	1 rad/ms	π rad/ms	8.6π rad/ms (adjust)
α_F	0.99	-	-
α_R	0.75	-	-
D	0.5	0.009	0.31 (adjust)
Transistor	Eqs. (2.5), (2.6), (3.1), (3.4)	2N3055	MJE13007

One problem of using bipolar junction transistors as a switch is that there is an undesirable delay when it switches between cutoff and saturation mode because of the reverse recovery time. This effect is identical to the effect in the circuit from Section 2.2 where chaos occurred because the diode does not immediately rectify after switching due to the presence of minority carriers. Likewise, in the bipolar junction transistor, a large concentration of minority carriers will keep the device saturated even after switching until they eventually decrease due to recombination. A delay also occurs when the bipolar junction transistor goes from cutoff to saturation because it takes time for the emitter-base junction to charge and become forward biased.

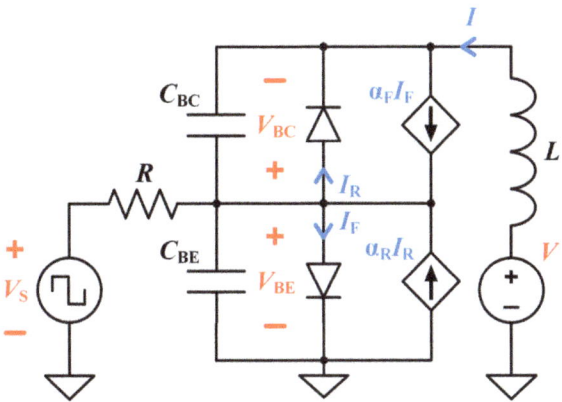

Fig. 3.35   Chaotic BJT switch with transistor model described by Eq. (3.15).

Chaos in this circuit was first observed by Ivanov (1991) and then briefly discussed by Hamill (1993). It appears to have received relatively less attention than other chaotic circuits possibly due to advances in the speed of bipolar junction transistors. Probably the best way to model this circuit is through the *charge control* model, used by Karadzinov et al. (1996) for this circuit, since it incorporates the reverse recovery time of the minority carriers, but this model is more advanced.

Instead, we use the Ebers–Moll model from Fig. 3.6 since it incorporates the cutoff and saturation mode. The forward-active region is also used because the bipolar junction transistor enters this mode as the concentration of minority carriers decreases. We modeled the delay using parasitic capacitances across the PN junctions in the Ebers–Moll model as shown in Fig. 3.35 and found that the capacitor across the collector-base junction must be nonlinear in order to obtain chaos.

The equations that describe the circuit in Fig. 3.35 are

$$\dot{V}_{BE} = (I_B - \alpha_F I_F + \alpha_R I_R + I)/C_{BE}$$
$$\dot{V}_{BC} = (-I - I_R + \alpha_F I_F)/C_{BC} \quad (3.15)$$
$$\dot{I} = (V - V_{BE} + V_{BC})/L,$$

where $I_B = (V_S - V_{BE})/R$ from Eq. (3.1) while $I_F = \max(0, V_{BE} - V_0)/R_D$ and $I_R = \max(0, V_{BC} - V_0)/R_D$ from Eq. (3.7).

The voltage source V_S is a square wave with frequency ω, a maximum value of $V_S = V_{S0}$, and a minimum value of $V_S = 0$ with a duty cycle D (the fraction of the cycle with $V_S > 0$). The factors α_F and α_R satisfy

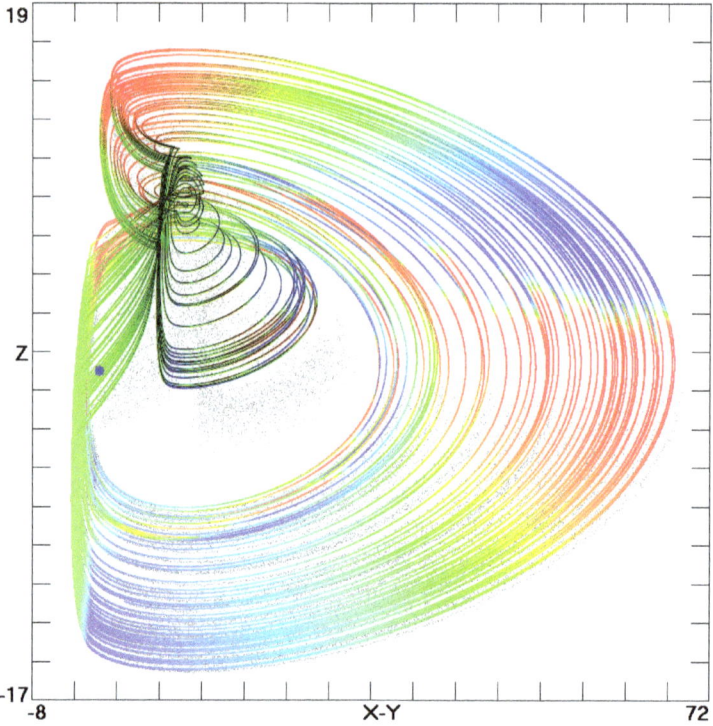

Fig. 3.36 Numerical solution of Eq. (3.16) with $a = 0.25$, $b = 10$, $c = 0.5$, $\alpha_F = 0.99$, and $\alpha_R = 0.75$.

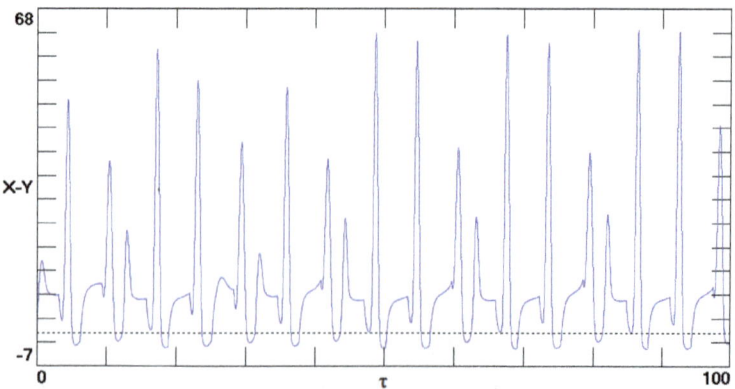

Fig. 3.37 Numerical waveform for $x - y$ versus τ from Eq. (3.16) with $a = 0.25$, $b = 10$, $c = 0.5$, $\alpha_F = 0.99$, and $\alpha_R = 0.75$.

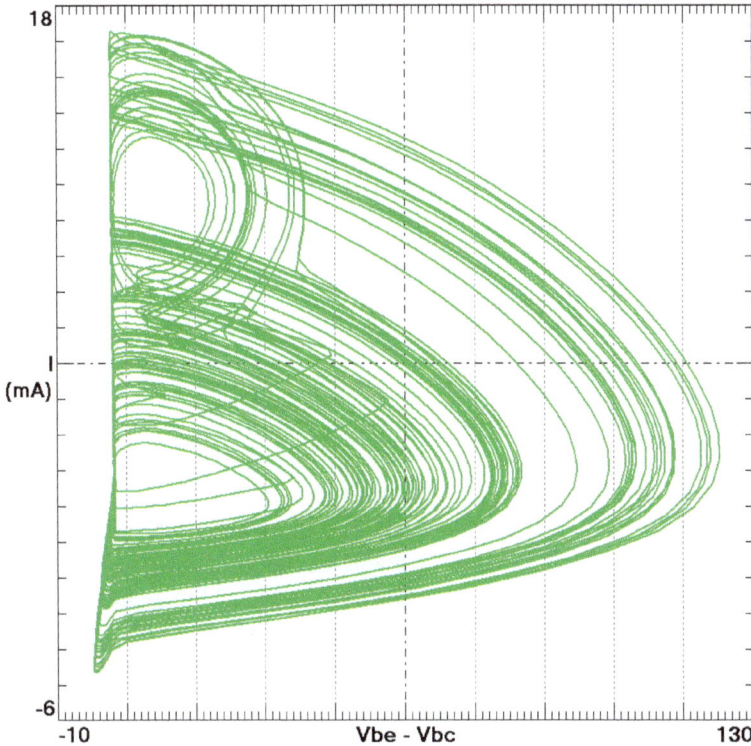

Fig. 3.38 Oscilloscope phase space plot of chaotic BJT switch from Fig. 3.34.

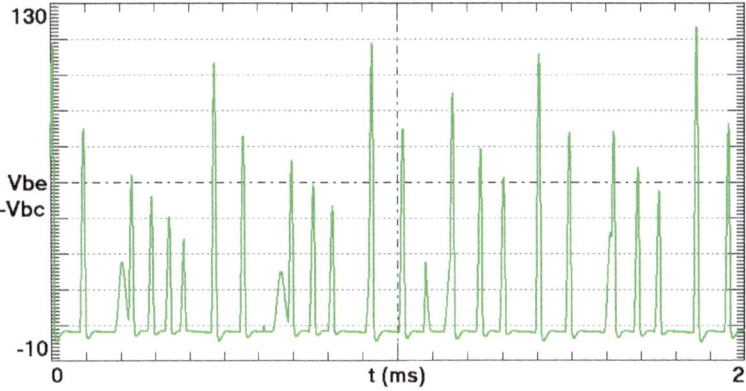

Fig. 3.39 Oscilloscope time plot of chaotic BJT switch from Fig. 3.34.

$0 < \alpha_R < \alpha_F < 1$. The base-collector capacitance C_{BC} is the diode capacitance given by $C_{BC} = C_{j0}/\sqrt{1 - V_{BC}/V_t}$ from Eq. (2.5) for $V_{BC} \leq 0$ and by $C_{BC} = C_{s0}e^{V_{BC}/V_t}$ from Eq. (2.6) for $V_{BC} > 0$ with $C_{j0} = C_{s0} = C_0$. The capacitance C_{BE} is assumed constant.

Elegant circuit values that give chaos are $R = 0.1$ kΩ, $R_D = 1$ kΩ, $C_{BE} = 4$ μF, $C_0 = 0.5$ μF, $L = 1$ H, $V = 10$ V, $V_0 = 1$ V, $V_{S0} = 10$ V, $V_t = 1$ V, $\omega = 1$ rad/ms, $\alpha_F = 0.99$, $\alpha_R = 0.75$, and $D = 0.5$. The dimensionless equations are

$$\dot{x} = a(\beta - \alpha_F \gamma + \alpha_R \delta + z)$$
$$\dot{y} = (-z - \delta + \alpha_F \gamma)/c \max[(1-y)^{-1/2}, e^y] \qquad (3.16)$$
$$\dot{z} = b - x + y,$$

where $\beta = b\,\mathrm{sgn}(\sin \tau, 0) - 1$, $\gamma = \max(0, x-1)$, and $\delta = \max(0, y-1)$.

The parameters $a = 0.25, b = 10, c = 0.5, \alpha_F = 0.99$, and $\alpha_R = 0.75$ give the attractor shown in Fig. 3.36 and a time series as shown in Fig. 3.37. The Lyapunov exponents are $(0.0735, 0, -0.3325, -2.5689)$ and the Kaplan–Yorke dimension is $D_{ky} = 2.2210$.

Agreement between the numerical, simulated, and experimental results is poor since the circuit relies on the parasitic capacitance of the bipolar junction transistor which varies among different models and may lead to chaotic parameters different from those listed in Table 3.7. Our simulation used the SPICE model for the 2N3055 since it had large capacitance parameters that made it easier to obtain chaos, although we had to use a different bipolar junction transistor model for the experiment. We also had to add a small parasitic capacitance across the inductor in the simulation, as Hamill (1993) also did, although it is unnecessary in the theoretical model.

You may have more success obtaining chaos by using a bipolar junction transistor designed for high power applications since the junction capacitances might be large enough to obtain chaos at lower frequencies. Some data sheets will also give the delay time, also called the 'storage time delay,' which should be on the order of a few microseconds.

The important parameters to adjust for chaos are the amplitude, duty cycle, and frequency of the square-wave source. Adjusting the offset of V_S may also help but was not required in our theoretical model or to obtain our experimental results. If you still have trouble observing chaos after adjusting these parameters, try reducing the inductance to see if chaos will occur at a higher frequency.

Once bifurcation is observed, try reducing R without making it too small since the bipolar junction transistor can be damaged by the large current entering the base. The bipolar junction transistor can also be damaged by large voltage spikes from the inductor when the current is abruptly interrupted. In practical applications, a diode would be added in parallel to the inductor, although doing so adds another nonlinear device to the circuit.

The experimental phase space plot of I versus $V_{BE} - V_{BC}$ is shown in Fig. 3.38, and the time plot of $V_{BE}(t) - V_{BC}(t)$ is shown in Fig. 3.39. Since probing the base for V_{BE} apparently stops the chaos, we opted instead for measuring $V_{BE} - V_{BC}$ by probing the collector. You can view I by adding a 50 Ω resistor between the inductor and V and measuring the current through this resistor. This circuit has one resistor, one inductor, and one bipolar junction transistor for a total of three components.

3.7 Lindberg–Murali–Tamasevicius Circuit

The discovery of the forced diode resonator (Section 2.2) prompted further interest in examining chaos in other semiconductor devices. Hoh and Yasuda (1994) were the first to find chaos using the *thyristor*, a semiconductor device that we will further describe in Chapter 5 but that can be modeled using transistors with parasitic capacitances. A simplification of the thyristor model in this circuit led to the transistor circuit developed by Irita et al. (1998). This circuit was then further simplified by Lindberg et al. (2005) by removing the DC source as shown in Fig. 3.40. The original circuit had an additional resistor in series with C_1, but this resistor can be neglected.

This chaotic circuit is unusual in that it operates in the reverse-active mode of the transistor, making it necessary to incorporate the Ebers–Moll model from Fig. 3.6 into the circuit as shown in Fig. 3.41. Initially, the transistor is in cutoff mode ($V_2 \approx 0$) until the sinusoidal voltage V_S enters its positive cycle, which charges C_2 and reverse-biases the CBJ due to the voltage across R. When V_2 reaches V_0, the diode corresponding to V_{BE} turns on and allows the forward current $\alpha_F I_F$ to flow causing the transistor to enter forward active mode and produce a large positive spike in I_C.

At some point, V_S will reach its maximum value and begin to decrease. The transistor will then enter the reverse-active mode since the voltage across C_2 reverses polarity during discharge (reverse-biasing the EBJ), and the current through R is reversed (forward-biasing the CBJ). The voltage across R turns on the diode corresponding to V_{BC} and allows the reverse

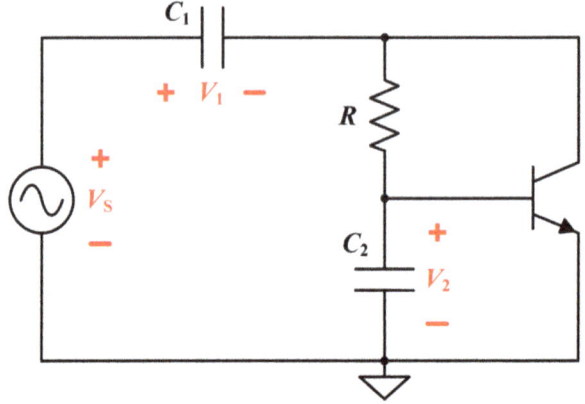

Fig. 3.40 Lindberg–Murali–Tamasevicius circuit.

Table 3.8 Component values for Fig. 3.40.

Component	Numerical	Simulated	Experimental
R	10 kΩ	10 kΩ	10 kΩ
R_D	0.05 kΩ	-	-
C_1	1 μF	1 μF	1 μF
C_2	1 μF	1 μF	1 μF
V_m	5 V	5 V	5 V
V_0	1 V	-	-
ω	1 rad/ms	0.636π rad/ms	0.636π rad/ms
α_F	0.9	-	-
α_R	0.4	-	-
Transistor	Eq. (3.4)	2N3904	2N3904

current $\alpha_R I_R$ to flow until V_2 drops below V_0. This creates a negative spike in I_C following the positive spike from the forward-active mode. This behavior was termed *disturbance of integration* by Lindberg et al. (2005) since R and C_2 form an *RC integrator* and C_2 integrates the current through R by charging, but it is interrupted when the transistor switches to reverse-active mode.

The equations that describe the circuit in Fig. 3.41 are

$$\dot{V}_1 = [(V_2 - V_1 - V_S)/R - I_C]/C_1$$
$$\dot{V}_2 = [I_B - (V_2 - V_1 - V_S)/R]/C_2, \tag{3.17}$$

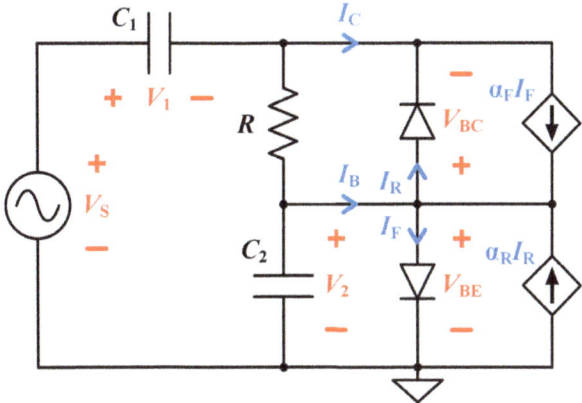

Fig. 3.41 Lindberg–Murali–Tamasevicius circuit with transistor model described by Eq. (3.17).

where $I_C = \alpha_F I_F - I_R$, $I_B = \alpha_R I_R + I_C - I_F$, $I_F = \max(0, V_2 - V_0)/R_D$, and $I_R = \max(0, V_2 - V_1 - V_S - V_0)/R_D$ from Eq. (3.7). The factors α_F and α_R satisfy $0 < \alpha_R < \alpha_F < 1$, and $V_S = V_m \sin \omega t$.

Elegant circuit values that give chaos are $R = 10$ kΩ, $R_D = 0.05$ kΩ, $C_1 = C_2 = 1$ μF, $V_m = 5$ V, $V_0 = 1$ V, $\omega = 1$ rad/ms, $\alpha_F = 0.9$, and $\alpha_R = 0.4$. The dimensionless equations are

$$\dot{x} = b(y - x - v) - \iota_c$$
$$\dot{y} = \iota_b - b(y - x - v), \quad (3.18)$$

where $\iota_c = \alpha_F \iota_f - \iota_r$, $\iota_b = \alpha_R \iota_r + \iota_c - \iota_f$, $\iota_f = c \max(0, y - 1)$, $\iota_r = c \max(0, y - x - v - 1)$, and $v = a \sin \tau$

The parameters $a = 5, b = 0.1, c = 20, \alpha_F = 0.9$, and $\alpha_R = 0.4$ give the attractor shown in Fig. 3.42 and a time series as shown in Fig. 3.43. The Lyapunov exponents are $(0.0374, 0, -7.5247)$, the Kaplan–Yorke dimension is $D_{ky} = 2.0050$, and the attractor has a global Class 1a basin of attraction. Because of the periodic forcing, the nonautonomous system does not have an equilibrium point, but the unforced autonomous system has a stable node at the origin with eigenvalues $(-0.1, -0.1)$.

Figure 3.44 shows the experimental phase space plot of V_2 versus V_1. You may need the instrumentation amplifier from Fig. 1.23 to obtain $V_1(t)$, whose time plot is shown in Fig. 3.45. None of the component values or forcing parameters in Table 3.8 are critical and can be adjusted as desired. The circuit has one resistor, two capacitors, and one bipolar junction transistor for a total of four components.

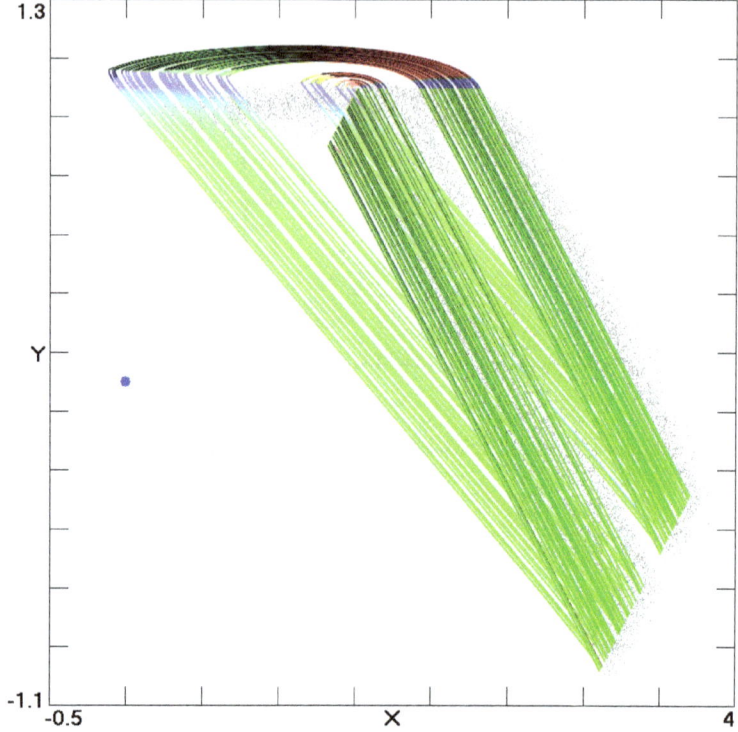

Fig. 3.42 Numerical solution of Eq. (3.18) with $a = 5$, $b = 0.1$, $c = 20$, $\alpha_F = 0.9$, and $\alpha_R = 0.4$.

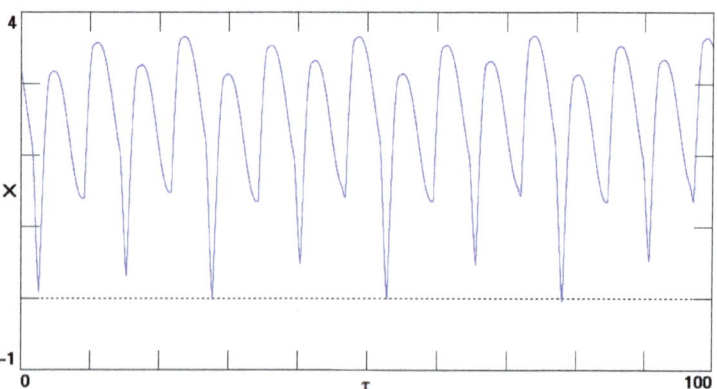

Fig. 3.43 Numerical waveform for $x(\tau)$ from Eq. (3.18) with $a = 5$, $b = 0.1$, $c = 20$, $\alpha_F = 0.9$, and $\alpha_R = 0.4$.

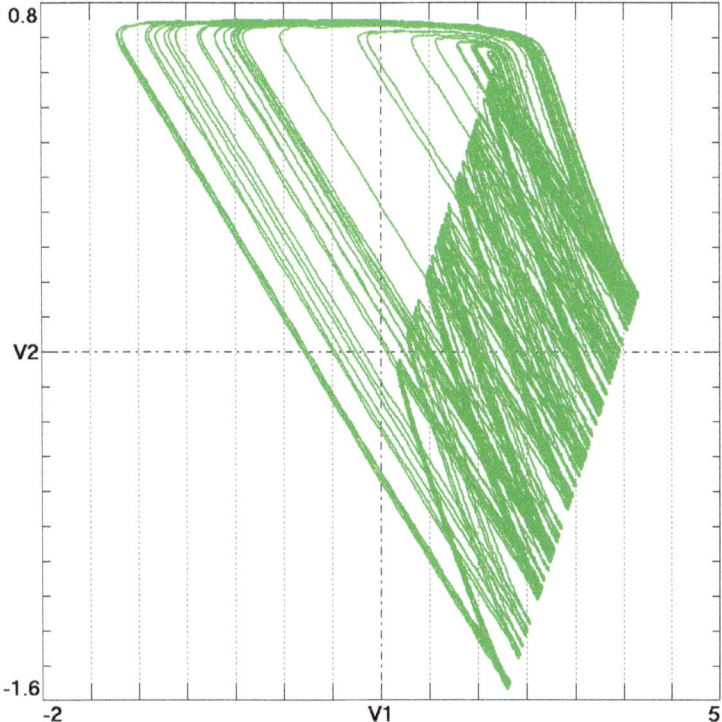

Fig. 3.44 Oscilloscope phase space plot of Lindberg–Murali–Tamasevicius circuit from Fig. 3.40.

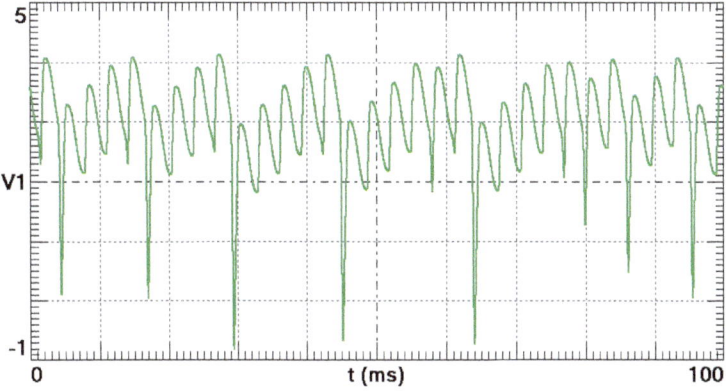

Fig. 3.45 Oscilloscope time plot of Lindberg–Murali–Tamasevicius circuit from Fig. 3.40.

3.8 Chaotic Hartley Oscillator

Another well known circuit is the *Hartley oscillator* which was invented by Ralph Hartley in 1915 (U.S patent 1,356,763) and later became the inspiration for the Colpitts oscillator from Section 3.2. This oscillator has an LC circuit that generates oscillations like those in the Colpitts oscillator, but it uses two inductors as a voltage divider rather than capacitors to sustain these oscillations through feedback. The frequency of oscillation is then adjusted by changing their inductance, making it less practical and thus less common than the Colpitts oscillator since it is easier to use a variable capacitor.

The first chaotic Hartley oscillator was proposed by Kvarda (2002), but Tchitnga *et al.* (2012) found a simpler implementation based on a junction field effect transistor as shown in Fig. 3.46. In their circuit, the capacitor for the LC circuit comes from the internal parasitic capacitance C_{GS} of the junction field effect transistor.

This circuit is arranged in the *common-drain amplifier* configuration, where the drain of the circuit is connected to a voltage source, the input is applied to the gate, and the output comes from the source for feedback. The junction field effect transistor in this configuration has a high input impedance (since the gate and source must be reverse-biased) and a low output impedance, making it ideal for sustaining oscillations in the LC circuit.

The equations that describe the equivalent circuit in Fig. 3.47 are

$$\begin{aligned}
\dot{V}_{DG} &= (I_2 - I_{DS})/C_{DG} \\
\dot{V}_{GS} &= (I_1 + I_2 - I_{DS} - I_{GS})/C_{GS} \\
\dot{I}_1 &= -V_{GS}/L_1 \\
\dot{I}_2 &= (V - V_{DG} - V_{GS})/L_2,
\end{aligned} \quad (3.19)$$

where $I_{GS} = \max(0, V_{GS} - V_0)/R_D$ from Eq. (3.5) and

$$I_{DS} = \begin{cases} 0 & V_{GS} \leq V_P \\ K(V_{GS} - V_P)^2 & V_{DS} \geq V_{GS} - V_P > 0 \\ K[2V_{DS}(V_{GS} - V_P) - V_{DS}^2] & V_{DS} < V_{GS} - V_P > 0 \end{cases} \quad (3.20)$$

from Eq. (3.6), where $V_{DS} = V_{DG} + V_{GS}$ and $V_P < 0$.

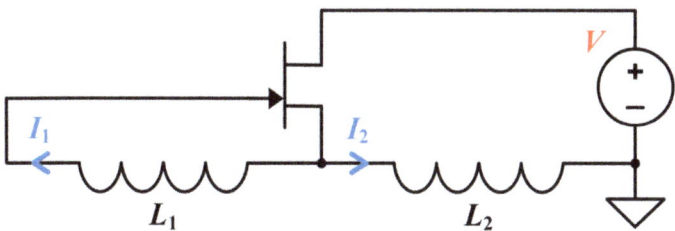

Fig. 3.46   Chaotic Hartley oscillator.

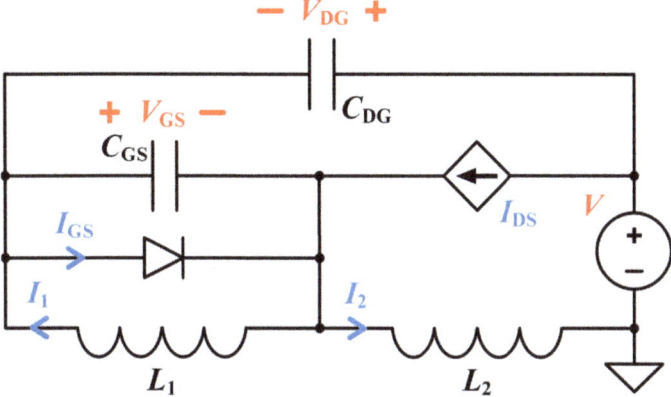

Fig. 3.47   Chaotic Hartley oscillator with transistor model described by Eq. (3.19).

Table 3.9   Component values for Fig. 3.46.

Component	Numerical	Simulated	Experimental
R_D	0.04 kΩ	-	-
C_{DG}	1 μF	-	-
C_{GS}	1 μF	-	-
L_1	1 H	1 H	1 H
L_2	0.1 H	0.1 H	0.1 H
V	5 V	5 V	8 V
V_P	−1 V	-	-
V_0	1 V	-	-
K	6 mA/V^2	-	-
Transistor	Eqs. (3.5), (3.6)	J201	J201

Elegant circuit values that give chaos are $R_D = 0.04$ kΩ, $C_{DG} = C_{GS} = 1$ μF, $L_1 = 1$ H, $L_2 = 0.1$ H, $V = 5$ V, $V_P = -1$ V, $V_0 = 1$ V, and $K = 6$ mA/V^2. The dimensionless equations are

$$\begin{aligned} \dot{x} &= u - \iota \\ \dot{y} &= z + u - \iota - a\max(0, y-1) \\ \dot{z} &= -y \\ \dot{u} &= b(c - x - y), \end{aligned} \quad (3.21)$$

where

$$\iota = \begin{cases} 0 & y \leq p \\ k(y-p)^2 & y > p \text{ and } x \geq -p \\ k[2(y-x)(y-p) - (y-x)^2] & y > p \text{ and } x < -p. \end{cases} \quad (3.22)$$

The parameters $a = 25, b = 10, c = 5, p = -1$, and $k = 6$ give the attractor shown in Fig. 3.48 and a time series as shown in Fig. 3.49. The Lyapunov exponents are $(0.0329, 0, -1.4506, -14.5576)$, the Kaplan–Yorke dimension is $D_{ky} = 2.0227$, and the attractor has a global Class 1a basin of attraction. The system has an unstable saddle focus at $(-5, 0, 0, 6)$ with eigenvalues $(0.1124 \pm 0.6451i, -2.3651, -9.8597)$.

The gate-source junction of the junction field effect transistor is usually considered to be an open circuit since this junction must be reverse-biased for the junction field effect transistor to operate. However, it is necessary to treat the junction as a diode to obtain chaos since the junction is both forward and reverse-biased in this circuit.

You will likely require different parameters than those in Table 3.9 since the parasitic capacitances of junction field effect transistors can vary among devices and are typically much less than 1 μF as evidenced by the higher frequency in the experimental data than in the numerical model. The most important parameter to adjust is the voltage V, which can help you obtain chaos even if you are using a different junction field effect transistor model. For example, we found chaos experimentally for the junction field effect transistor model J204 for $V = 10$ V and the same inductor values in Table 3.9.

If you are still having trouble obtaining chaos, another method is to add a resistor in series with the source or gate, or in parallel with L_1 or L_2. This resistor is necessary for the junction field effect transistor model J111 where we experimentally added a 100 Ω ($\pm 1\%$) resistor in series with the source and used $V = 17.8$ V. The chaos is more delicate for this particular junction field effect transistor model.

The experimental phase space plot of I_2 versus V_{DG} is shown in Fig. 3.50, and the time plot of $V_{DG}(t)$ is shown in Fig. 3.51. The voltage V_{GS} appeared to be less consistent among the numerical, simulated, and experimental results, and I_1 has a small amplitude since the gate-source junction is mostly reverse-biased. For this reason, V_{DG} and I_2 were chosen for the phase space plot.

You can measure V_{DG} by probing the gate of the junction field effect transistor and taking the difference from the DC voltage V. We measured the current I_2 through a grounded 50 Ω resistor in series with the inductor. This circuit has two inductors and one junction field effect transistor for a total of three components.

3.9 JFET-based Wien Bridge Oscillator

The Wien bridge oscillator, previously introduced in Chapter 2, is one of the most useful oscillators since it does not require an inductor to generate periodic oscillations. The most elegant chaotic Wien bridge oscillator was described by Namajunas and Tamasevicius (1995) who inserted a junction field effect transistor into one of the RC networks of the circuit to generate chaos. This circuit later inspired the various chaotic Wien bridge oscillators described by Elwakil and Soliman (1997a) and the conjecture of Elwakil and Kennedy (2000a) for creating chaotic oscillators by simply adding a junction field effect transistor and a capacitor to existing periodic oscillators. Elwakil and Kennedy (1999a) were able to simplify the original circuit using an inductor, but this makes the circuit more difficult to implement. Their modified Wien bridge oscillator is shown in Fig. 3.52.

The junction field effect transistor in this circuit will operate as a current source when the gate is connected to the source ($V_{GS} = 0$). As a result, if the junction field effect transistor is in saturation mode, the maximum current (I_{DSS}) will flow through the junction field effect transistor. The current through the junction field effect transistor is maintained at this value, but it becomes nonlinear when it enters the triode region.

The junction field effect transistor in this configuration is called a *constant-current diode*, and it can be used to protect devices by limiting the current through them. However, it is not robust since I_{DSS} and V_P vary among junction field effect transistors even of the same model, making the junction field effect transistor obsolete for this application.

The *I-V* characteristic of the constant-current diode is similar to the diode biased by a current source introduced in Chapter 2 and described

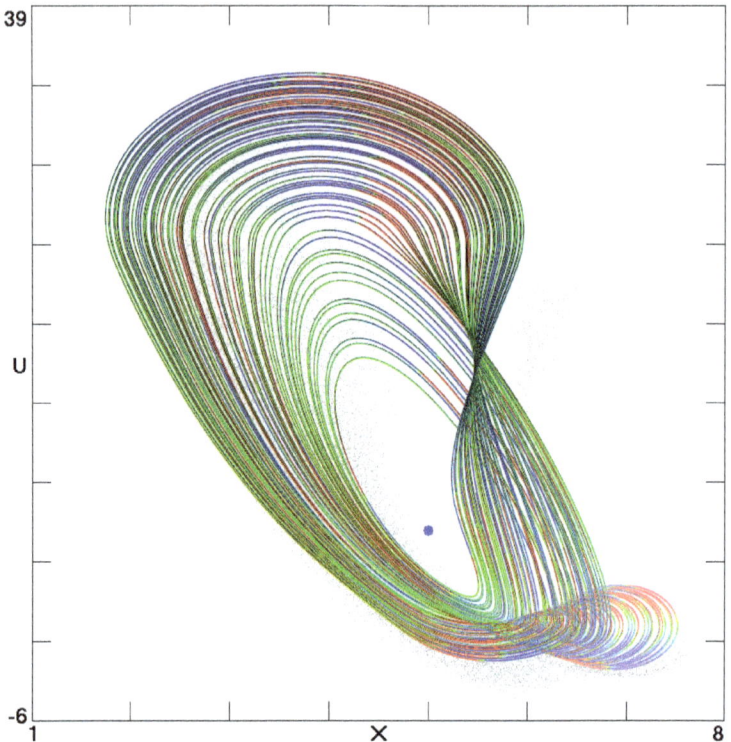

Fig. 3.48 Numerical solution of Eq. (3.21) with $a = 25$, $b = 10$, $c = 5$, $p = -1$, and $k = 6$.

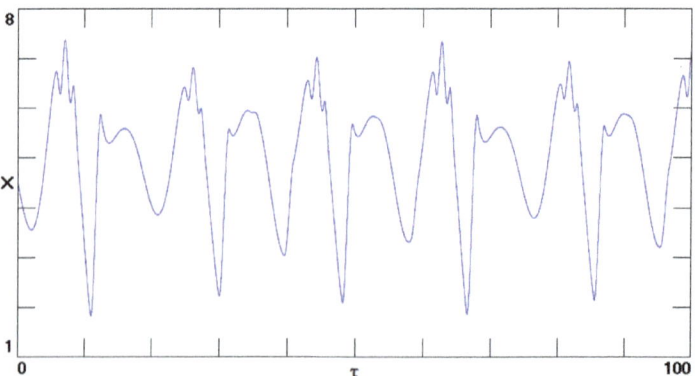

Fig. 3.49 Numerical waveform for $x(\tau)$ from Eq. (3.21) with $a = 25$, $b = 10$, $c = 5$, $p = -1$, and $k = 6$.

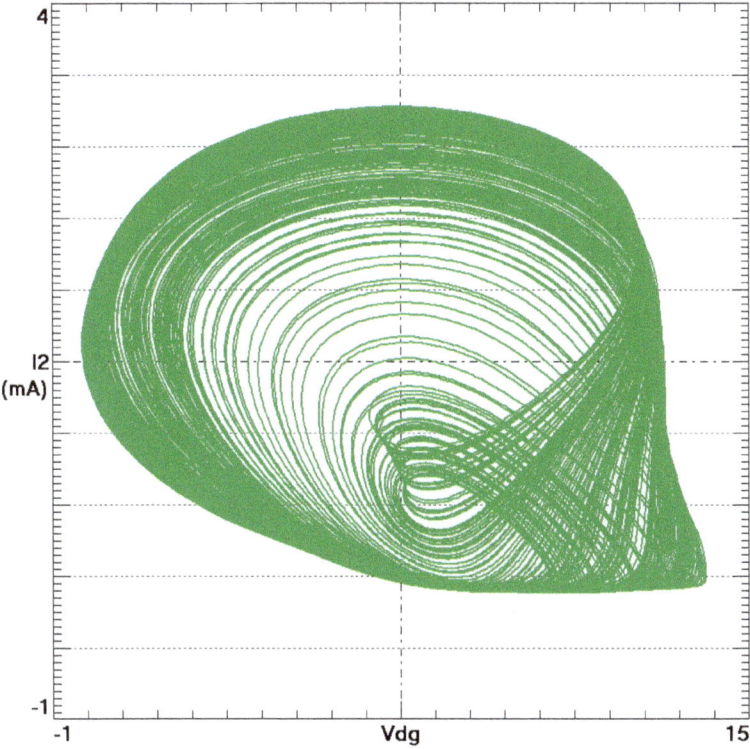

Fig. 3.50   Oscilloscope phase space plot of chaotic Hartley oscillator from Fig. 3.46.

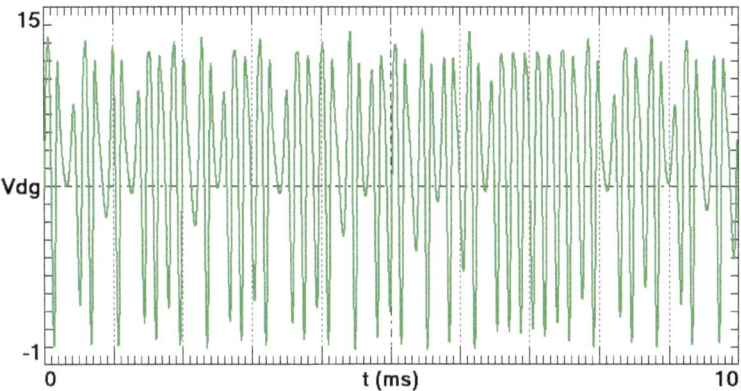

Fig. 3.51   Oscilloscope time plot of chaotic Hartley oscillator from Fig. 3.46.

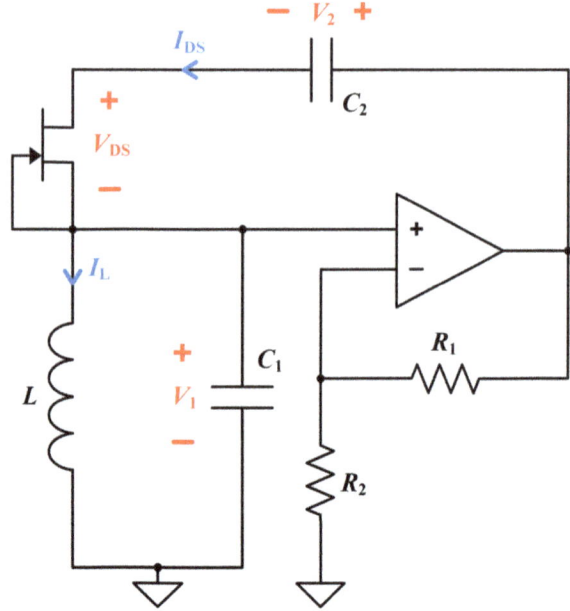

Fig. 3.52 JFET-based Wien bridge oscillator described by Eq. (3.23).

Table 3.10 Component values for Fig. 3.52.

Component	Numerical	Simulated	Experimental
R_1	1 kΩ	1 kΩ	1 kΩ (±1%)
R_2	1 kΩ	1 kΩ	1 kΩ (±1%)
R_L	0.5 kΩ	739 Ω	470 Ω (adjust)
R_S	0.2 kΩ	110 Ω	-
C_1	1 μF	1 μF	1 μF
C_2	1 μF	1 μF	1 μF
L	1 H	1 H	1 H
V_P	−1 V	-	−1.1 V
Op amp	-	Ideal	LM741
Transistor	Eq. (3.6)	J204	J204

by Munmuangsaen and Srisuchinwong (2013), which allows the diode and current source in several of the circuits in Chapter 2 to be replaced with a junction field effect transistor.

For example, Tchitnga et al. (2016) showed that the biased diode in the Banlue–Buncha diode circuit from Section 2.5 can be replaced by a junction field effect transistor. The same is true for the Elwakil–Kennedy diode oscillator from Section 2.7 which originally used a junction field effect transistor instead of a biased diode.

The equations that describe the circuit in Fig. 3.52 are

$$\dot{V}_1 = (I_{DS} - I_L)/C_1$$
$$\dot{V}_2 = I_{DS}/C_2 \qquad (3.23)$$
$$\dot{I}_L = (V_1 - I_L R_L)/L,$$

where I_{DS} from Eq. (3.6) can be simplified to

$$I_{DS} = \begin{cases} I_{DSS} & V_{DS} > -V_P \\ V_{DS}/R_S & V_{DS} \leq -V_P, \end{cases} \qquad (3.24)$$

where $V_{DS} = V_1 R_1/R_2 - V_2$ and $I_{DSS} = -V_P/R_S$.

Elegant circuit values that give chaos are $R_1 = R_2 = 1$ kΩ, $R_L = 0.5$ kΩ, $R_S = 0.2$ kΩ, $C_1 = C_2 = 1$ μF, $L = 1$ H, and $V_P = -1$ V. The dimensionless equations are

$$\dot{x} = \iota - z$$
$$\dot{y} = \iota \qquad (3.25)$$
$$\dot{z} = x - az,$$

where $\iota = b \min(p, x - y)$.

The parameters $a = 0.5, b = 5$, and $p = 1$ give the attractor shown in Fig. 3.53 and a time series as shown in Fig. 3.54. The Lyapunov exponents are $(0.0314, 0, -0.5314)$, the Kaplan–Yorke dimension is $D_{ky} = 2.0591$, and the attractor has a global Class 1a basin of attraction. The system has an unstable saddle focus at the origin with eigenvalues $(0.5893 \pm 1.6221i, -1.6787)$.

The experimental phase space plot of V_2 versus V_1 is shown in Fig. 3.55, and the time plot of $V_1(t)$ is shown in Fig. 3.56. Since R_1/R_2 is an adjustable parameter, it is probably easier to probe V_1 and use the instrumentation amplifier from Fig. 1.23 to measure V_2.

The most critical values in the junction field effect transistor are I_{DSS} and V_P, but unfortunately, they vary among junction field effect transistors of the same type. You may need to measure both values in your chosen junction field effect transistor to determine whether it will work with our parameters. Our J204 had the experimental parameters $I_{DSS} = 0.92$ mA and $V_P = -1.1$ V. It is preferable to choose a junction field effect transistor with a small V_P to avoid saturating the operational amplifier.

If I_{DSS} of your junction field effect transistor is larger than ours, it can be reduced by adding a resistor (R_S) in series with the source. Increasing

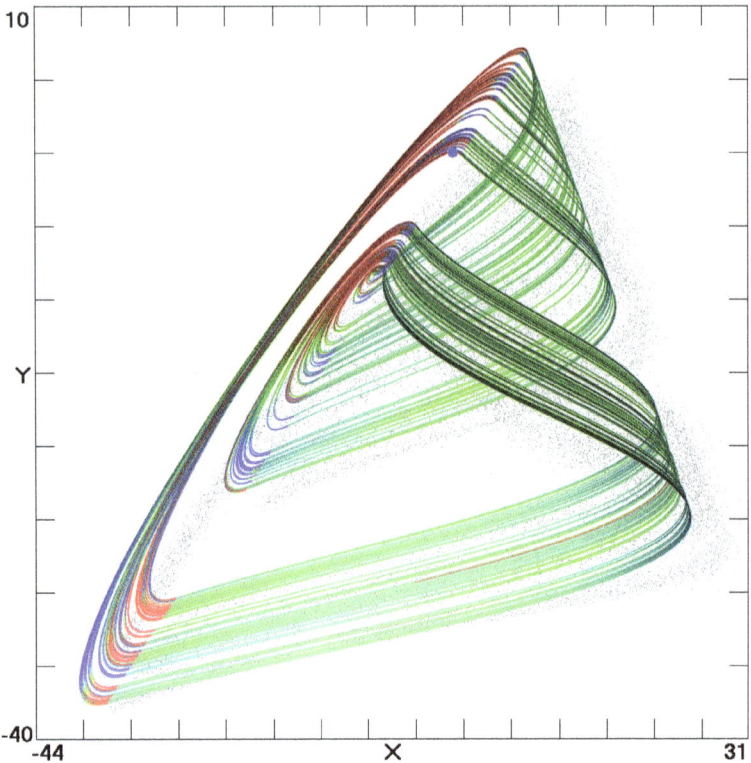

Fig. 3.53 Numerical solution of Eq. (3.25) with $a = 0.5$, $b = 5$, and $p = 1$.

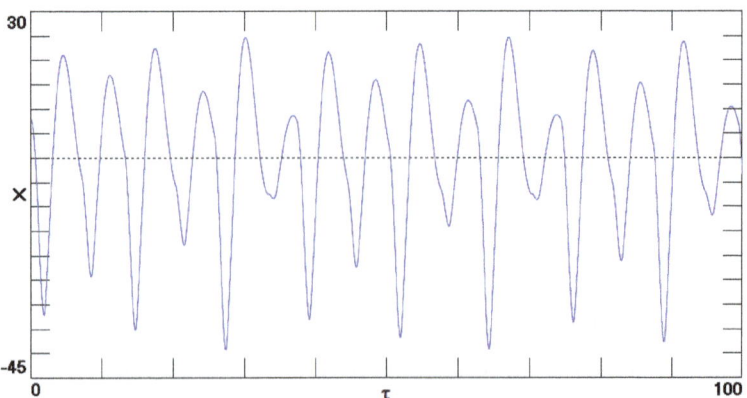

Fig. 3.54 Numerical waveform for $x(\tau)$ from Eq. (3.25) with $a = 0.5$, $b = 5$, and $p = 1$.

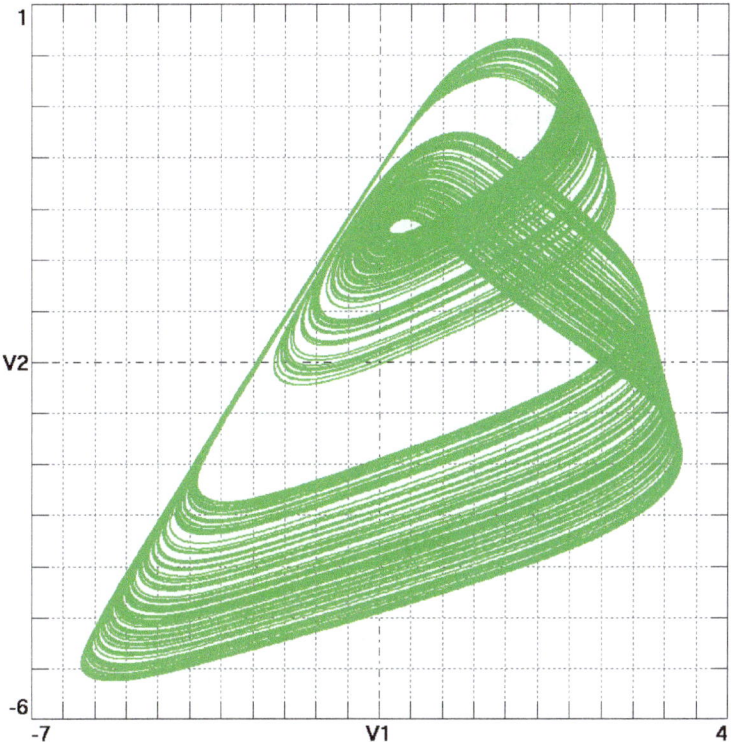

Fig. 3.55 Oscilloscope phase space plot of JFET-based Wien bridge oscillator from Fig. 3.52.

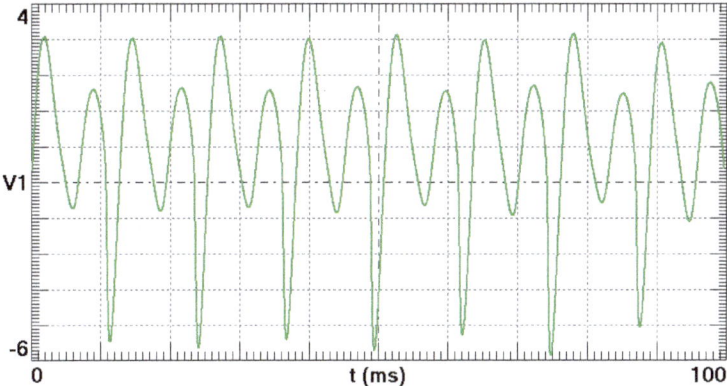

Fig. 3.56 Oscilloscope time plot of JFET-based Wien bridge oscillator from Fig. 3.52.

this resistance will allow you to reduce I_{DS} to any desired value below I_{DSS}. This is done in the simulation since the I_{DSS} of the simulated J204 is slightly larger than our experimental J204. On the other hand, I_{DSS} may be low. For example, another J204 that we tested had $I_{DSS} = 0.2$ mA and did not exhibit chaos for our parameters. In such a case, you should choose another junction field effect transistor.

Component values are given in Table 3.10. This circuit is sensitive to variations in the parameters (a 1.7% change in parameters is likely to destroy the chaos), and so you should use ±1% resistors and adjust R_L until chaos is observed. You will likely require a different value for R_L than ours due to the different parameters in your junction field effect transistor. The circuit has two resistors, two capacitors, one inductor, one operational amplifier, and a junction field effect transistor for a total of seven components.

3.10 Chaotic MOS Amplifier

One of the critical technologies that allowed the emergence of the computer industry was the development of the integrated circuit, which contains a large number of miniaturized transistors to function as computer memory or micro-processors. These transistors are mainly MOSFETs since they are easier to fabricate than the bipolar junction transistor and improve in performance when their size is scaled down. Both of these factors allow for a large number of MOSFETs to be easily implanted on an integrated circuit.

On the other hand, it is difficult to implement large resistances and capacitances in an integrated circuit due to their size. A solution to this problem is to use *switched capacitors*, which consist of a capacitor C between two MOSFET switches. These switches are controlled by a square wave voltage (called a *clock* or CLK) that has two states, *high* and *low*. One switch will turn on when CLK is high and turn off when CLK is low. The other switch is controlled by an inverted clock signal ($\overline{CLK}$) so that it switches opposite to the first one. When the switching occurs at some frequency f, the circuit is equivalent to a resistor with a value of $R = 1/fC$, which allows any resistance value to be obtained by simply changing the frequency.

Switched capacitors in chaotic circuits were proposed by Rodriguez-Vazquez et al. (1985) and later outlined by Rodriguez-Vazquez et al. (1987) as a way to build circuits modeling chaotic map functions. They also form the basis of the *field programmable analog array* (FPAA), which contains sections of switched capacitors and operational amplifiers. Circuits can then be rapidly created by programming the connections between these components. Most chaotic circuits built using FPAAs are considered to be a type of analog computer, which will be discussed further in Chapter 7.

The most elegant chaotic switched-capacitor circuit was designed by Pham et al. (1996). This circuit, shown in Fig. 3.57 with its equivalent circuit in Fig. 3.58, is based on a *MOS amplifier* formed by the MOSFET and resistor R. Sweeping the input voltage at V_{GS} of this amplifier from ground to V_{DD} causes the MOSFET to access the cutoff, saturation, and triode mode. Of these regions, the saturation region is the most suited for amplification since the output of the amplifier taken at V_{DS} is approximately $V_{DS} = -GV_{GS}$ where G is the voltage gain that is directly proportional to the value of R. Thus the MOSFET is typically biased so that its Q-point is in this mode.

In this chaotic circuit, no bias is applied, which makes the MOSFET nonlinear since the output signal is distorted by the cutoff region and fed back through a switched capacitor. When CLK is high, the left switch is turned on to store the output of the amplifier in the capacitor C while the right switch is turned off through $\overline{CLK}$. When the clock voltage is low, the stored value in C is transferred back to the amplifier input and stored in the small gate capacitance of the MOSFET.

When SW_1 is ON and SW_2 is OFF,

$$\dot{V}_{DS} = (V_{DD} - RI_{DS} - V_{DS})/RC$$
$$\dot{V}_{GS} = 0, \qquad (3.26)$$

where I_{DS} is given by

$$I_{DS} = \begin{cases} 0 & V_{GS} \leq V_T \\ K(V_{GS} - V_T)^2 & V_{DS} \geq V_{GS} - V_T > 0 \\ K[2V_{DS}(V_{GS} - V_T) - V_{DS}^2] & V_{DS} < V_{GS} - V_T > 0 \end{cases} \qquad (3.27)$$

from Eq. (3.6) with $V_T > 0$ and $V_{DD} = 5$ V.

When SW_1 is OFF and SW_2 is ON, the capacitor is directly connected to V_{GS}, whose value quickly equilibrates to V_{DS} in a short time Δt according to

$$\dot{V}_{DS} = 0$$
$$\dot{V}_{GS} = (V_{DS} - V_{GS})/\Delta t. \qquad (3.28)$$

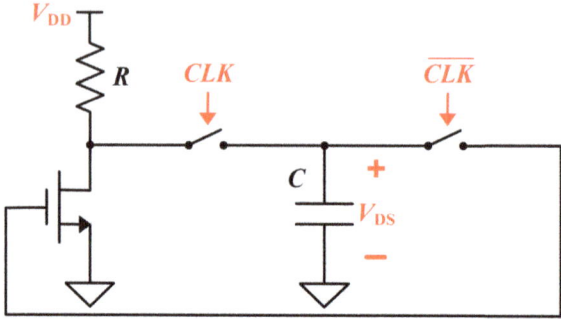

Fig. 3.57 Chaotic MOS amplifier.

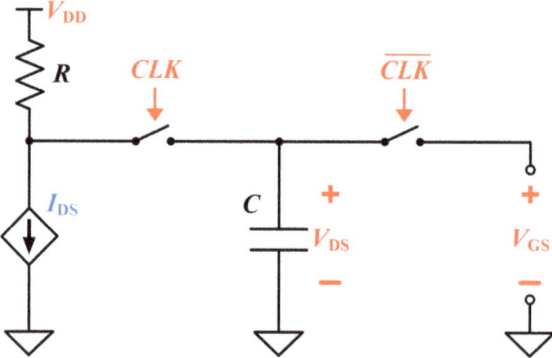

Fig. 3.58 Chaotic MOS amplifier with transistor model described by Eq. (3.26).

Table 3.11 Component values for Fig. 3.58.

Component	Numerical	Simulated	Experimental
R	9 kΩ	150 Ω	1 kΩ
R_G	-	-	62 kΩ (adjust)
C	1 μF	1 nF	-
C_1	-	-	10 nF (PP)
C_2	-	-	10 nF (PP)
V_{DD}	5 V	5 V	5 V
V_T	1 V	-	-
K	1 mA/V^2	-	-
Δt	0.1 ms	-	-
f	-	400 kHz	20 kHz
Sample-and-hold	-	Ideal	LF398
Transistor	Eq. (3.6)	TN0702	TN0702

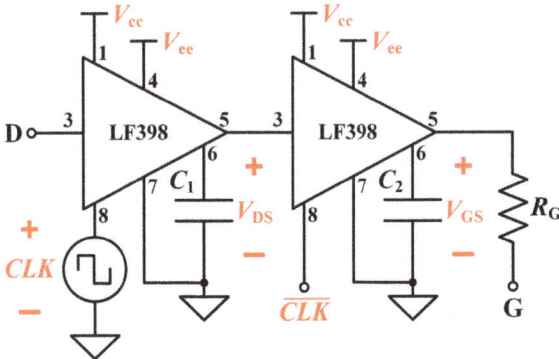

Fig. 3.59 Switched capacitor implementation using sample-and-hold circuits.

Elegant circuit values that give chaos are $R = 9$ kΩ, $C = 1$ μF, $V_{DD} = 5$ V, $V_T = 1$ V, $K = 1$ mA/V^2, and $\Delta t = 0.1$ ms. The dimensionless equations are

$$\dot{x} = \begin{cases} (v-x)/a - \iota & \sin z > 0 \\ 0 & \sin z \leq 0 \end{cases}$$
$$\dot{y} = \begin{cases} 0 & \sin z > 0 \\ b(x-y) & \sin z \leq 0, \end{cases} \quad (3.29)$$

where

$$\iota = \begin{cases} 0 & y \leq 1 \\ (y-1)^2 & y > 1 \text{ and } x \geq y-1 \\ 2x(y-1) - x^2 & y > 1 \text{ and } x < y-1. \end{cases} \quad (3.30)$$

The parameters $a = 5, b = 10$, and $v = 5$ give the attractor shown in Fig. 3.60 and a time series as shown in Fig. 3.61. The Lyapunov exponents are $(0.0292, 0, -5.1494)$, the Kaplan–Yorke dimension is $D_{ky} = 2.0057$, and the attractor has a global Class 1a basin of attraction. This system is unusual since it does not have any equilibrium points, and so the strange attractor is hidden even though it is globally attracting.

We implemented the switched capacitor using *sample-and-hold* circuits (LF398) as shown in Fig. 3.59 that can be connected between the gate of the MOSFET labeled 'G' and the drain labeled 'D.' This circuit will initially follow a signal until it is given a command to store a constant value of

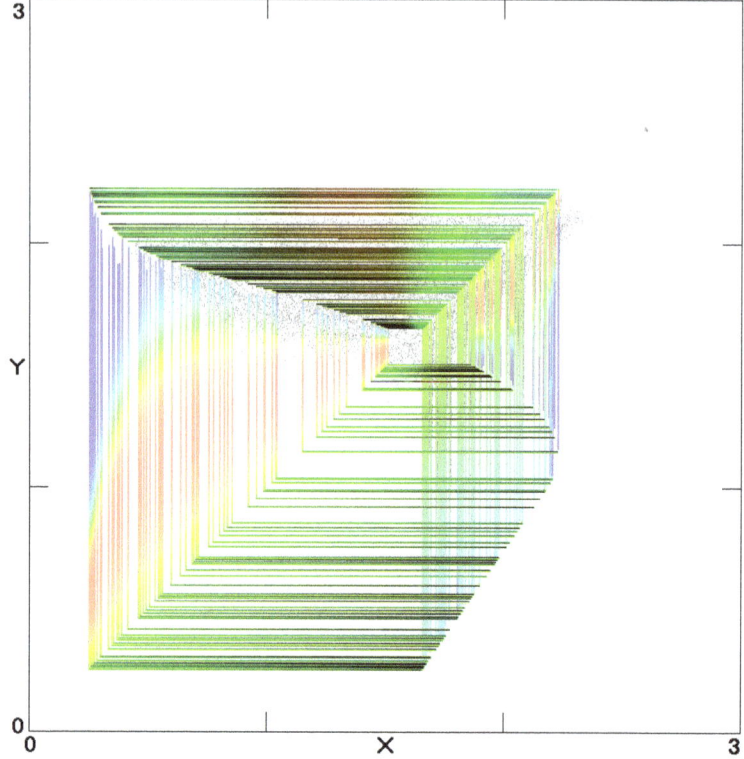

Fig. 3.60 Numerical solution of Eq. (3.29) with $a = 5$, $b = 10$, and $v = 5$.

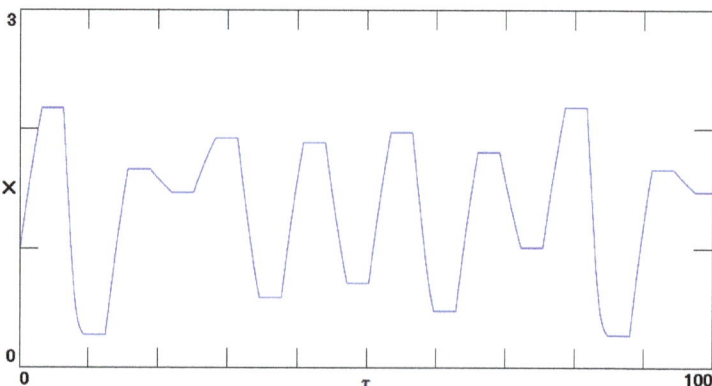

Fig. 3.61 Numerical waveform for $x(\tau)$ from Eq. (3.29) with $a = 5$, $b = 10$, and $v = 5$.

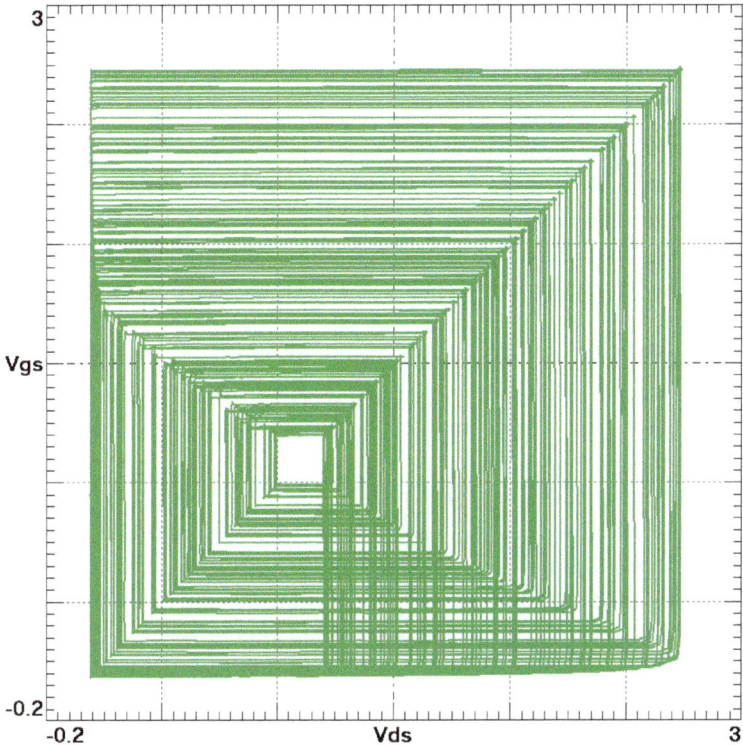

Fig. 3.62 Oscilloscope phase space plot of chaotic MOS amplifier from Fig. 3.57.

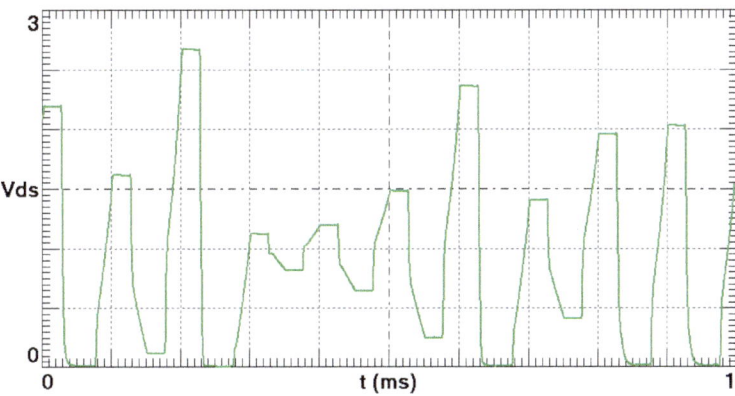

Fig. 3.63 Oscilloscope time plot of chaotic MOS amplifier from Fig. 3.57.

the input at the instant the command is given. They are mainly used to temporarily hold a recorded value of the input so that an analog-to-digital converter has enough time to process it.

The value of the input is stored in a capacitor, which can be connected externally to an LF398 through pin 6 and read from the output at pin 5. You must minimize the leakage current on this capacitor to provide an accurate reading of the input. In addition, capacitors used in sample-and-hold circuits should be polypropylene, since ceramic capacitors will often incompletely discharge and cause a 1-2% difference in voltage at the next reading.

The exact value of this capacitance is not critical, but it should not be too large since an excessive charging time will delay acquisition of the input signal. On the other hand, small capacitances are prone to leakage currents that degrade accuracy.

Interestingly, chaos can still be obtained if pin 6 is left disconnected without a capacitor, possibly because of the small parasitic capacitance from this unused pin. This suggests that acquiring accurate signals is not particularly important for this circuit. However, these unused pins will pick up external noise and distort the output and cannot be grounded since that will eliminate the chaos. Thus we used 10 nF polypropylene (PP) capacitors for both sample-and-hold circuits.

Like the switched capacitor, the sample-and-hold circuits are controlled by CLK to determine its mode of operation. When CLK is low, the capacitor is connected to the input. When CLK is high, a switch is opened so that the last input is stored in the capacitor. You can implement this switching by using a square wave signal with a duty cycle of 50%, an amplitude of +5 V for the high state, and a minimum of 0 V for the low state as required for digital signals. One sample-and-hold circuit should be controlled by $\overline{CLK}$ which is the same signal but 180° out of phase with CLK.

The most important parameter of the MOSFET is the threshold voltage, labeled as $V_{GS(th)}$ in most data sheets. We tested several MOSFETs but only found chaos for those with a threshold voltage below 2 V such as the TN0702, TN0604N3, and ALD21290. MOSFETs are also easily damaged by static electricity and are best handled on an anti-static mat with a wrist strap. Some MOSFETs also have an additional connection to their substrate called a *body* that is usually directly connected to the source. If this connection is available on your MOSFET, you should manually connect it to the source.

This circuit is designed to operate as a digital circuit, and so it only oscillates between 0 V to 5 V with the voltage supply of the amplifier set at $V_{DD} = 5$ V. However, a dual voltage supply is required for the LF398 and was set to $V_{cc} = 15$ V and $V_{ee} = -15$ V to improve the quality of the signal. Thus you must set V_{DD} using a voltage divider and buffer from Fig. 1.29.

The experimental phase space plot of V_{GS} versus V_{DS} is shown in Fig. 3.62, and the time plot of $V_{DS}(t)$ is shown in Fig. 3.63. Both V_{DS} and V_{GS} are best viewed by probing the sample-and-hold capacitors C_1 and C_2, respectively.

The main parameters to adjust for chaos in Table 3.11 are the frequency of the clock f, the resistance R, and an additional resistor in series with the gate R_G. The chaos should occur in the kHz range since a clock frequency in the MHz range apparently introduces additional parasitics unrelated to the circuit operation.

After a limit cycle is obtained, you should adjust R_G until chaos occurs. The value of R is less critical and can be kept at 1 kΩ and adjusted last. Be careful not to adjust these resistances too low since this can draw too much current from the power supply and change V_{DD}, which you may want to check with a voltmeter as you adjust the resistances.

All the components in this circuit can be replaced with MOSFETs if an integrated circuit implementation is required. The switch can be constructed using a *transmission gate* that uses two MOSFETs to conduct or block a signal, while R can be replaced with a MOSFET whose drain is connected to its gate. In our implementation, the circuit has two sample-and-hold circuits, two capacitors, one NMOS transistor, and two resistors for a total of seven components.

Chapter 4

Tunnel Diode Circuits

The tunnel diode is one of the most fascinating of the semiconductor devices, invented by Physics Nobel laureate (1973) Leo Esaki. This two-terminal device exploits quantum tunneling to produce a negative resistance and hysteresis, allowing it to be used as an active device in chaotic circuits.

4.1 Tunnel Diode Junction

The earlier Chapter 2 on conventional diodes described the PN junction and its characteristics that allow rectification. Those devices rely on the semiconductor properties of silicon whose conductivity can be changed by increasing the electron and hole concentration in the n-type and p-type regions, respectively. This concentration is typically kept low in a conventional diode.

A *tunnel diode*, shown schematically in Fig. 4.1, can be constructed by making this concentration very high, which leads to some unusual properties. The large concentration causes the depletion width to become very thin since the holes and electrons cannot diffuse very far into the opposing regions. Then, if a forward bias is applied to the diode, a quantum physics phenomenon called *tunneling* occurs where the electrons pass through the thin barrier rather than requiring a potential V_0 to jump over it. It is as if you could walk through a wall, which is impossible in classical mechanics, but common in the quantum world.

A small forward-bias voltage is sufficient for electrons to tunnel through this barrier from the n-type region to the p-type region. This flow of electrons is called a *tunneling current*, and it increases with the forward-bias voltage but eventually reaches a peak and begins to decrease. Since the

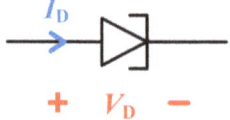

Fig. 4.1   Tunnel diode.

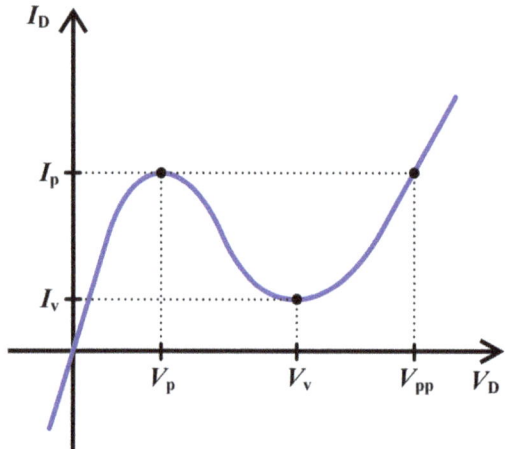

Fig. 4.2 I_D-V_D curve of a tunnel diode.

tunneling current decreases with increasing voltage, this creates a *negative resistance* region. This negative resistance is a unique property of the tunnel diode, which allows it to become an active device and deliver power when biased in this region. Further increase in the voltage causes the tunneling current to disappear completely, and the main forward current is a diffusion current as in a normal PN junction diode. No rectification occurs under reverse-bias conditions, unlike in a conventional diode.

4.1.1   Tunnel diode I-V characteristic

The tunnel diode *I-V* characteristic in Fig. 4.2 has the shape of an 'N' and thus is commonly called an *N-type* negative resistance. With an applied forward bias, the tunneling current reaches a peak at the point (I_p, V_p) before decreasing rapidly in the negative resistance region. This negative resistance continues until the tunneling current ceases in a valley at point

(I_v, V_v). With a further increase in voltage beyond V_{pp}, the tunnel diode behaves like an ordinary diode. The parameters of a tunnel diode will vary for different models. For example, a 1N3712 tunnel diode has typical parameters of $I_p = 1$ mA, $V_p = 65$ mV, $I_v = 0.12$ mA, and $V_v = 350$ mV.

There are two main currents in this device, the normal diode current I_{diode} that can be modeled by Eq. (2.1) and the tunneling current I_{tunnel}. This allows the tunnel diode to be modeled by $I = I_{diode} + I_{tunnel}$ [Sze and Kwok (2006)]. Usually an excess tunneling current I_{excess} is added in the model to determine V_v, but that is not essential. Thus the I-V characteristic of the tunnel diode is given approximately by

$$I(V) = I_S(e^{V/V_t} - 1) + I_p(V/V_p)e^{1-V/V_p} \qquad (4.1)$$

with the condition $I_S \ll I_p e^{-V_p/V_t}$.

The shape of the I-V curve in Fig. 4.2 implies that for a given current in the range $I_v < I < I_p$, there are three possible voltages, the middle of which is unstable and occurs only transiently. Thus when the current is slowly increased from a small value, the voltage remains small until the current reaches a value of I_p, whereupon the voltage abruptly increases from V_p to V_{pp}. The voltage remains large until the current falls to I_v, whereupon it drops to a value near zero. Hence the voltage depends on the past history of the current and thus is an example of hysteresis.

Hysteresis can be modeled in a system of ordinary differential equations by considering a variable voltage V whose time derivative $\dot{V}$ is zero unless $I > I_p$ and $V < V_{pp}$, in which case it is assigned a large positive value $\dot{V} = \dot{V}_m$ or unless $I_p < I_v$ and $V > 0$, in which case it is assigned a large negative value $\dot{V} = -\dot{V}_m$, while the other variables are held constant. In practice, a value of $\dot{V}_m = 100$ V/ms is adequate and reasonably approximates the finite switching time of a tunnel diode.

A slightly more realistic hysteresis model is to assume the voltage approaches its new value V_{new} exponentially according to $\dot{V} = (V_{new} - V)/\Delta t$, where Δt is a small switching time that we typically take as 2 microseconds, although the value is not critical. This is the model that we will use for most of the circuits that follow.

4.1.2 Tunnel diode emulator

Tunnel diodes were used in high-frequency applications in the past but have largely been replaced by other technologies, making the tunnel diode mostly obsolete and unavailable from electronics suppliers. An alternative

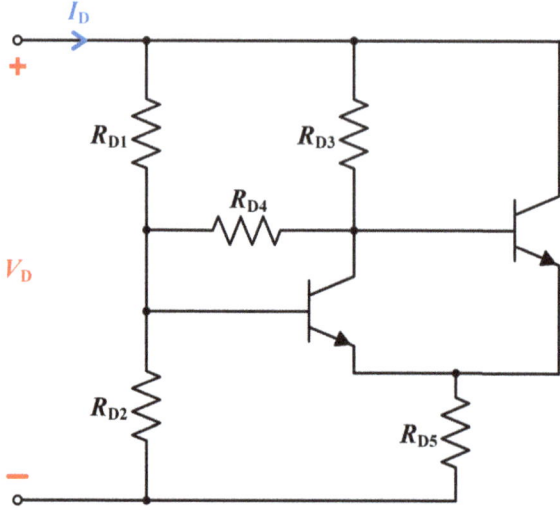

Fig. 4.3 Tunnel diode emulator.

is to construct an *emulator* circuit that synthesizes the N-type *I-V* characteristics of the tunnel diode. Various N-type emulators based on bipolar junction transistors were explored by Chua *et al.* (1983), one of which we chose to use in this chapter and is shown in Fig. 4.3.

Most of the circuits described by Chua *et al.* (1983) implement the negative resistance using two bipolar junction transistors and at least three resistors. The components R_{D1}, R_{D2}, and R_{D3} and the left bipolar junction transistor in Fig. 4.3 form an *inverter* whose voltage output increases as V_D increases but eventually decreases. This decrease in voltage at the base-emitter junction of the right-hand transistor will decrease its collector current and leads to the negative resistance in the *I-V* characteristic. Additional resistors R_{D4} and R_{D5} can be added in the emulator to smooth the negative resistance region or to implement the subsequent normal diode behavior.

This emulator, although a poor fit to Eq. (4.1), qualitatively emulates the tunneling current, and the resistors of this circuit can be adjusted to obtain parameters V_p, V_v, I_p, and I_v characteristic of a particular tunnel diode. For example, in one of the circuits the emulator is constructed with $R_{D1} = 80.4$ kΩ, $R_{D2} = 19.9$ kΩ, $R_{D3} = 38$ kΩ, $R_{D4} = 5$ kΩ, and $R_{D5} = 50$ Ω with 2N3904 transistors to approximately implement the tunnel diode parameters $V_p = 2$ V, $V_v = 4$ V, $I_p = 0.4$ mA, and $I_v = 0.2$ mA.

However, these exact resistances may yield different parameters because of the varying characteristics of bipolar junction transistors. You can check your emulator by applying a 100 Hz triangular-wave voltage across V_D and plotting this voltage against the current measured with a series 50 Ω resistor or I-V converter from Fig. 1.24.

Unfortunately, each resistance value will affect several parameters, so you may need some trial and error to obtain the desired I-V characteristic. For the circuits in this chapter, we used resistance values to give elegant tunnel diode parameters rather than to model a particular tunnel diode. Although this emulator requires two bipolar junction transistors and five resistors, we consider it to be a single component since it is meant to emulate a single, albeit now rare, solid-state device.

4.2 Forced Relaxation Oscillator

Perhaps the simplest chaotic circuit using a tunnel diode is the forced relaxation oscillator in Fig. 4.4. This is just a simplified modern implementation of van der Pol's periodically-forced relaxation oscillator in Fig. 1.16 but using an inductor with its parasitic resistance in place of the RC circuit and with the tunnel diode replacing the neon lamp. It can also be viewed as a variant of the forced diode resonator in Fig. 2.10 but using hysteresis as the nonlinearity rather than the nonlinear parasitic capacitance of the diode.

The equations that describe the circuit in Fig. 4.4 are

$$\dot{I} = (V_S - V_D - IR)/L$$

$$\dot{V}_D = \begin{cases} (V_p - V_D)/\Delta t & I < I_v \text{ or } \dot{V}_D < 0 \\ (V_v - V_D)/\Delta t & I > I_p \text{ or } \dot{V}_D > 0, \end{cases} \quad (4.2)$$

where $V_S = V_0 + V_1 \sin \omega t$. The time Δt is the switching time for the hysteresis loop and is typically a few microseconds. The hysteresis loop is assumed to be bounded by $V_p = 2$ V, $V_v = 4$ V, $I_p = 0.4$ mA, and $I_v = 0.2$ mA.

Elegant circuit values that give chaos are $R = 1$ kΩ, $L = 1$ H, $V_0 = 3$ V, $V_1 = 0.5$ V, and $\omega = 5$ rad/ms. The dimensionless equations are

$$\dot{x} = a + b \sin c\tau - y - x$$

$$\dot{y} = \begin{cases} d(2-y) & x < 0.2 \text{ or } \dot{y} < 0 \\ d(4-y) & x > 0.4 \text{ or } \dot{y} > 0. \end{cases} \quad (4.3)$$

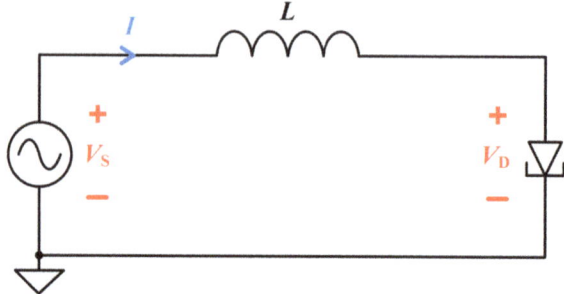

Fig. 4.4   Forced relaxation oscillator described by Eq. (4.2).

Table 4.1   Component values for Fig. 4.4.

Component	Numerical	Simulated	Experimental
R	1 kΩ	1 kΩ	1 kΩ
L	1 H	1 H	1 H
V_0	3 V	3.1 V	3 V (adjust)
V_1	0.5 V	0.9 V	0.5 V (adjust)
V_p	2 V	2 V	2 V
V_v	4 V	4 V	4 V
I_p	0.4 mA	0.4 mA	0.4 mA
I_v	0.2 mA	0.2 mA	0.2 mA
ω	5 rad/ms	5 rad/ms	5 rad/ms
Δt	0.002 ms	-	-
Tunnel Diode	-	Emulator	Emulator

The parameters $a = 3$, $b = 0.5, c = 5$, and $d = 500$ give the attractor shown in Fig. 4.5 and a time series as shown in Fig. 4.6. The x-y plot in Fig. 4.5 shows the nearly rectangular hysteresis loop of the diode, and the $x(\tau)$ waveform in Fig. 4.6 shows that the chaos occurs in the phase of the oscillation with a nearly constant amplitude. The Lyapunov exponents and Kaplan–Yorke dimension cannot be easily determined because of the discontinuity in the diode voltages. The attractor has a global Class 1a basin of attraction.

Because the Lyapunov exponents cannot be easily and reliably calculated, it is necessary to have another way to verify that the solution of the model equations is chaotic rather than, for example, quasiperiodic. For that purpose, the plot in Fig. 4.5 was calculated twice with slightly different initial conditions (typically differing by about one percent the size of a pixel on the computer screen), and the two plots were compared pixel by pixel to see how closely they overlay. For a periodic or quasiperiodic solution, none or just a few pixels will differ, whereas most (thousands) will differ if the solution is chaotic. This method was used for all the subsequent cases where Lyapunov exponents are hard to calculate.

This circuit uses the tunnel diode emulator from Fig. 4.3 with the parameters $V_p = 2$ V, $V_v = 4$ V, $I_p = 0.4$ mA, and $I_v = 0.2$ mA. These parameters were experimentally obtained by setting the resistances to $R_{D1} = 90.6$ kΩ, $R_{D2} = 28.5$ kΩ, $R_{D3} = 21.0$ kΩ, $R_4 = 4.7$ kΩ, and $R_5 = 50$ Ω with 2N3904 transistors. You will need to check your own tunnel diodes using the method in Section 4.1.2 to see if your parameters are close to ours, but the circuit is not sensitive to these parameters, and resistances close to ours should suffice.

The experimental phase space plot of V_D versus I for this circuit is shown in Fig. 4.7 and should be a nearly rectangular loop showing the hysteresis of the tunnel diode. However, if you were to measure I through an I-V converter circuit from Fig. 1.24, you may notice spikes in the current corresponding to when the tunnel diode switches states. These spikes are caused by the parasitic capacitance in the inductor whose current briefly rises to a large value due to the nearly instantaneous change in the inductor voltage. We added a 10 pF capacitor in the simulation to demonstrate this effect and to show how these spikes appear in I and distort the phase space plot.

Interestingly, the spikes are less noticeable if I is measured through an instrumentation amplifier (Fig. 1.23) across R only when it is between the frequency generator and the inductor. This arrangement results in less distortion in the phase space plot. The phase space plot also shows that the currents at which the tunnel diode switches voltages appear to shift slightly from their originally measured values of $V_p = 2$ V, $V_v = 4$ V, $I_p = 0.4$ mA, and $I_v = 0.2$ mA. However, the circuits in this chapter using the hysteretic behavior of the tunnel diode are not sensitive to these parameters and can still be used to obtain chaos.

The time plot of $I(t)$ in Fig. 4.8 shows a higher frequency of about 3100 Hz compared to the numerical waveform of about 1600 Hz. This difference is fairly sensitive to V_0 and V_1. A frequency closer to 1600 Hz can be experimentally obtained by setting $V_0 = 2.5$ V and $V_1 = 0.16$ V.

Component values are given in Table 4.1 where both the amplitude V_1 and offset voltage V_0 of the sinusoidal waveform can be adjusted through the function generator to obtain chaos. This circuit consists of an inductor and tunnel diode for a total of two components.

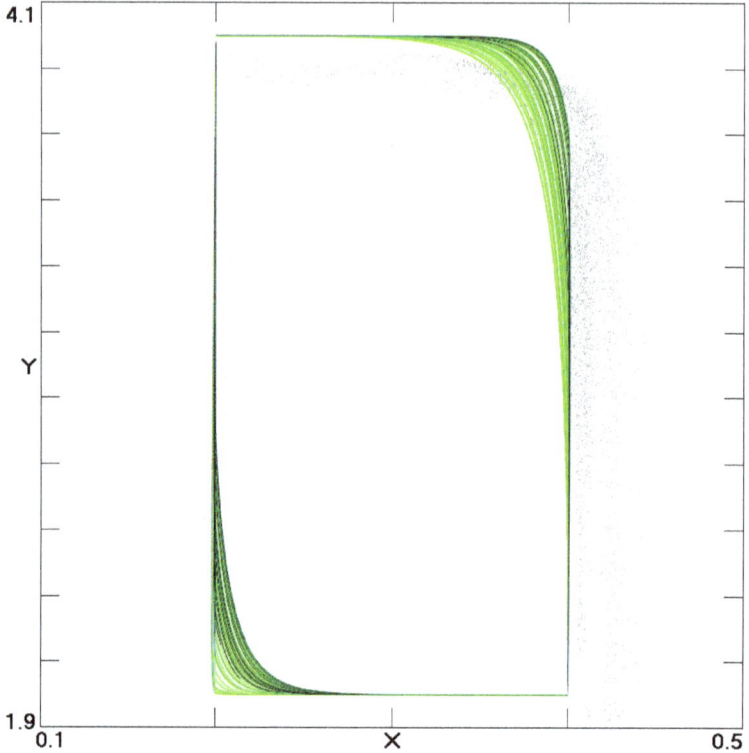

Fig. 4.5 Numerical solution of Eq. (4.3) with $a = 3$, $b = 0.5$, $c = 5$, and $d = 500$.

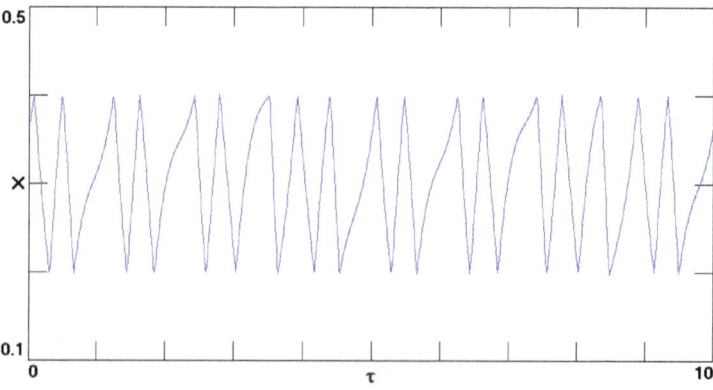

Fig. 4.6 Numerical waveform for $x(\tau)$ from Eq. (4.3) with $a = 3$, $b = 0.5$, $c = 5$, and $d = 500$.

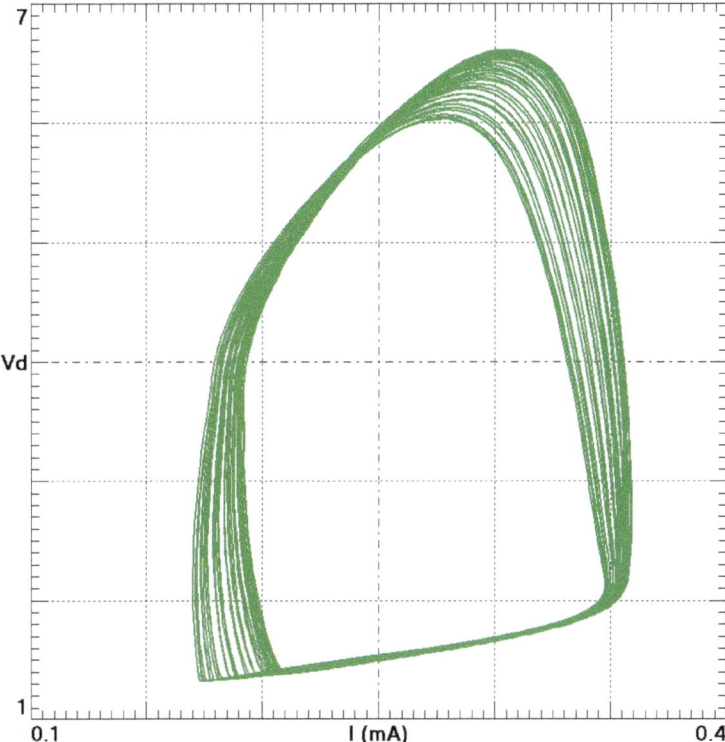

Fig. 4.7 Oscilloscope phase space plot of the forced relaxation oscillator from Fig. 4.4.

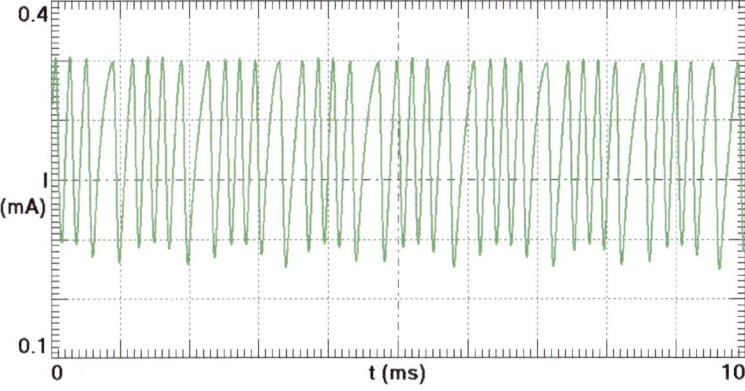

Fig. 4.8 Oscilloscope time plot of the forced relaxation oscillator from Fig. 4.4.

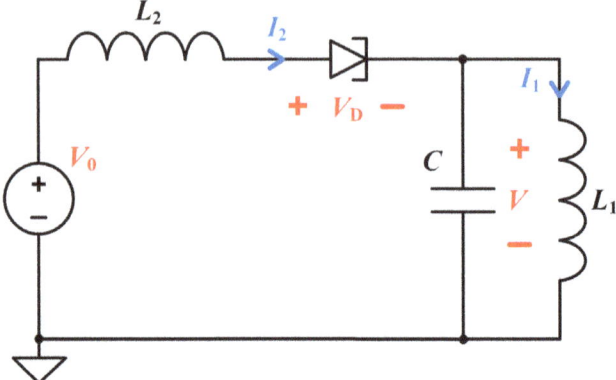

Fig. 4.9 Autonomous relaxation oscillator described by Eq. (4.4).

Table 4.2 Component values for Fig. 4.9.

Component	Numerical	Simulated	Experimental
R	0.5 kΩ	0.5 kΩ	0.5 kΩ
C	0.2 µF	0.2 µF	0.2 µF
L_1	0.2 H	0.2 H	0.2 H (gyrator)
L_2	1 H	1 H	1 H
V_0	3 V	2.3 V	2.3 V (adjust)
V_p	2 V	2 V	2 V
V_v	4 V	4 V	4 V
I_p	0.4 mA	0.4 mA	0.4 mA
I_v	0.2 mA	0.2 mA	0.2 mA
Δt	0.002 ms	-	-
Tunnel Diode	-	Emulator	Emulator

4.3 Autonomous Relaxation Oscillator

Following a suggestion of Bernhardt (1991), we replaced the AC source in Fig. 4.4 with a high-Q LC circuit as shown in Fig. 4.9 and found that the resulting autonomous circuit can also exhibit chaos. This circuit is the electrical analog of the chaotic dripping faucet [Shaw (1984)] in which the release of a water drop corresponds to the tunnel diode switching states, causing the water left behind to oscillate, thereby affecting when the next drip occurs. The switching transient excites oscillations in the LC circuit, which would otherwise decay exponentially.

The equations that describe the circuit in Fig. 4.9 are

$$\dot{V} = (I_2 - I_1)/C$$
$$\dot{I}_1 = V/L_1$$
$$\dot{I}_2 = (V_0 - V_D - I_2 R - V)/L_2 \quad (4.4)$$
$$\dot{V}_D = \begin{cases} (V_p - V_D)/\Delta t & I_2 < I_v \text{ or } \dot{V}_D < 0 \\ (V_v - V_D)/\Delta t & I_2 > I_p \text{ or } \dot{V}_D > 0. \end{cases}$$

The time Δt is the switching time for the hysteresis loop and is typically a few microseconds. The hysteresis loop is assumed to be bounded by $V_p = 2$ V, $V_v = 4$ V, $I_p = 0.4$ mA, and $I_v = 0.2$ mA.

Elegant circuit values that give chaos are $R = 0.5$ kΩ, $C = 0.2$ μF, $L_1 = 0.2$ H, $L_2 = 1$ H, and $V_0 = 3$ V. The dimensionless equations are

$$\dot{x} = a(z - y)$$
$$\dot{y} = bx$$
$$\dot{z} = c - u - dz - x \quad (4.5)$$
$$\dot{u} = \begin{cases} e(2 - u) & z < 0.2 \text{ or } \dot{u} < 0 \\ e(4 - u) & z > 0.4 \text{ or } \dot{u} > 0. \end{cases}$$

The parameters $a = b = 5, c = 3, d = 0.5$, and $e = 500$ give the attractor shown in Fig. 4.10 and a time series as shown in Fig. 4.11. The chaos occurs over a narrow range of the parameters near the onset of oscillation, with most of the solutions being stable equilibria, limit cycles, or tori. The Lyapunov exponents and Kaplan–Yorke dimension cannot be easily determined because of the discontinuity in the diode voltage and current, but chaos was confirmed by the method described in the previous section. The attractor has a global Class 1a basin of attraction.

The circuit uses the tunnel diode emulator from Fig. 4.3 with the parameters $V_p = 2$ V, $V_v = 4$ V, $I_p = 0.4$ mA, and $I_v = 0.2$ mA. These parameters were experimentally obtained by setting the resistances to $R_{D1} = 90.7$ kΩ, $R_{D2} = 20.9$ kΩ, $R_{D3} = 21.25$ kΩ, $R_4 = 3.8$ kΩ, and $R_5 = 50$ Ω with 2N3904 transistors.

It is necessary to implement L_1 using the gyrator from Fig. 1.25 to provide an LC circuit with sufficiently high Q, and we provide both an inductor and gyrator version of L_1 in the circuit simulation, where L_1 required a small resistance. On the other hand, you should add an extra 0.5 kΩ resistor in series with L_2 for R.

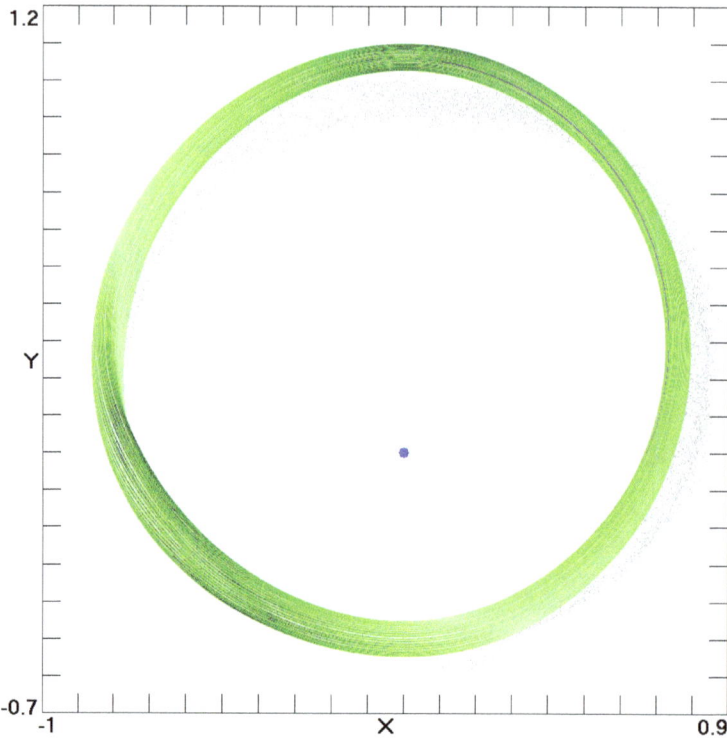

Fig. 4.10 Numerical solution of Eq. (4.5) with $a = b = 5$, $c = 3$, $d = 0.5$, and $e = 500$.

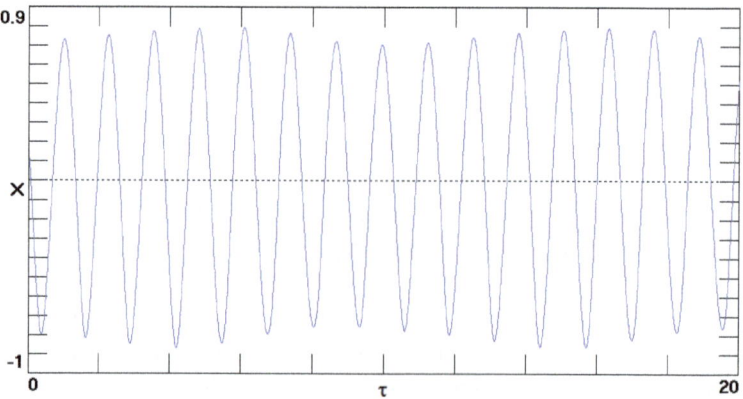

Fig. 4.11 Numerical waveform for $x(\tau)$ from Eq. (4.5) with $a = b = 5$, $c = 3$, $d = 0.5$, and $e = 500$.

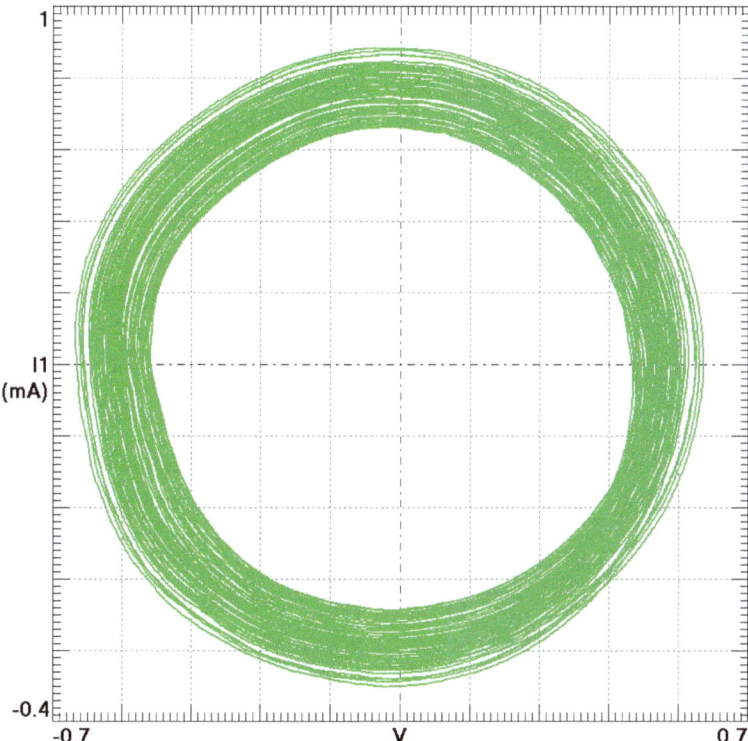

Fig. 4.12 Oscilloscope phase space plot of the autonomous relaxation oscillator from Fig. 4.9.

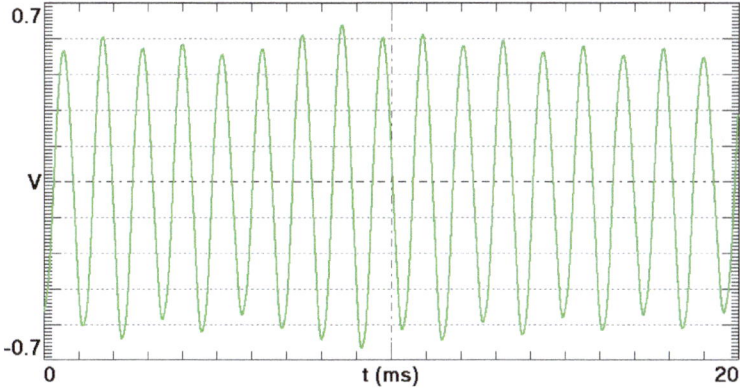

Fig. 4.13 Oscilloscope time plot of the autonomous relaxation oscillator from Fig. 4.9.

The voltage V_0 is set by a buffer and voltage divider from Fig. 1.29 since a dual voltage supply is required for the gyrator. The main parameter in Table 4.2 to vary for chaos is V_0, but the chaos is not sensitive to this value, and high precision components are also not required in this circuit.

The experimental phase space plot of I_1 versus V is shown in Fig. 4.12, and the time plot of $V(t)$ is shown in Fig. 4.13, where the current I_1 can be found by measuring the current through R_1 of the gyrator with V already at one end. This circuit has one capacitor, two inductors, and one tunnel diode for a total of four components.

4.4 Chua Tunnel Diode Oscillator

Ever since Matsumoto (1984) described Chua's circuit, there have been countless published variants with different configurations and nonlinear devices. Abdomerovic *et al.* (2000) proposed the first variant based on a tunnel diode that acts as an active device and requires a voltage bias V to operate in the negative resistance region. Without this bias, the oscillation will inevitably damp out.

However, our investigation of this circuit and its model revealed no indication of chaos even using the parameters given by Abdomerovic *et al.* (2000). We note that they did not actually build the circuit but only modeled it using a questionable numerical integrator, and the load line is nearly tangent to the I-V curve, so that their attractor is tiny. We believe it would be very hard to make their circuit work as designed, and that at best it would be very delicate.

Shi and Ran (2004) modified this circuit by using a blocking capacitor and a large inductor to prevent a resistive voltage divider used to adjust V from affecting the tunnel diode. We found that their circuit is still chaotic if a capacitor and resistor are eliminated, leading to the circuit shown in Fig. 4.14.

Pikovsky and Rabinovich (1981) reported another tunnel diode circuit based on a circuit by van der Pol (1920) which used a vacuum tube and transformer to act as a linear active resistor. However, the chaos in that circuit does not depend on the properties of the tunnel diode, which can be replaced by a conventional diode, and thus their circuit is a member of the Saito family described in Section 2.8.

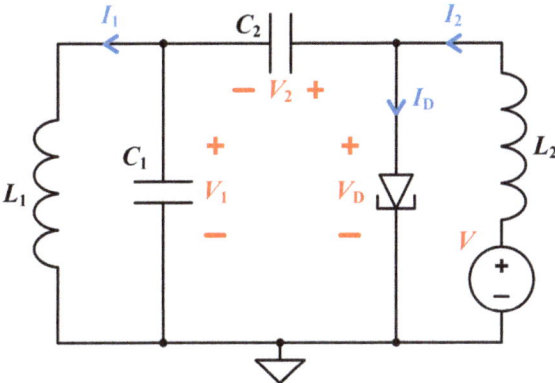

Fig. 4.14   Chua tunnel diode oscillator described by Eq. (4.6).

Table 4.3   Component values for Fig. 4.14.

Component	Numerical	Simulated	Experimental
C_1	0.3 μF	0.3 μF	0.3 μF (PP, ±3%)
C_2	1 μF	1 μF	1 μF (PP, ±3%)
L_1	0.8 H	0.8 H	0.8 H (Gyrator)
L_2	1 H	1.5 H	1.5 H (Gyrator)
V	1.6 V	2.5 V	1.6 V
V_P	1 V	2.2 V	1.8 V
V_t	1 V	-	-
I_P	1 mA	6.7 mA	4.8 mA
I_S	0.001 mA	-	-
Tunnel Diode	Eq. (4.1)	Emulator	Emulator

The equations that describe the circuit in Fig. 4.14 are

$$\begin{aligned}
\dot{V}_1 &= (I_2 - I_D - I_1)/C_1 \\
\dot{V}_2 &= (I_2 - I_D)/C_2 \\
\dot{I}_1 &= V_1/L_1 \\
\dot{I}_2 &= (V - V_2 - V_1)/L_2,
\end{aligned} \qquad (4.6)$$

where I_D is given by Eq. (4.1) with $V_D = V_1 + V_2$.

Elegant circuit values that give chaos are $C_1 = 0.3$ μF, $C_2 = 1$ μF, $L_1 = 0.8$ H, $L_2 = 1$ H, $V = 1.6$ V, $I_S = 0.001$ mA, $V_t = 1$ V, $I_p = 1$ mA, and $V_p = 1$ V. The dimensionless equations are

$$\dot{x} = (u - \iota - z)/a$$
$$\dot{y} = u - \iota$$
$$\dot{z} = x/b \qquad (4.7)$$
$$\dot{u} = c - y - x,$$

where $\iota = 0.001(e^{x+y} - 1) + (x+y)e^{1-x-y}$.

The parameters $a = 0.3$, $b = 0.8$, and $c = 1.6$ give the attractor shown in Fig. 4.15 and a time series as shown in Fig. 4.16. The Lyapunov exponents are (0.0238, 0, −0.1117, −1.4336), the Kaplan–Yorke dimension is $D_{ky} = 2.2127$, and the attractor has a Class 3 basin of attraction with $P \approx 2/r^{0.5}$. The system has an unstable focus at (0, 1.6, 0, 0.88205) with eigenvalues $(0.6614 \pm 2.7245i, 0.0414 \pm 0.7269i)$.

This circuit is particularly challenging to construct since the chaos is sensitive to the value of the capacitors and the inductors listed in Table 4.3 (a 1.5% variation in their values is likely to destroy the chaos). We used polypropylene capacitors for C_1 and C_2 since they are available at a higher precision ($\pm 3\%$) compared to ceramic capacitors (usually $\pm 10\%$).

The circuit is also sensitive to the value of the parasitic resistances in the inductors and must be emulated using the gyrator from Fig. 1.25. Note that the grounded end of the gyrator corresponding to L_2 should instead be connected to the voltage source V, which you will need to set using the buffer and voltage divider from Fig. 1.29 since the gyrators require a dual power supply.

The main parameters to adjust for chaos are the resistors in the tunnel diode emulator. We experimentally obtained $V_p = 1.8$ V and $I_p = 4.8$ mA through $R_{D1} = 35.9$ kΩ, $R_{D2} = 19.1$ kΩ, and $R_{D3} = 30.4$ kΩ with 2N3904 transistors. You should be able to obtain a limit cycle similar to the attractor for similar resistance values, and then you should observe chaos by varying each of the resistances slightly. This particular tunnel diode chaotic circuit relies on the negative resistance region of I_{tunnel} and does not require I_{diode}, so you can replace R_{D4} with an open circuit and R_{D5} with a short circuit.

The experimental plot of V_2 versus V_1 is shown in Fig. 4.17, and the time series of $V_1(t)$ is in Fig. 4.18. The differences between the experimental and numerical plots are attributed to the tunnel diode emulator which does not well fit Eq. (4.1). The circuit has two capacitors, two inductors, and a tunnel diode for a total of five components.

4.5 Coupled Relaxation Oscillator

A simple (nonchaotic) relaxation oscillator can be created by using a tunnel diode in series with an inductor L_1 powered by a DC source V. This DC source must bias the tunnel diode so that the operating point is in its negative resistance region. When the circuit is energized, the current through the inductor begins to rise until it reaches I_p in Fig. 4.2, whereupon the diode voltage abruptly increases from V_p to V_{pp}. Since the polarity of the inductor is reversed at this point ($V_D > V$), the inductor current begins to decrease until it reaches I_v, whereupon it jumps again to the left-hand side of the curve to repeat the oscillation in a hysteresis loop.

A single such oscillator cannot exhibit chaos, but Gollub et al. (1978) added a second relaxation oscillator and coupled them with a resistor R as shown in Fig. 4.19. A variation of this circuit was also described by Gollub et al. (1980) where R instead couples the two relaxation oscillators directly, but we only found quasiperiodic solutions and no chaos for that circuit.

The equations that describe the circuit in Fig. 4.19 are

$$\dot{I}_1 = [V - R(I_1 + I_2) - V_{D1}]/L_1$$
$$\dot{I}_2 = [V - R(I_1 + I_2) - V_{D2}]/L_2$$
$$\dot{V}_{D1} = \begin{cases} (V_p - V_{D1})/\Delta t & I_1 < I_v \text{ or } \dot{V}_{D1} < 0 \\ (V_v - V_{D1})/\Delta t & I_1 > I_p \text{ or } \dot{V}_{D1} > 0 \end{cases} \quad (4.8)$$
$$\dot{V}_{D2} = \begin{cases} (V_p - V_{D2})/\Delta t & I_2 < I_v \text{ or } \dot{V}_{D2} < 0 \\ (V_v - V_{D2})/\Delta t & I_2 > I_p \text{ or } \dot{V}_{D2} > 0. \end{cases}$$

The time Δt is the switching time for the hysteresis loop and is typically a few microseconds. The hysteresis loop is assumed to be bounded by $V_p = 2$ V, $V_v = 4$ V, $I_p = 0.4$ mA, and $I_v = 0.2$ mA.

Elegant circuit values that give chaos are $R = 1$ kΩ, $L_1 = 1$ H, $L_2 = 1.6$ H, and $V = 2.8$ V. The dimensionless equations are

$$\dot{x} = a - x - y - u$$
$$\dot{y} = (a - x - y - v)/b$$
$$\dot{u} = \begin{cases} c(2 - u) & x < 0.2 \text{ or } \dot{u} < 0 \\ c(4 - u) & x > 0.4 \text{ or } \dot{u} > 0 \end{cases} \quad (4.9)$$
$$\dot{v} = \begin{cases} c(2 - v) & y < 0.2 \text{ or } \dot{v} < 0 \\ c(4 - v) & y > 0.4 \text{ or } \dot{v} > 0. \end{cases}$$

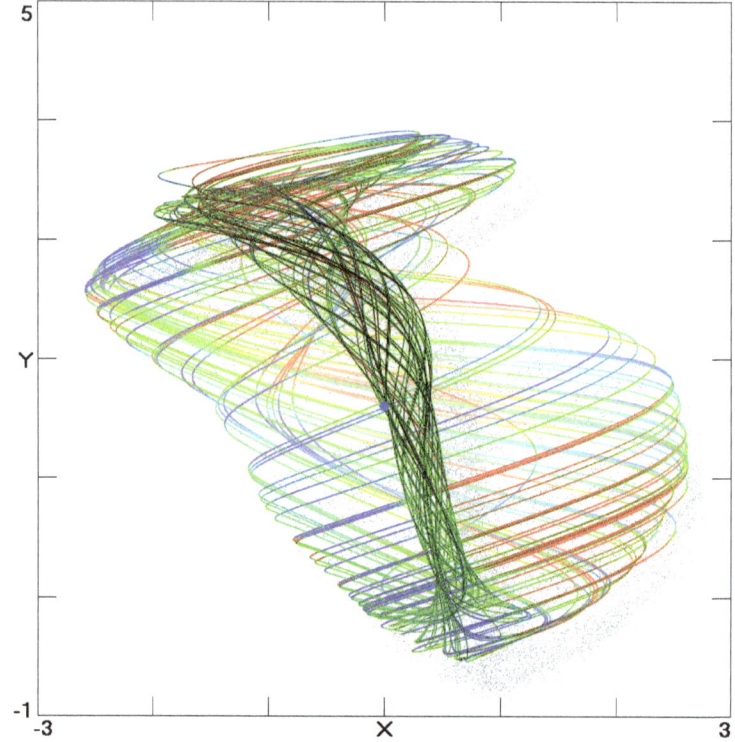

Fig. 4.15 Numerical solution of Eq. (4.7) with $a = 0.3$, $b = 0.8$, and $c = 1.6$.

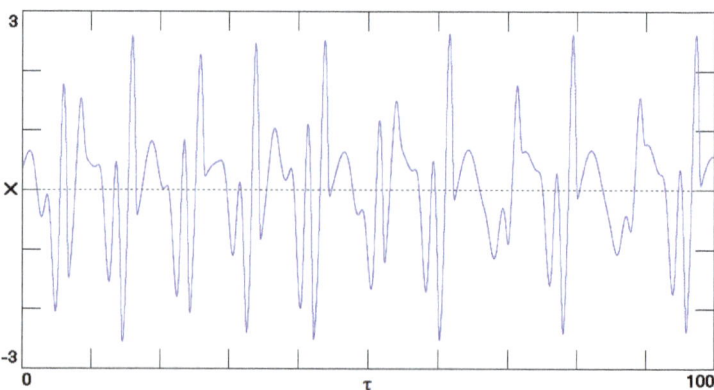

Fig. 4.16 Numerical waveform for $x(\tau)$ from Eq. (4.7) with $a = 0.3$, $b = 0.8$, and $c = 1.6$.

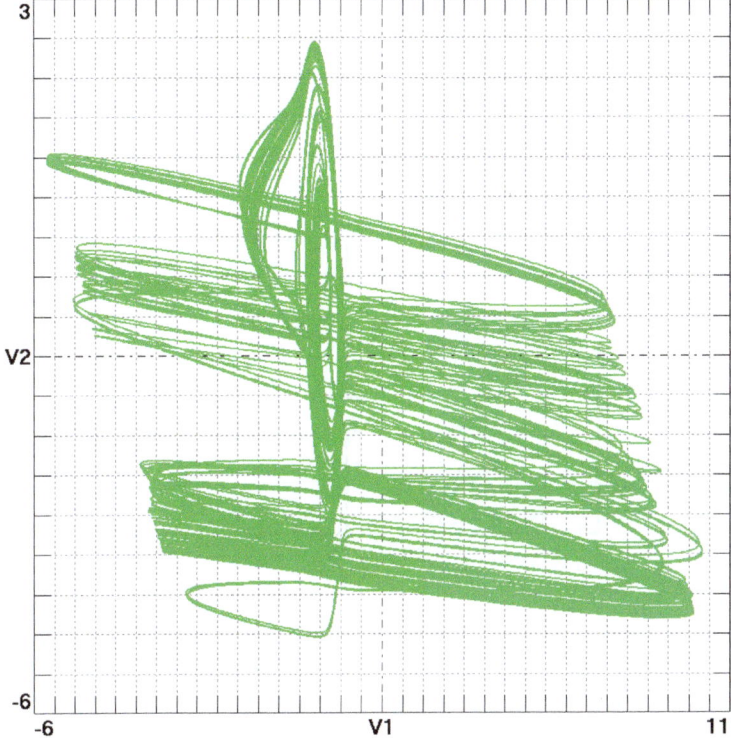

Fig. 4.17 Oscilloscope phase space plot of the Chua tunnel diode oscillator from Fig. 4.14.

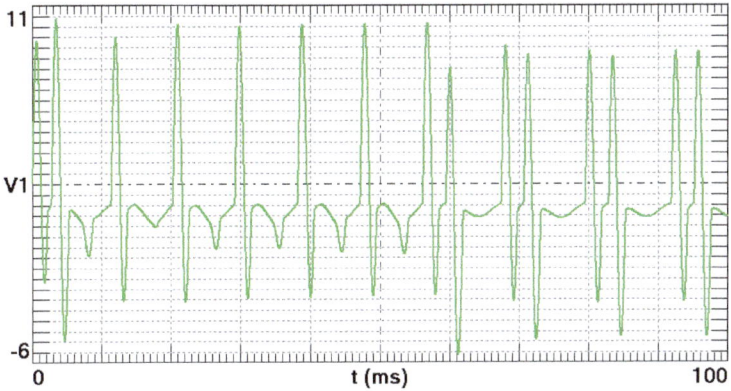

Fig. 4.18 Oscilloscope time plot of the Chua tunnel diode oscillator from Fig. 4.14.

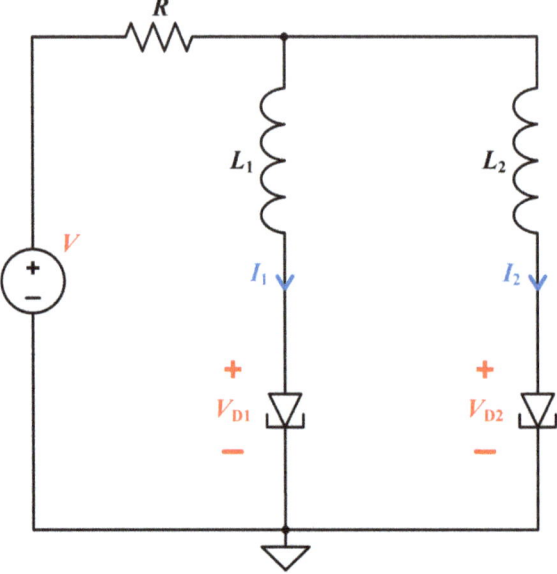

Fig. 4.19 Coupled relaxation oscillator described by Eq. (4.8).

Table 4.4 Component values for Fig. 4.19.

Component	Numerical	Simulated	Experimental
R	1 kΩ	1 kΩ	1 kΩ
L_1	1 H	1 H	1 H
L_2	1.6 H	1.6 H	1.6 H
V	2.8 V	2.8 V	2.8 V (adjust)
V_p	2 V	2 V	2 V
V_v	4 V	4 V	4 V
I_p	0.4 mA	0.4 mA	0.4 mA
I_v	0.2 mA	0.2 mA	0.2 mA
Δt	0.002 ms	-	-
Tunnel Diode	-	Emulator	Emulator

The parameters $a = 2.8$, $b = 1.6$, and $c = 500$ give the attractor shown in Fig. 4.20 and a time series as shown in Fig. 4.21. The chaos occurs over a narrow range of the parameters near the onset of oscillation, with most of the solutions being stable equilibria, limit cycles, or tori. The Lyapunov exponents and Kaplan–Yorke dimension cannot be easily determined because of the discontinuity in the diode voltages, but chaos was confirmed by the method previously described in Section 4.2. The attractor has a global Class 1a basin of attraction.

This circuit uses two tunnel diode emulators from Fig. 4.3, both of which have the parameters $V_p = 2$ V, $V_v = 4$ V, $I_p = 0.4$ mA, and $I_v = 0.2$ mA. We experimentally obtained these parameters by setting the resistances in one of our emulators to $R_{D1} = 80.4$ kΩ, $R_{D2} = 19.9$ kΩ, $R_{D3} = 38$ kΩ, $R_4 = 5$ kΩ, and $R_5 = 50$ Ω with 2N3904 transistors. The second emulator used slightly different resistances due to the varying parameters of the transistors and were chosen as $R_{D1} = 77.8$ kΩ, $R_{D2} = 18.4$ kΩ, $R_{D3} = 33.3$ kΩ, $R_4 = 4.9$ kΩ, and $R_5 = 50$ Ω also with 2N3904 transistors. The parameters of these emulators do not need to be exactly matched, although we did so to have better agreement with the numerical results.

This circuit has mostly quasi-periodic behavior, and it may be difficult to distinguish this from chaos. The most important parameter to adjust in Table 4.4 is V, and you should see chaos when it is adjusted to the point where the oscillations nearly cease. The phase space plot of I_2 versus I_1 in Fig. 4.22 shows that the values of current at which the tunnel diode emulator switches appear to shift from their originally measured values of $I_p = 0.4$ mA and $I_v = 0.2$ mA as previously mentioned in Section 4.2. The time series of I_1 in Fig. 4.23 also shows a higher frequency than the numerical value, which is due to the frequency of oscillation depending sensitively on the value of V.

The amplitudes of I_1 and I_2 are small and difficult to measure using a current-sensing resistor, so we instead used the I-V converter circuit from Fig. 1.24 with $R = 1$ kΩ and $C = 10$ nF. As previously explained in Section 4.2, the parasitic capacitance of the inductor leads to spikes in the inductor current, which can be removed by the I-V converter but still results in some distortion of the phase space plot. These spikes can be reproduced in the simulation by adding 10 pF capacitors in parallel with the inductors. The parasitic resistances in our inductors were negligible with 39 Ω for L_1 and 50 Ω for L_2. This circuit has a resistor, two inductors, and two tunnel diodes for a total of five components.

This circuit is particularly suited for extension to a higher-dimensional circuit with additional parallel inductors and tunnel diodes to simulate a complex coupled dynamical network.

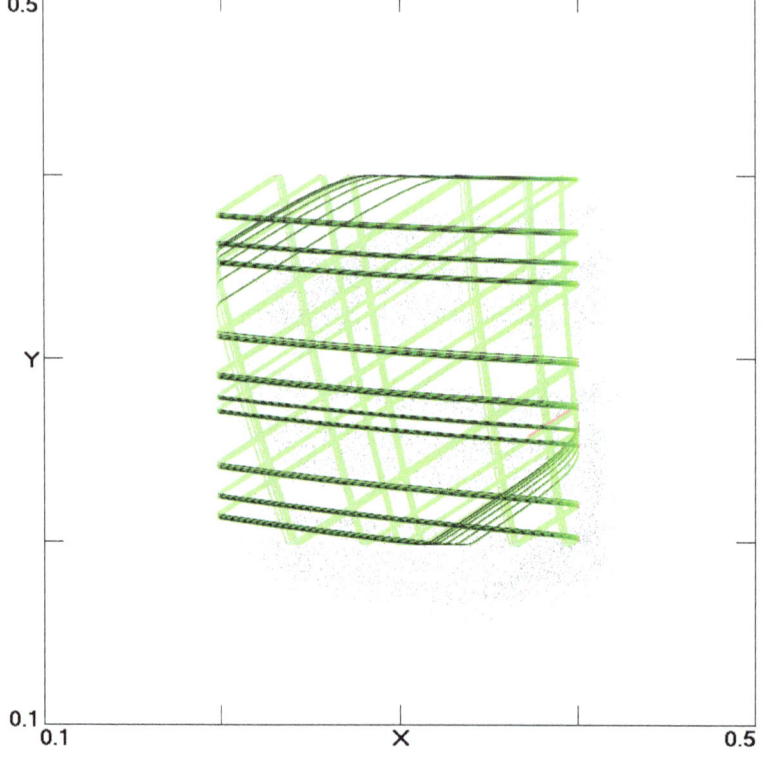

Fig. 4.20 Numerical solution of Eq. (4.9) with $a = 2.8$, $b = 1.6$, and $c = 500$.

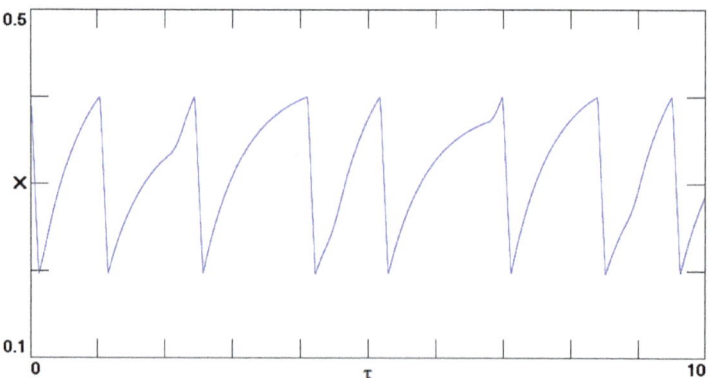

Fig. 4.21 Numerical waveform for $x(\tau)$ from Eq. (4.9) with $a = 2.8$, $b = 1.6$, and $c = 500$.

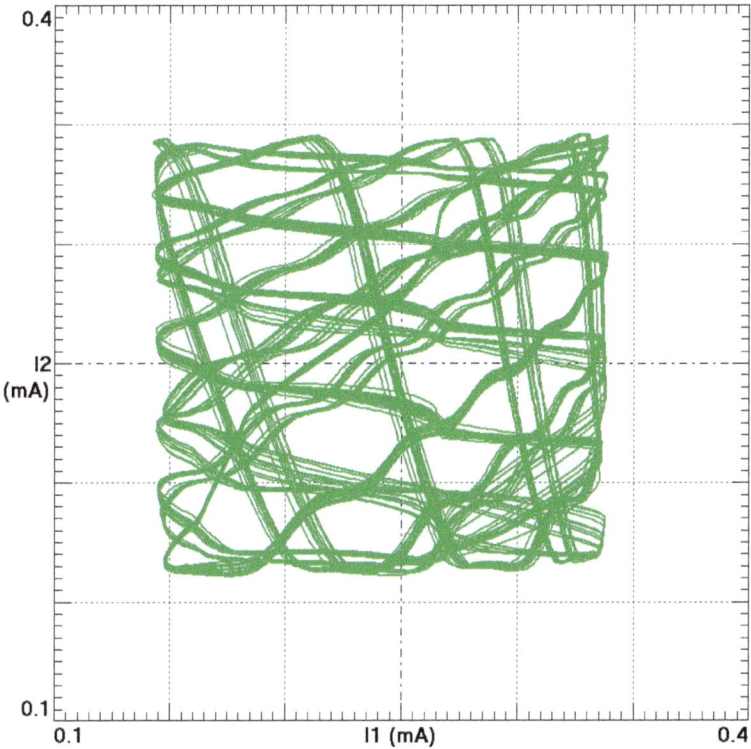

Fig. 4.22 Oscilloscope phase space plot of the coupled relaxation oscillator from Fig. 4.19.

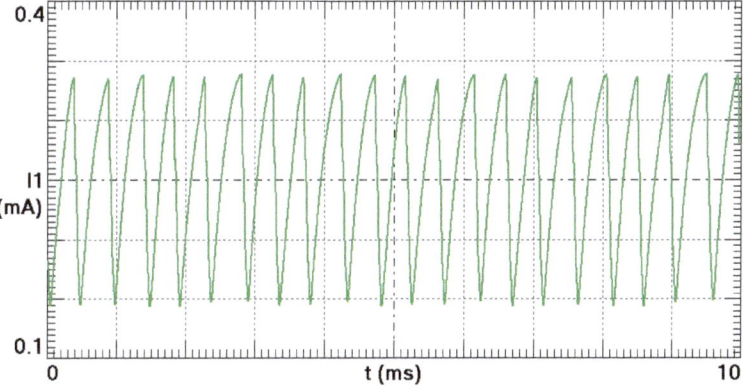

Fig. 4.23 Oscilloscope time plot of the coupled relaxation oscillator from Fig. 4.19.

Chapter 5

Thyristor Circuits

The thyristor is a semiconductor switch that is often used to control high-voltage and high-current circuits. Since the thyristor exhibits hysteresis, it can also be used as the nonlinear element in a wide range of chaotic circuits, several elegant examples of which are shown in this chapter.

5.1 Thyristor Characteristics

Thyristors encompass a large family of devices that operate as a switch with an 'on' and an 'off' state. They initially exist in the 'off' state by blocking the current and only begin to conduct when a voltage threshold called the *breakover voltage* (V_{BO}) is reached. Then they turn 'on' and act like a short circuit, carrying whatever current is applied. They will 'latch' in the 'on' state until the voltage goes to zero, thereby exhibiting hysteresis.

Thyristors have several advantages over mechanical switches, especially when controlling high-power circuits. In particular, electro-mechanical relay switches have moving parts that eventually wear out and can be destroyed by the large voltages that occur when abruptly opening a circuit with a large inductive load. This problem is circumvented by using semiconductor devices whose operation we now describe.

5.1.1 *Silicon controlled rectifier*

The majority of thyristors are based on the *silicon controlled rectifier* (SCR), which has three terminals, respectively called the anode (A), cathode (K), and gate (G) as shown in Fig. 5.1. In a practical application, a motor or high-power lamp would be in series with the anode and cathode and would initially be 'off' until V_{BO} is reached or until a small current pulse is applied to the gate.

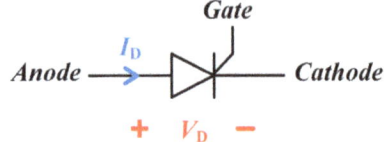

Fig. 5.1 Silicon controlled rectifier.

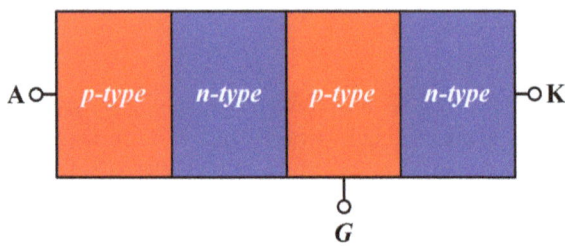

Fig. 5.2 PN-PN junction.

The silicon controlled rectifier as shown in Fig. 5.2 consists of a *PN-PN junction* whose operation can be understood by considering the equivalent circuit with two bipolar junction transistors as shown in Fig. 5.3. Each transistor controls the base current entering the other transistor. When a voltage below V_{BO} is initially applied from anode to cathode, both transistors are in cutoff mode since no base current can flow through either one.

When the voltage reaches V_{BO}, one of the transistors will conduct some collector current. If the NPN transistor were to begin conducting in this case, current will flow out of the base of the PNP transistor and cause it to enter saturation mode. This in turn allows current to enter the base of the NPN transistor and causes it also to enter saturation mode. The PN-PN junction is turned on when both transistors are in saturation and will continue to conduct current from anode to cathode until the applied voltage is low enough so that both transistors return to cutoff mode.

The thyristor can also be switched on when a small current enters the gate, which leads to the base of the NPN transistor. We now show how to use this terminal to construct a more practical thyristor for use in our chaotic circuits.

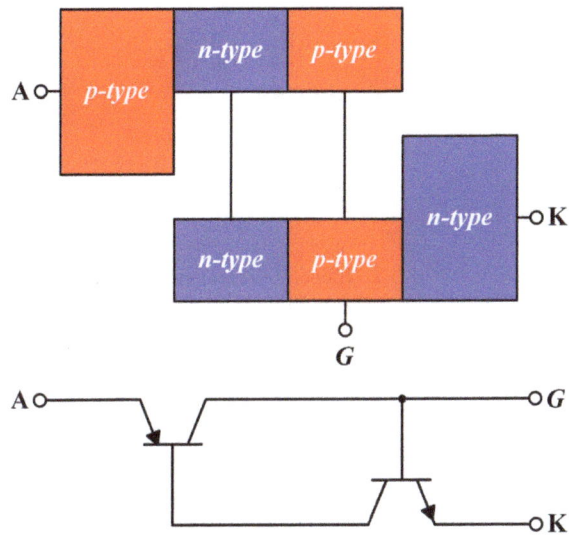

Fig. 5.3 PN-PN junction with transistor model consisting of two bipolar junction transistors.

5.1.2 *Silicon bilateral switch*

Most silicon controlled rectifiers have a breakover voltage of several hundred volts, which makes them impractical for our circuits. A type of thyristor with a much lower breakover voltage is the *silicon bilateral switch* (SBS), whose schematic is shown in Fig. 5.4. This device is based on two parallel silicon controlled rectifiers connected opposite to one another as shown in Fig. 5.5, allowing current to flow in either direction. Note that these silicon controlled rectifiers have their gates connected to the anodes rather than to the cathode in contrast to Fig. 5.1, which means that it switches on when a current pulse flows *out* of the gate.

A special type of diode called a *Zener diode* (labeled Z_1 and Z_2 in Fig. 5.5) can be used to lower the breakover voltage. This diode has the same forward-bias properties as a normal diode but will also conduct at a relatively low reverse-bias voltage V_Z. Such Zener diodes are connected to the gates of the silicon controlled rectifiers as shown in the equivalent circuit in Fig. 5.5. Any voltage applied to the silicon bilateral switch must reach a voltage $V_D = V_Z + V_0$ in order to turn on the silicon controlled rectifiers, where V_0 is the forward voltage drop of the diode from Eq. (2.2). The

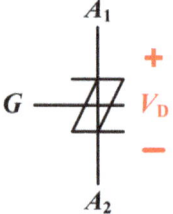

Fig. 5.4   Silicon bilateral switch.

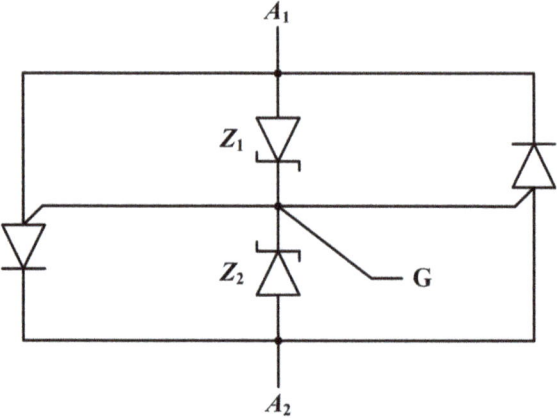

Fig. 5.5   Silicon bilateral switch equivalent circuit consisting of two SCRs and two Zener diodes.

silicon controlled switch we will be using (BS08D-T112) has a maximum switching voltage of 9 V, and switches off around 0.6 V.

The switching voltage of the silicon bilateral switch can be further reduced either by additional external Zener diodes or by resistors as shown in Fig. 5.6. In the latter case, the silicon bilateral switch turns on when the current through the resistor reaches the gate turn-on current. If the resistance is sufficiently low, the silicon bilateral switch will turn on for voltages smaller than V_{BO}. For example, in our device with $R_D = 100$ kΩ, the switching voltage is around 4 V. Note that this voltage will vary among devices, and so your results may vary.

To keep the circuits simple, we will assume a switching voltage of 4 V, but in most cases, the voltages and currents in the circuit will scale linearly with whatever switching voltage your device has. We also assume that the device will switch off at 0 V rather than the more typical 0.6 V to

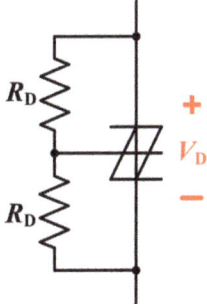

Fig. 5.6 Silicon bilateral switch with a voltage divider.

simplify the equations, although this approximation leads to some differences between the numerical and experimental results.

The silicon bilateral switch can also be made to switch at two different bidirectional breakover voltages if the two R_D resistors in Fig. 5.6 are different. Since the circuits in this chapter only require unidirectional switching, only one R_D is used to reduce V_{BO} across V_D, whereas the maximum breakover voltage is still required to switch in the opposite direction. Since the purpose of the resistors is simply to scale down the voltages and currents, we consider them as part of the thyristor and do not include them in our component count.

5.1.3 Thyristor I-V characteristic

The thyristor is called an *S-type* negative resistance device because of the I-V characteristic shown in Fig. 5.7 in contrast to the N-type negative resistance of the tunnel diode shown in Fig. 4.2. Chua *et al.* (1983) described several S-type negative resistance emulators that can be used as a substitute if desired.

With two states, the model for a thyristor is relatively straightforward. When $V_D < V_{BO}$, the device starts in the 'off' (nonconducting) state, $I_D = 0$, with a small leakage current that we will neglect. When $V_D \geq V_{BO}$, the device enters the 'on' (conducting) state and becomes a short circuit forcing V_D to near zero. Increasing the gate current I_G will lower the breakover voltage V_{BO} so that the thyristor can turn on for smaller applied voltages.

There are some chaotic circuits where the thyristor is less than a perfect switch and should be modeled instead as two coupled transistors from

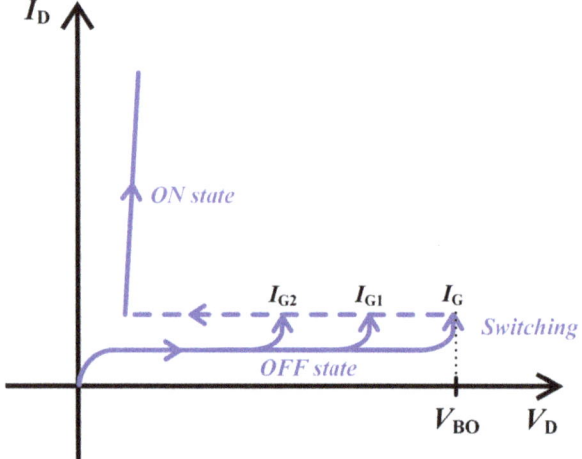

Fig. 5.7 Thyristor I_D-V_D curve with $I_{G2} > I_{G1} > I_G = 0$.

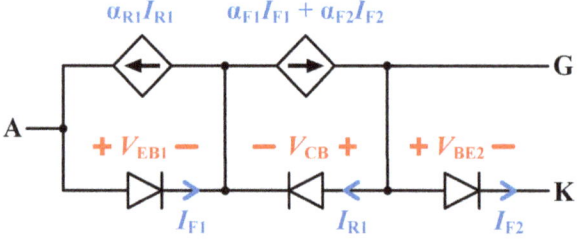

Fig. 5.8 Thyristor with Ebers–Moll model.

Fig. 5.3. For this reason, we adopt the thyristor model in Fig. 5.8 developed by Nienhaus *et al.* (1976) which uses the Ebers–Moll model from Fig. 3.6 to represent both bipolar junction transistors. Here V_{EB1} and V_{CB} represent the voltages across the PNP transistor junctions, and V_{BE2} is across the NPN transistor junction. Additionally, $\alpha_{R1}I_{R1}$ refers to the reverse active current of the PNP transistor, and $\alpha_{F1}I_{F1}+\alpha_{F2}I_{F2}$ refers to the sum of the forward active currents of both bipolar junction transistors. The reverse-active current of the NPN transistor can be neglected since it does not have a significant effect on the silicon controlled rectifier. The equations for this model are

$$I_{F1} = \max(0, V_{EB1} - V_0)/R_D$$
$$I_{R1} = \max(0, V_{CB} - V_0)/R_D \quad (5.1)$$
$$I_{F2} = \max(0, V_{BE2} - V_0)/R_D,$$

where V_0 is the forward voltage drop of the respective junction (typically 0.6 V), and R_D is the forward resistance (typically small but not zero).

5.2 Forced Thyristor Circuit

You can obtain chaos with a thyristor by simply driving a silicon controlled rectifier with a sinusoidal voltage source as shown in Fig. 5.9. This was first done by Hoh and Yasuda (1994) who were inspired by early investigations of chaos in physical phenomenon. In one of these studies, Aoki and Yamamoto (1989) found chaos in a material that exhibited an S-type negative resistance when an alternating voltage was applied to it.

Hoh and Yasuda (1994) suggested that a thyristor could be used instead and proposed a model that treated the thyristor as a switch with a junction capacitance from Eq. (2.5) and a delay time to turn on. However, this model had poor agreement with the experimental results. Irita et al. (1995) found that chaos instead arises from the coupled transistors described in Section 4.1.1, which gives a more accurate but less elegant model as shown in Fig. 5.10. Irita et al. (1998) later found a circuit where only the NPN bipolar junction transistor was needed for chaos, which Lindberg et al. (2005) subsequently simplified as we explained in Section 3.7.

This circuit works by charging C_{BE2} from V_2 through R_2 while the bipolar junction transistors switch between reverse-active mode when $V_S < 0$ and cutoff when $V_S > 0$. The voltage V_{BE2} gradually rises to V_0 during these cycles. If V_{BE2} reaches V_0 when $V_S > 0$, both bipolar junction transistors enter forward-active mode, and the thyristor begins conducting. The process repeats when C_{BE2} is discharged below V_0 for the following negative cycle.

The equations that describe the circuit in Fig. 5.10 are

$$\dot{V}_{EB1} = (I_D - I_{F1} + \alpha_{R1}I_{R1})/C_{EB1}$$
$$\dot{V}_{CB} = -(I_D + I_{R1} - \alpha_{F1}I_{F1} - \alpha_{F2}I_{F2})/C_{CB} \quad (5.2)$$
$$\dot{V}_{BE2} = [(I_D + (V_2 - V_{BE2})/R_2 - I_{F2}]/C_{BE2},$$

where $I_{F1} = \max(0, V_{EB1} - V_0)/R_D$, $I_{R1} = \max(0, V_{CB} - V_0)/R_D$ and

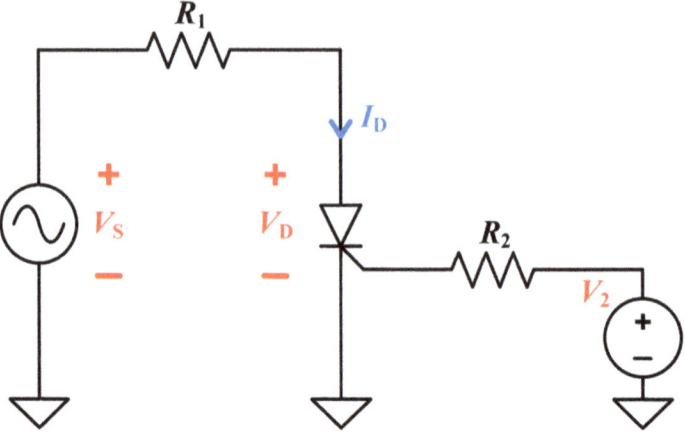

Fig. 5.9 Forced thyristor circuit.

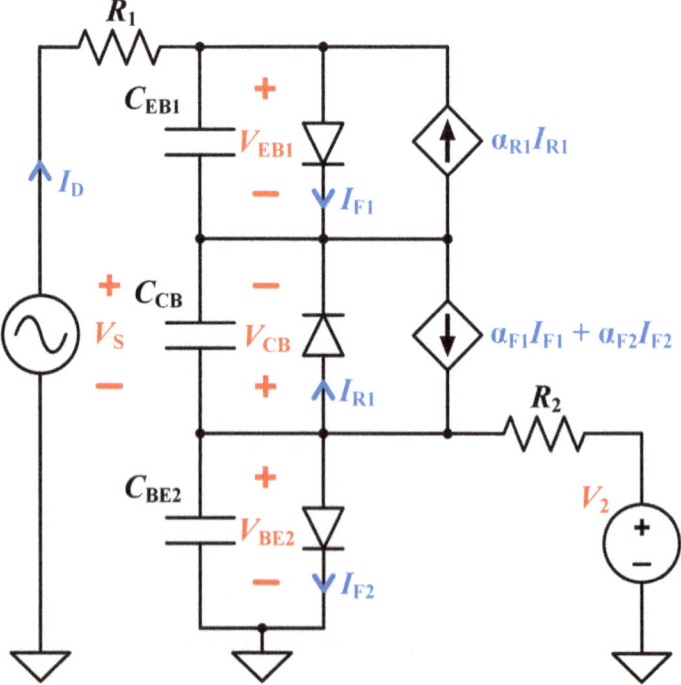

Fig. 5.10 Forced thyristor circuit with the thyristor model described by Fig. 5.8.

Table 5.1 Component values for Fig. 5.9.

Component	Numerical	Simulated	Experimental
R_1	4 kΩ	4 kΩ	4 kΩ (adjust)
R_2	50 kΩ	1 MΩ	48.5 kΩ (adjust)
R_D	0.1 kΩ	-	-
C_{EB1}	0.1 μF	-	-
C_{CB}	0.01 μF	-	-
C_{BE2}	0.1 μF	-	-
V_0	0.2 V	-	-
V_1	10 V	10 V	10 V (adjust)
V_2	2 V	10 V	2 V (adjust)
ω	1 rad/ms	62.8 rad/ms	100 rad/ms (adjust)
α_{R1}	0.6	-	-
α_{F1}	0.99	-	-
α_{F2}	0.99	-	-
Thyristor	-	Coupled BJTs	EC103M1

$I_{F2} = \max(0, V_{BE2} - V_0)/R_D$ from Eq. (5.1). Additionally, $I_D = (V_S - V_{EB1} + V_{CB} - V_{BE2})/R_1$ and $V_S = V_1 \sin \omega t$. The factors α_F and α_R satisfy $0 < \alpha_R < \alpha_F < 1$.

Elegant circuit values that give chaos are $R_1 = 4$ kΩ, $R_2 = 50$ kΩ, $R_D = 0.1$ kΩ, $C_{EB1} = C_{BE2} = 0.1$ μF, $C_{CB} = 0.01$ μF, $V_0 = 0.2$ V, $V_1 = 10$ V, $V_2 = 2$ V, $\omega = 1$ rad/ms, $\alpha_{R1} = 0.6$, and $\alpha_{F1} = \alpha_{F2} = 0.99$. The dimensionless equations are

$$\begin{aligned}\dot{x} &= a(\iota - \iota_1 + \alpha_R \iota_2) \\ \dot{y} &= -b(\iota + \iota_2 - \alpha_F \iota_1 - \alpha_F \iota_3) \\ \dot{z} &= a(\iota + e(c - z) - \iota_3),\end{aligned} \quad (5.3)$$

where $\iota = f(d \sin \tau - x + y - z)$, $\iota_1 = g \max(0, x - h)$, $\iota_2 = g \max(0, y - h)$, and $\iota_3 = g \max(0, z - h)$.

The parameters $a = 10$, $b = 100$, $c = 2$, $d = 10$, $e = 0.02$, $f = 0.25$, $g = 10$, $h = 0.2$, $\alpha_R = 0.6$, and $\alpha_F = 0.99$ give the attractor shown in Fig. 5.11 and a time series as shown in Fig. 5.12. The Lyapunov exponents are (0.0467, 0, 0, −0.9785), the Kaplan–Yorke dimension is $D_{ky} = 2.0477$, and the attractor has a global Class 1a basin of attraction. Because the system is periodically forced, it does not have an equilibrium point, but the unforced system has a stable node close to the origin.

The fact that the system has two zero Lyapunov exponents suggests that there is a constant of the motion and that the system can in principle be reduced to a two-dimensional system with a sinusoidal forcing. This is confirmed by a *Poincaré section* that is approximately one-dimensional and

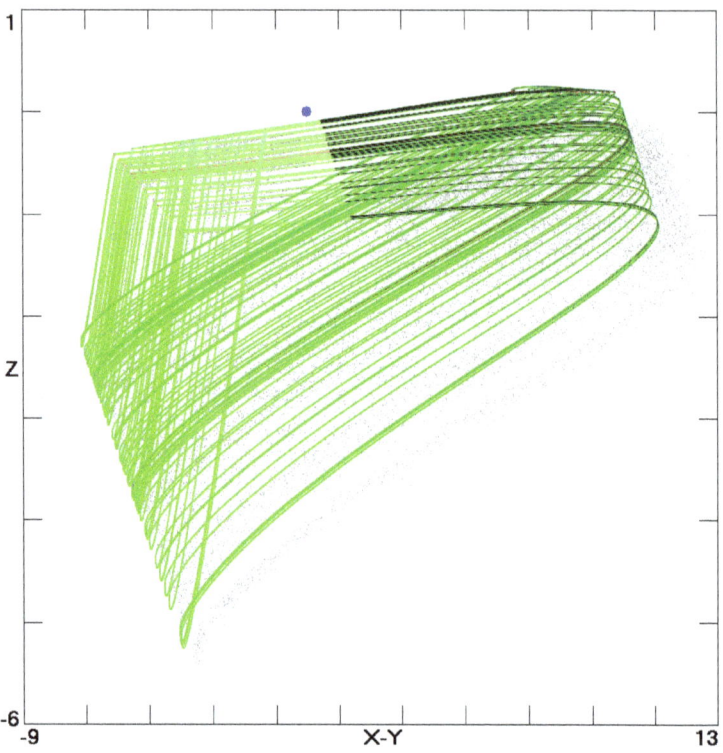

Fig. 5.11 Numerical solution of Eq. (5.3) with $a = 10$, $b = 100$, $c = 2$, $d = 10$, $e = 0.02$, $f = 0.25$, $g = 10$, $h = 0.2$, $\alpha_R = 0.6$, and $\alpha_F = 0.99$.

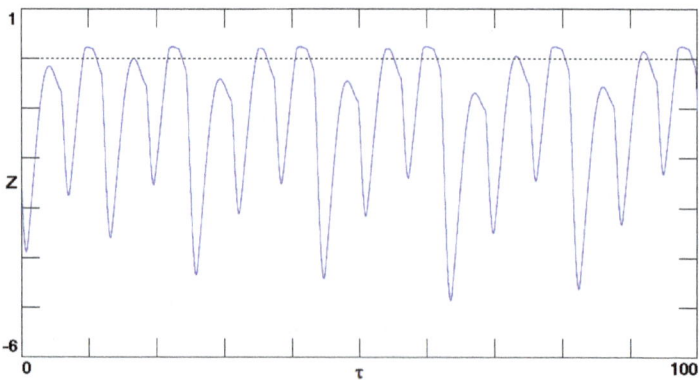

Fig. 5.12 Numerical waveform for $z(\tau)$ from Eq. (5.3) with $a = 10$, $b = 100$, $c = 2$, $d = 10$, $e = 0.02$, $f = 0.25$, $g = 10$, $h = 0.2$, $\alpha_R = 0.6$, and $\alpha_F = 0.99$.

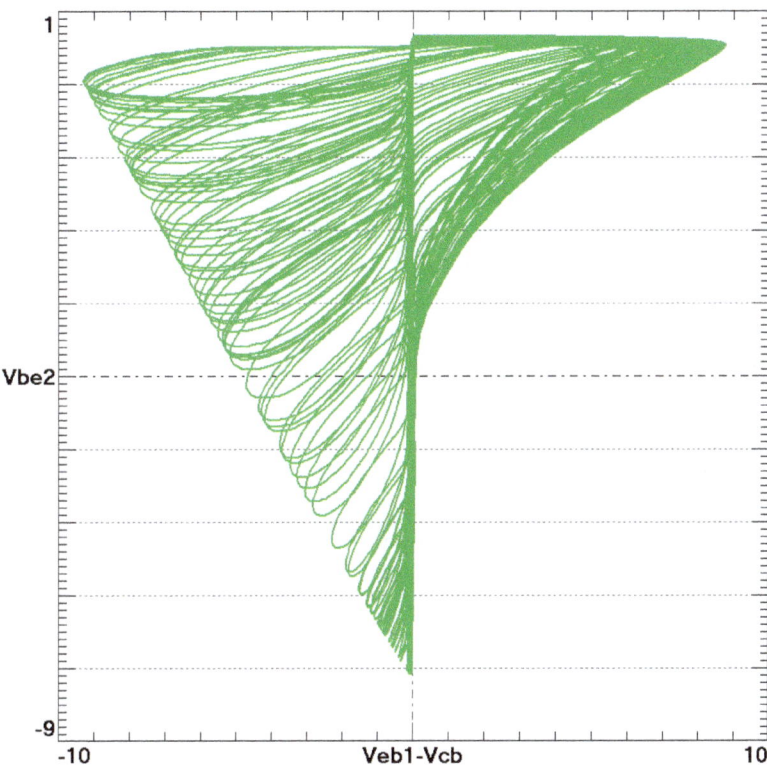

Fig. 5.13 Oscilloscope phase space plot of forced thyristor circuit from Fig. 5.9.

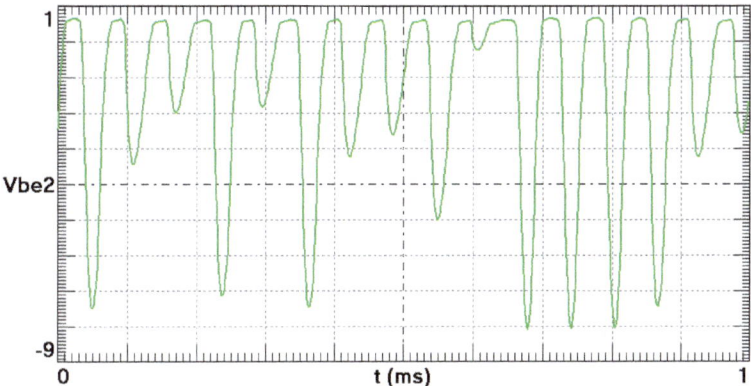

Fig. 5.14 Oscilloscope time plot of forced thyristor circuit from Fig. 5.9.

a *correlation dimension* that is approximately two-dimensional as is consistent with the Kaplan–Yorke dimension ignoring one of the zero Lyapunov exponents.

The experimental phase space plot of V_{BE2} versus $V_{EB1} - V_{CB}$ is shown in Fig. 5.13. Since it is not possible to view V_{EB1} or V_{CB} individually, you can instead measure $V_{EB1} - V_{CB}$ by taking the difference between the voltage at the anode and gate of the thyristor. The time plot of $V_{BE2}(t)$ is shown in Fig. 5.14 and is mostly negative while it charges to V_0. The thyristor then turns on when V_0 is reached.

Agreement between the numerical, simulated, and experimental results is poor since chaos is dependent on the nonlinear parasitic capacitances of the thyristor. In the numerical equations, we consider these capacitances to be constant, although using the nonlinear junction and diffusion capacitances described by Eq. (2.5) and Eq. (2.6) could lead to better agreement between the numerical and experimental results.

The simulation used the bipolar junction transistor parameters described by Irita *et al.* (1995) for the 2SB561 PNP and 2SC367 NPN bipolar junction transistors. These were coupled together as shown in Fig. 5.3 to model the thyristor.

Chaos is found more easily in a thyristor with a gate trigger current (I_{gt}) of 25 µA or less, which is the current required to turn on the silicon controlled rectifier. Other devices that we tested with this rated current that gave chaos were the P0118MA-2AL3 and EC103D1.

You can look for chaos by measuring the voltage at the anode of the thyristor and adjusting R_1, R_2, and V_2 until you observe a decrease in voltage at the positive cycle which corresponds to the thyristor turning on. The particular values we found using our device for these parameters are listed in Table 5.1. Note that V_2 should be close to V_0 so that V_{BE2} can reach the threshold voltage.

You can then adjust the angular frequency ω until you see period-doubling as an indication of impending chaos. Some devices may also require adjusting the offset voltage of V_S by +1 to +2 V, but this was not required with the EC103M1. This circuit has two resistors and a thyristor for a total of three components.

5.3 van der Pol Relaxation Oscillator

A nonlinear device with behavior similar to the thyristor is the neon lamp, which was used in many of the early chaotic circuits such as Fig. 1.16.

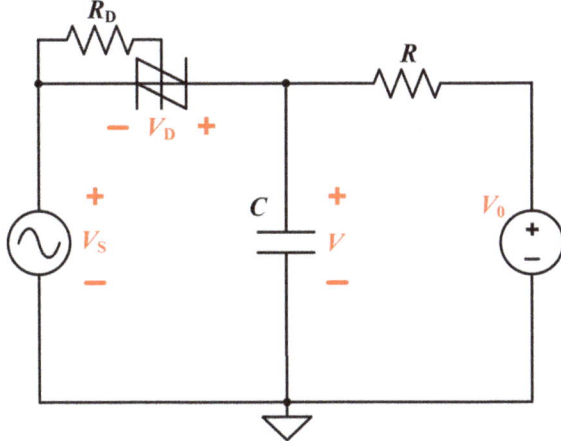

Fig. 5.15 van der Pol relaxation oscillator.

Table 5.2 Component values for Fig. 5.15.

Component	Numerical	Simulated	Experimental
R	360 kΩ	400 kΩ	348 kΩ (adjust)
R_D	-	-	100 kΩ
C	0.01 μF	0.01 μF	0.01 μF
V_0	20 V	20 V	25 V
V_1	1 V	5 V	0.6 V
V_{BO}	4 V	4 V	4 V
ω	1 rad/ms	1 rad/ms	1 rad/ms
Thyristor	-	Ideal switch	BS08D-T112

Like the thyristor, this lamp initially blocks current since it contains an insulating neon gas. When a sufficiently large voltage is reached, the neon gas is ionized, and a conducting path forms and persists until the applied voltage is lowered sufficiently.

The similarity of the neon lamp and the thyristor allows us to create a modern implementation of many of the early chaotic circuits that were first proposed using neon lamps. For example, the circuit in Fig. 5.15 is described by the following simple equations,

$$\begin{aligned} \dot{V} &= (V_0 - V)/RC \quad &\text{SBS off} \\ V &= V_S \quad &\text{SBS on}, \end{aligned} \quad (5.4)$$

where $V_S = V_1 \sin \omega t$. The current through the thyristor is zero until $V_D \geq$

V_{BO}, whereupon the device becomes a short circuit, forcing V_D to near zero.

There are a number of ways to model hysteresis of this sort, one of simplest of which is to instantly reset the capacitor voltage to V_S whenever $V - V_S$ exceeds V_{BO} and to assume the thyristor ceases conducting immediately thereafter. This assumption implies a large but brief current in the thyristor. Such a model will fail to describe the circuit if the thyristor continues conducting after the capacitor voltage has been brought to V_S, which can occur if the current does not also rapidly fall to a small value.

Elegant circuit values that give chaos are $R = 360$ kΩ, $C = 0.01$ μF, $V_0 = 20$ V, $V_1 = 1$ V, $V_{BO} = 4$ V, and $\omega = 1$ rad/ms. Note that for a given set of voltages, the only parameter in the model is the product ωRC, and the circuit may require a large value of R to make the thyristor turn off after discharging the capacitor. The chaos occurs in narrow ranges of the parameter surrounded by quasiperiodic solutions. The dimensionless equations are

$$\dot{x} = (a - x)/b$$
$$x = \sin \tau \text{ when } x - \sin \tau \geq c.$$
(5.5)

The parameters $a = 20$, $b = 3.6$, and $c = 4$ give the attractor shown in Fig. 5.16 and a time series as shown in Fig. 5.17. Because of the discontinuity in the equations, it is not possible to calculate the Lyapunov exponents, although we confirmed the chaos using the method described in Section 4.2. The attractor has a global Class 1a basin of attraction.

The component values for this circuit are given in Table 5.2 where the main parameter to adjust is R. It is important to confirm that your silicon bilateral switch has a similar V_{BO} using the method Section 5.1.2 since this parameter can vary among devices for the same R_D. We found chaos to occur for a sufficiently large R and small C.

As you adjust the parameters, avoid making R too small since the resulting current can damage the thyristor and function generator. Making R too small or V_1 too large can also cause the thyristor to become stuck in the conducting state, which causes $V = V_S$. You can fix this problem by reducing V_1 and restarting the power supply.

The experimental phase space plot of V versus V_S is shown in Fig. 5.18, and the time plot of $V(t)$ is shown in Fig. 5.19 where the difference from the numerical and simulated results is attributed to the ideal voltage-controlled switch with hysteresis used to model the silicon bilateral switch. The circuit has a resistor, a capacitor, and a thyristor for a total of three components.

5.4 Autonomous Relaxation Oscillator

You can also replace the sinusoidal voltage source in Fig. 5.15 with a passive LC circuit as shown in Fig. 5.20. Bernhardt (1991) first described such a circuit using a neon lamp as the switch which we replaced with a tunnel diode in Section 4.3. Elwakil and Kennedy (2000c) described several other variants of this circuit in which they converted a periodic oscillator [Elwakil (2000)] into a chaotic one using coupled bipolar junction transistors to emulate the switching behavior. The equations for this circuit are

$$\begin{aligned}\dot{V}_1 &= -I/C_1 & \text{SBS off} \\ \dot{V}_2 &= (V - V_2)/RC_2 & \text{SBS off} \\ \dot{I} &= V_1/L & \text{SBS off} \\ V_1 &= V_2 & \text{SBS on.}\end{aligned} \quad (5.6)$$

To implement the condition $V_1 = V_2$ when the thyristor turns on, we assume the total charge on the two capacitors is conserved and both capacitors instantly acquire a voltage $V_0 = (C_1 V_1 + C_2 V_2)/(C_1 + C_2)$, whereupon the thyristor turns off. This implies a large but brief current through the thyristor and a loss of electrical energy through parasitic resistance in the thyristor.

Elegant circuit values that give chaos are $R = 200$ kΩ, $C_1 = C_2 = 0.1$ μF, $L = 0.1$ H, $V = 5$ V, and $V_{BO} = 4$ V. The dimensionless equations are

$$\begin{aligned}\dot{x} &= -az \\ \dot{y} &= (d-y)/b \\ \dot{z} &= cx \\ x &= y = (x+y)/2 \text{ when } y - x \geq 4.\end{aligned} \quad (5.7)$$

The parameters $a = 10$, $b = 20$, $c = 10$, and $d = 5$ give the attractor shown in Fig. 5.21 and a time series as shown in Fig. 5.22. Because of the discontinuity in the equations, it is not possible to calculate the Lyapunov exponents, although we confirmed the chaos using the method described in Section 4.2. The attractor has a global Class 1a basin of attraction.

It is critical that the LC circuit composed of L and C_1 has a high quality factor; otherwise, it will be difficult to obtain chaos. You can look at the time plot of $V_1(t)$ in Fig. 5.24 as an example, where the LC circuit oscillates with minimal damping. Since $Q = R\sqrt{C/L}$ for a parallel resonant circuit, you will need a large value for C_1 and a small value for L.

Since the quality factor will be reduced by the parasitic resistance of a real inductor, we instead used the gyrator from Fig. 1.25 with the parameters $R_1 = R_2 = R_3 = R_4 = 1$ kΩ and $C = 1$ nF to implement $L = 0.001$ H.

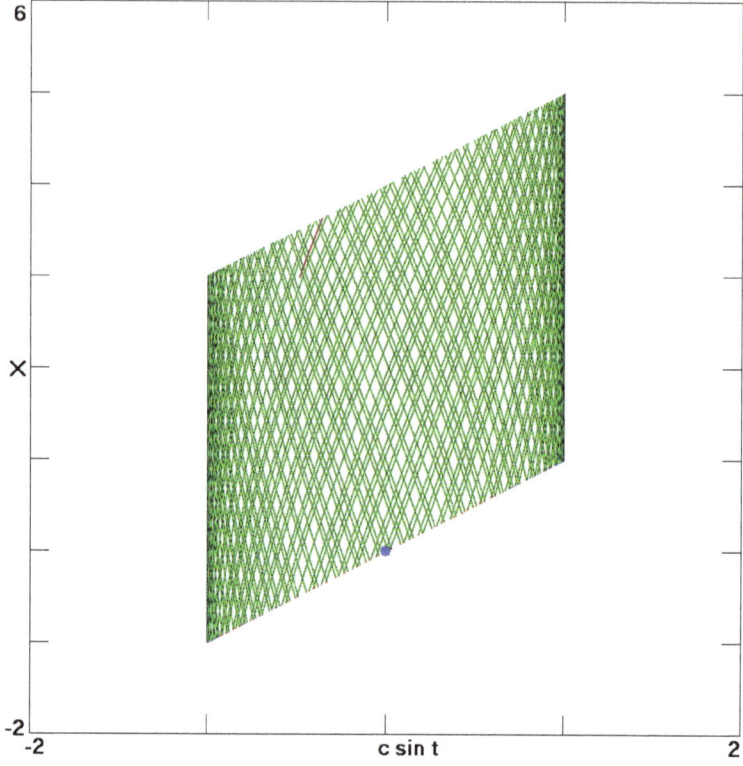

Fig. 5.16 Numerical solution of Eq. (5.5) with $a = 20$, $b = 3.6$, and $c = 4$.

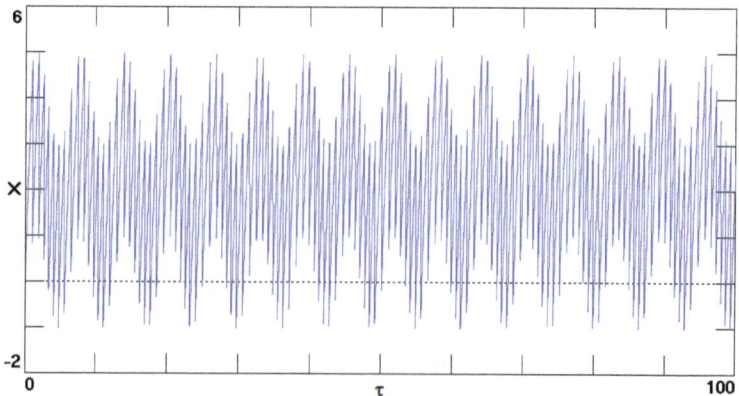

Fig. 5.17 Numerical waveform for $x(\tau)$ from Eq. (5.5) with $a = 20$, $b = 3.6$, and $c = 4$.

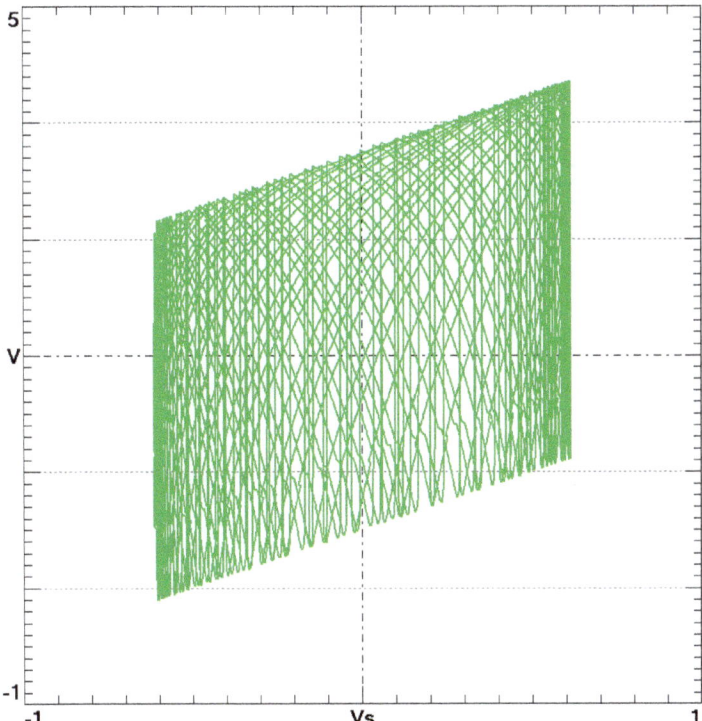

Fig. 5.18 Oscilloscope phase space plot of van der Pol relaxation oscillator circuit from Fig. 5.15.

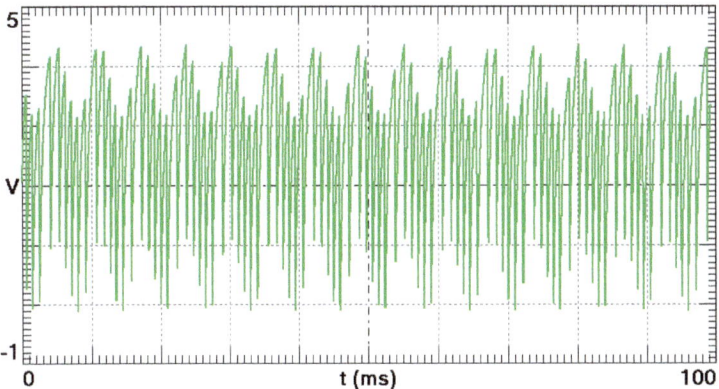

Fig. 5.19 Oscilloscope time plot of van der Pol relaxation oscillator circuit from Fig. 5.15.

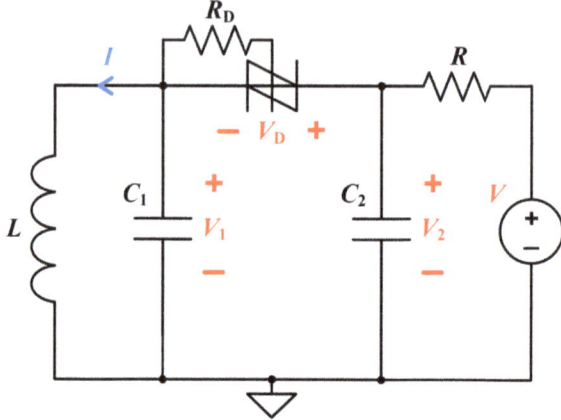

Fig. 5.20 Autonomous relaxation oscillator.

Table 5.3 Component values for Fig. 5.20.

Component	Numerical	Simulated	Experimental
R	200 kΩ	100 kΩ	100 kΩ
R_D	-	-	50 kΩ
C_1	0.1 μF	1 μF	1 μF (PP, ±3%)
C_2	0.1 μF	0.1 μF	0.1 μF (PP, ±3%)
L	0.05 H	0.001 H	0.001 H (gyrator)
V	5 V	5 V	13.4 V (adjust)
V_{BO}	4 V	2.3 V	2.3 V
Thyristor	-	Ideal switch	BS08D-T112

This inductance can be further reduced by changing either the resistors or the capacitors, although we noticed that the gyrator will become active (a negative parasitic resistance) if the resistances are too low.

The voltage V in Table 5.3 is the main parameter to adjust for chaos, and you can set it using the buffer circuit from Fig. 1.29. Since the circuit is chaotic for very specific values of V, you may want to use a potentiometer with a large resistance to fine-tune the voltage. If you increase V too much, the thyristor will become stuck in the 'off' state, and you must reduce V to near zero to restart the oscillations (or reduce V slightly and restart the circuit).

We could only obtain chaos by using high-tolerance polypropylene capacitors for C_1 and C_2 and by reducing V_{BO} of the thyristor to 2.4 V by reducing R_D. The phase space plot of V_2 versus V_1 is shown in Fig. 5.23, where the diagonal lines correspond to when the thyristor briefly conducts and feeds energy into the LC circuit. The circuit has one resistor, two capacitors, one inductor, and one thyristor for a total of five components.

5.5 Coupled Relaxation Oscillators

You can also observe chaos in an autonomous circuit consisting of two coupled relaxation oscillators, one example of which using tunnel diodes was shown in Section 4.5. Rolf Landauer proposed such a circuit using neon lamps in a 1977 IBM internal memorandum and called it 'poor man's chaos.' Figure 5.25 shows a simplified version of his circuit using silicon bilateral switches as described by the equations

$$\begin{aligned} \dot{V}_1 &= (V - V_1 - V_2)/RC_1 & SBS_1 \text{ off} \\ V_1 &= 0 & SBS_1 \text{ on} \\ \dot{V}_2 &= (V - V_1 - V_2)/RC_2 & SBS_2 \text{ off} \\ V_2 &= 0 & SBS_2 \text{ on}. \end{aligned} \quad (5.8)$$

Elegant circuit values that give chaos are $R = 100$ kΩ, $C_1 = 0.01$ μF, $C_2 = 0.0079$ μF, $V = 12$ V, and $V_{BO} = 4$ V. Note that only the product RC appears in the equations, but it is necessary to make R sufficiently large that the thyristor ceases conducting when the voltage across it reaches zero. The chaos occurs in a narrow range of the parameters surrounded by quasiperiodic solutions. The dimensionless equations are

$$\begin{aligned} \dot{x} &= (a - x - y) \\ x &= 0 \text{ when } x \geq c \\ \dot{y} &= (a - x - y)/b \\ y &= 0 \text{ when } y \geq c. \end{aligned} \quad (5.9)$$

The parameters $a = 12$, $b = 0.79$, and $c = 4$ give the attractor shown in Fig. 5.26 and a time series as shown in Fig. 5.27. In the latter figure, the quantity $x + y$ is plotted versus time since it shows the chaos more clearly than x or y alone, whose values oscillate between the limits of 0 and 4. Because of the discontinuity in the equations, it is not possible to calculate the Lyapunov exponents, although we confirmed the chaos using the method described in Section 4.2. The attractor has a Class 2 basin of attraction that fills about 75% of the space. The remaining 25% attract to the line of equilibria at $x + y = a$.

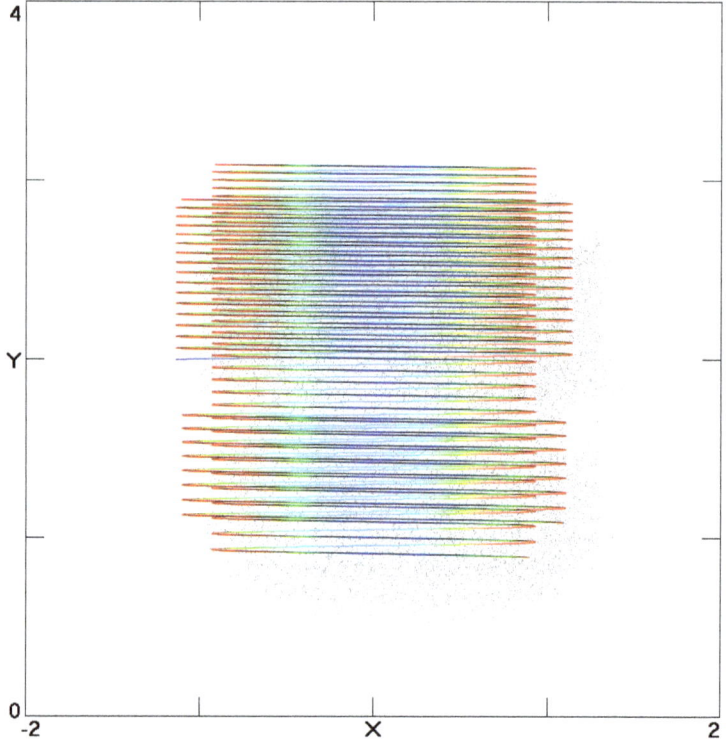

Fig. 5.21 Numerical solution of Eq. (5.7) with $a = 10$, $b = 20$, $C = 10$, and $d = 5$.

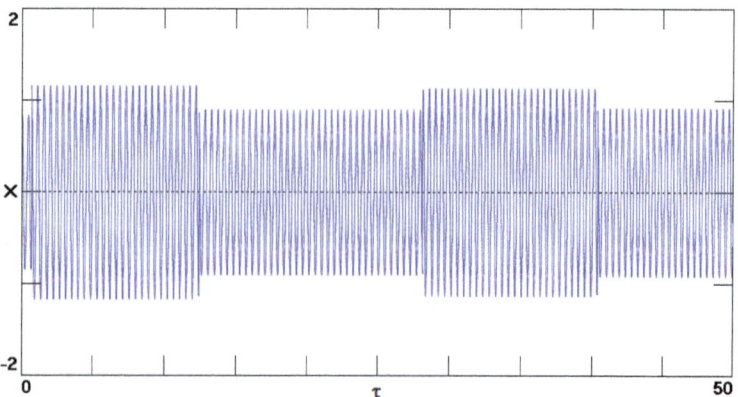

Fig. 5.22 Numerical waveform for $x(\tau)$ from Eq. (5.7) with $a = 10$, $b = 20$, $C = 10$, and $d = 5$.

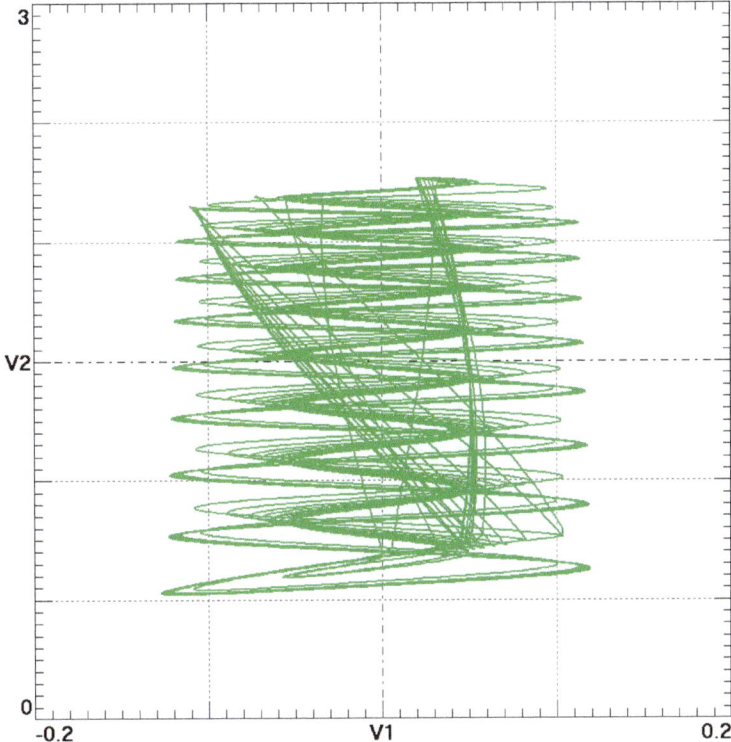

Fig. 5.23 Oscilloscope phase space plot of the autonomous relaxation oscillator from Fig. 5.20.

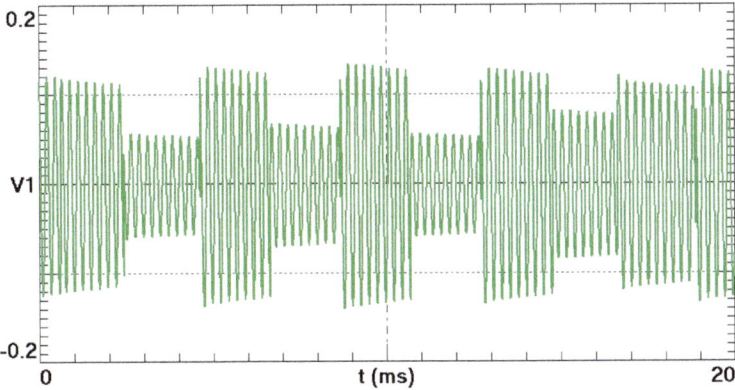

Fig. 5.24 Oscilloscope time plot of the autonomous relaxation oscillator from Fig. 5.20.

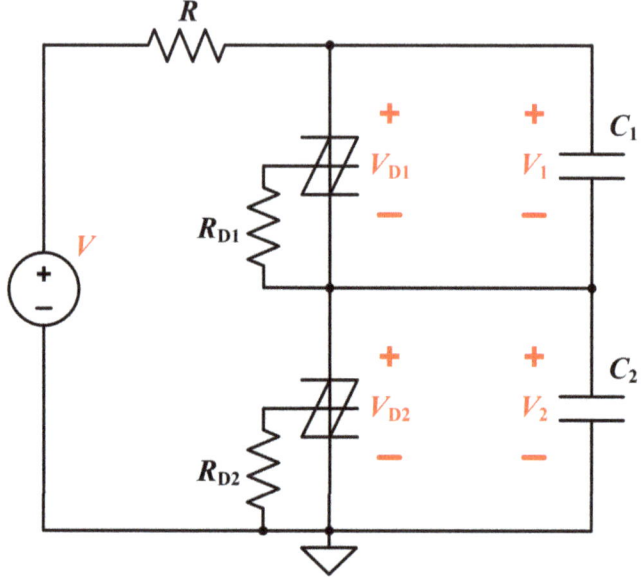

Fig. 5.25   Coupled relaxation oscillators.

Table 5.4   Component values for Fig. 5.25.

Component	Numerical	Simulated	Experimental
R	100 kΩ	174 kΩ	174 kΩ (adjust)
R_{D1}	-	-	100 kΩ
R_{D2}	-	-	100 kΩ
C_1	0.01 µF	0.01 µF	0.01 µF (PP, ±3%)
C_2	0.0079 µF	0.008 µF	0.008 µF
V	12 V	12 V	21 V (adjust)
V_{BO}	4 V	4 V	4 V
Thyristor	-	Ideal switch	BS08D-T112

The components are listed in Table 5.4, the phase space plot of V_2 versus V_1 is in Fig. 5.28, and the time plot of $V_1(t) + V_2(t)$ is shown in Fig. 5.29. You can measure V_1 by taking the potential difference across C_1. It is easier to obtain chaos by using a high-tolerance polypropylene capacitor for C_1. The value of C_2 does not need to be as precise, but it should be a few nanofarads less than C_1. You can then adjust V and R to observe chaos. This circuit has a resistor, two capacitors, and two thyristors for a total of five components.

5.6 Many Coupled Oscillators

For some applications, it is useful to have more than two coupled relaxation oscillators, for example as a way to simulate complex high-dimensional chaotic systems that are ubiquitous in nature. The circuit in Fig. 5.25 can be extended in a simple way by adding more thyristors and capacitors in series with the two shown in the figure. However, that necessitates a proportionally larger source voltage and makes it difficult to measure the individual capacitor voltages.

A better arrangement is to put the relaxation oscillators in parallel as shown in Fig. 5.30 which shows the case of three such oscillators. In this circuit, rather than the thyristors conducing briefly and then turning off, they remain in conduction, but only one at a time. For example, if D_1 is conducting, the equations that describe the circuit are

$$\dot{V}_1 = (V_2 - V_1 - V)/R_2 C_1 + (V_3 - V_1 - V)/R_3 C_1$$
$$\dot{V}_2 = (V - V_2 + V_1)/R_2 C_2 \qquad (5.10)$$
$$\dot{V}_3 = (V - V_3 + V_1)/R_3 C_3,$$

and similarly for the other two thyristors whose voltages are respectively $V_{D2} = V_2 - V_1$ and $V_{D3} = V_3 - V_1$. When either of these voltages exceeds V_{BO}, the corresponding thyristor begins conducting, dropping its voltage to zero while thyristor D_1 ceases conducting.

Elegant circuit values that give chaos are $R = 1$ kΩ, $R_2 = 2$ kΩ, $R_3 = 5$ kΩ, $C_1 = 1$ μF, $C_2 = 2$ μF, $C_3 = 3$ μF, $V = 22$ V, and $V_{BO} = 4$ V. The dimensionless equations are

D_1 conducting :
$$\dot{x} = (y - x - a)/b + (z - x - a)/e$$
$$\dot{y} = (a - y + x)/d$$
$$\dot{z} = (a - z + x)/h$$

D_2 conducting :
$$\dot{x} = a - x + y$$
$$\dot{y} = (x - y - a)/b + (z - y - a)/g \qquad (5.11)$$
$$\dot{z} = (a - z + y)/h$$

D_3 conducting :
$$\dot{x} = a - x + z$$
$$\dot{y} = (a - y + z)/d$$
$$\dot{z} = (x - z - a)/c + (y - z - a)/f.$$

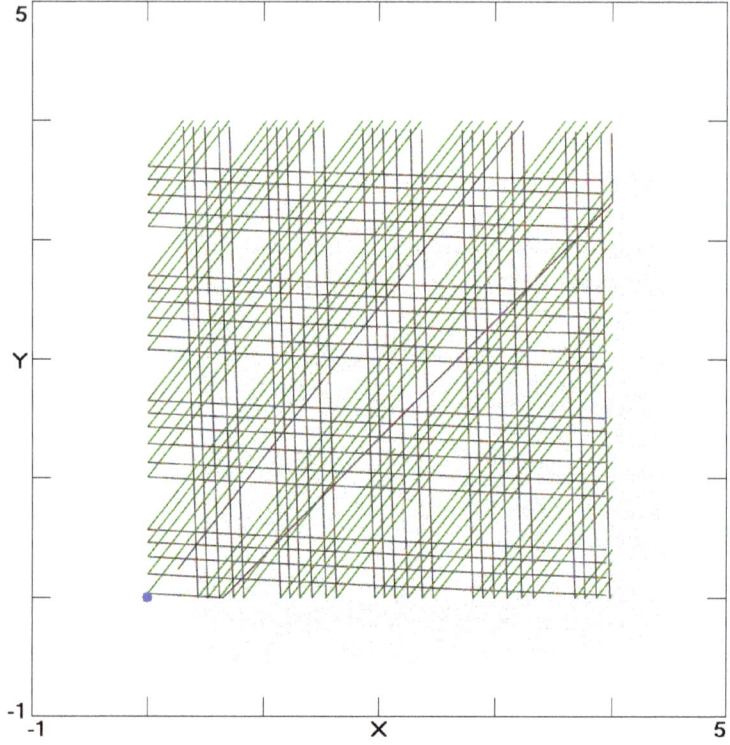

Fig. 5.26 Numerical solution of Eq. (5.9) with $a = 12$, $b = 0.79$, and $c = 4$.

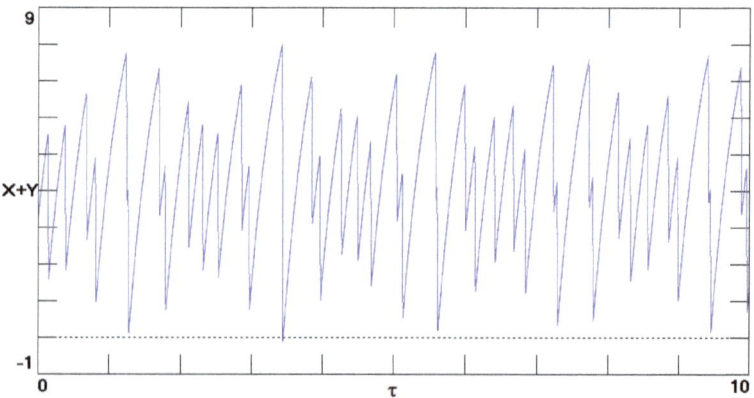

Fig. 5.27 Numerical waveform for $x + y$ versus τ from Eq. (5.9) with $a = 12$, $b = 0.79$, and $c = 4$.

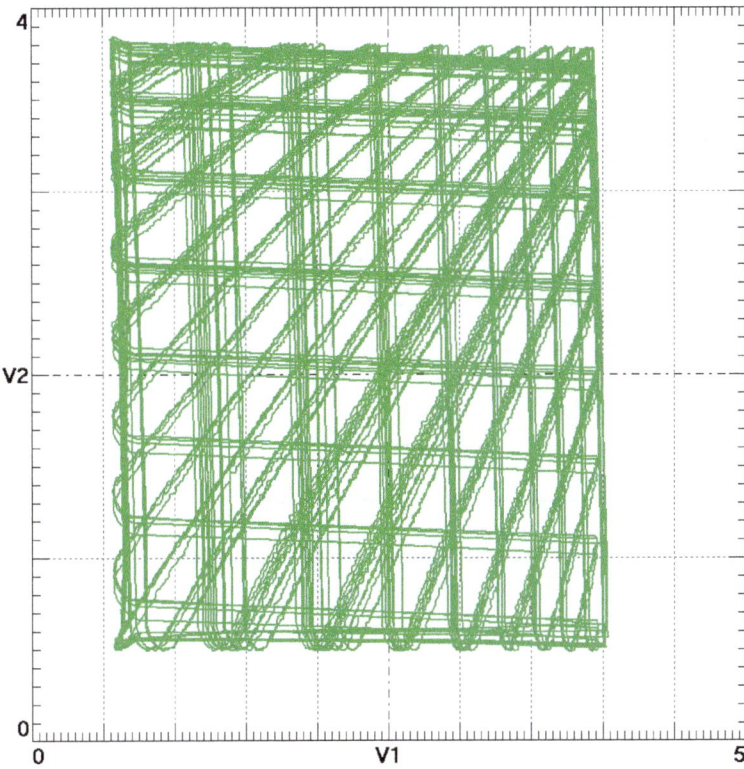

Fig. 5.28 Oscilloscope phase space plot of coupled relaxation oscillators from Fig. 5.25.

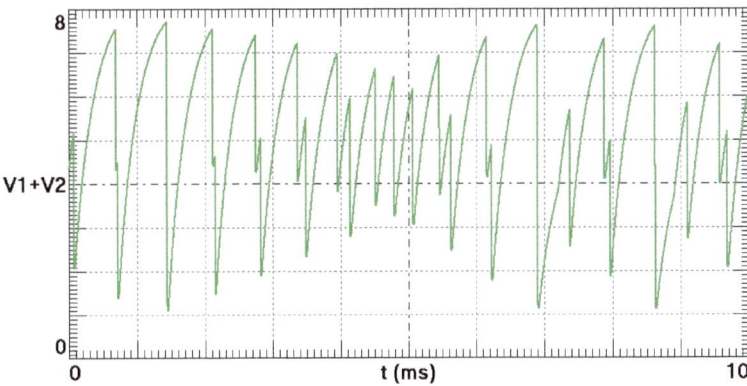

Fig. 5.29 Oscilloscope time plot of coupled relaxation oscillators from Fig. 5.25.

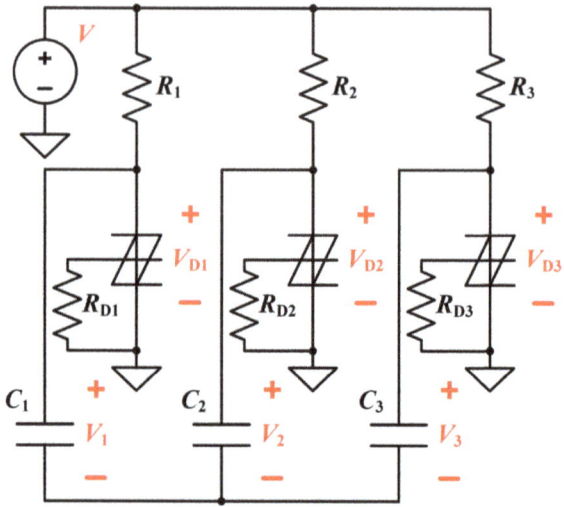

Fig. 5.30   Many coupled oscillators.

Table 5.5   Component values for Fig. 5.30.

Component	Numerical	Simulated	Experimental
R_1	1 kΩ	1 kΩ	1 kΩ
R_2	2 kΩ	2 kΩ	2 kΩ
R_3	5 kΩ	5 kΩ	5 kΩ
R_{D1}	-	-	100 kΩ
R_{D2}	-	-	100 kΩ
R_{D3}	-	-	100 kΩ
C_1	1 μF	1 μF	1 μF
C_2	2 μF	2 μF	2 μF
C_3	3 μF	3 μF	3 μF
V	22 V	22 V	22 V (adjust)
V_{BO}	4 V	4 V	4 V
Thyristor	-	Ideal switch	BS08D-T112

The parameters $a = 22$, $b = 2$, $c = 3$, $d = 4$, $e = 5$, $f = 6$, $g = 10$, and $h = 15$ give the attractor shown in Fig. 5.31 and a time series as shown in Fig. 5.32. Because of the discontinuity in the equations, it is not possible to calculate the Lyapunov exponents, although we confirmed the chaos using the method described in Section 4.2. The attractor has a global Class 1a basin of attraction and no equilibrium points. Hence it cannot fail to oscillate for the given parameters.

A curious property of the circuit is that a constant voltage can be added to all the capacitors without altering the behavior of the circuit, for example by charging one or more of the capacitors prior to turning on the power supply. This implies that the system is actually two-dimensional with one of the capacitor voltages linearly dependent on the other two. The numerical plots show the case where the capacitors are initially discharged so that the initial condition is at the origin. However, chaos is still possible because of the hysteresis.

You can obtain chaos easily in this circuit by only adjusting V. Component values are given in Table 5.5, and the phase space plot of V_2 versus V_1 is shown in Fig. 5.33. The time plot of $V_1(t)$ shown in Fig. 5.34 has a higher frequency than the numerical value primarily due to the model neglecting the 0.6 V forward voltage drop of the thyristors when they turn off.

You can measure V_1 and V_2 by connecting the ground of the oscilloscope probe (usually a black alligator clip) to the common point of the capacitors and then measuring the signal at the positive end of C_1 and C_2. The circuit has three resistors, three capacitors, and three thyristors for a total of nine components.

5.7 Saito Family Thyristor Circuit

The Saito family of chaotic circuits is based on a combination of reactive elements with an active resistor and a nonlinear device. A large number of such circuits exist including the example from Section 2.8 based on a diode. Saito (1988) also explored chaotic circuits where the nonlinear device has hysteresis generated by saturating operational amplifiers. Elwakil and Kennedy (1999b) further investigated this circuit using a transistor circuit from Chua et al. (1983) with an S-type negative resistance as the nonlinear device. A variation of this circuit using a silicon bilateral switch is shown in Fig. 5.35 with the equations

$$\begin{aligned} \dot{V} &= (V/R - I)/C \quad &SBS \text{ off} \\ V &= 0 \quad &SBS \text{ on} \\ \dot{I} &= V/L. \end{aligned} \quad (5.12)$$

Elegant circuit values that give chaos are $R = 1.5$ kΩ, $C = 0.1$ μF, $L = 0.01$ H, and $V_{BO} = 4$ V. The dimensionless equations are

$$\begin{aligned} \dot{x} &= a(x/b - y) \\ x &= 0 \quad &\text{when } x \geq 4 \\ \dot{y} &= cx. \end{aligned} \quad (5.13)$$

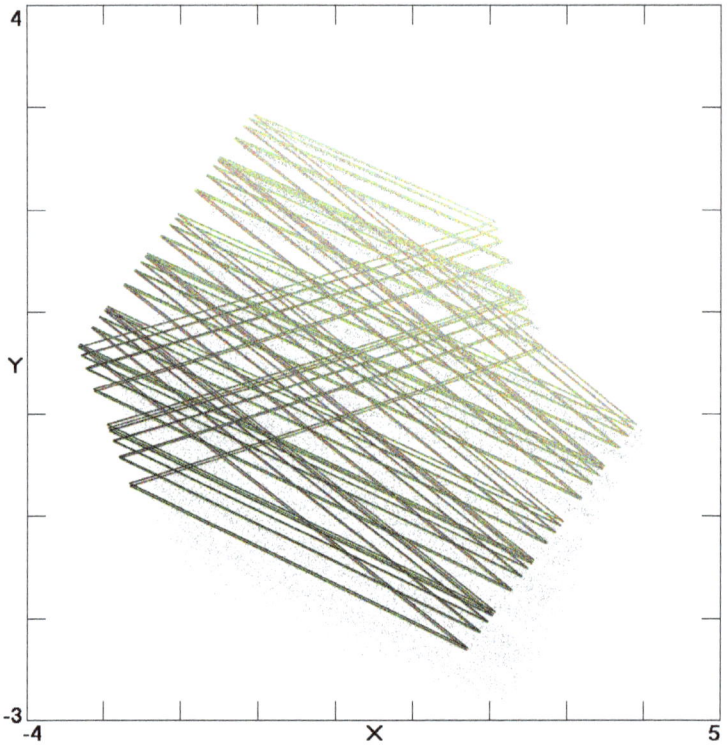

Fig. 5.31 Numerical solution of Eq. (5.11) with $a = 22$, $b = 2$, $c = 3$, $d = 4$, $e = 5$, $f = 6$, $g = 10$, and $h = 15$.

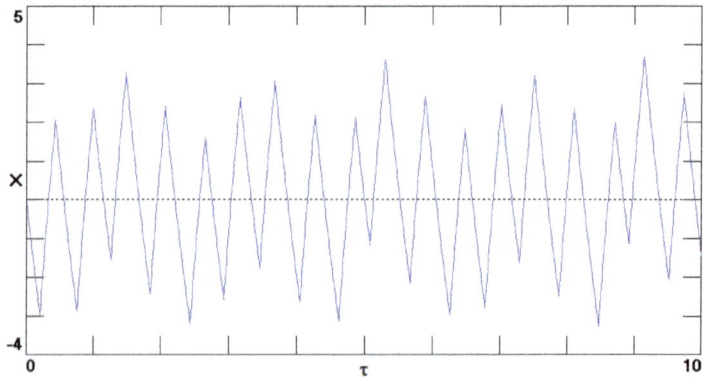

Fig. 5.32 Numerical waveform for $x(\tau)$ from Eq. (5.11) with $a = 22$, $b = 2$, $c = 3$, $d = 4$, $e = 5$, $f = 6$, $g = 10$, and $h = 15$.

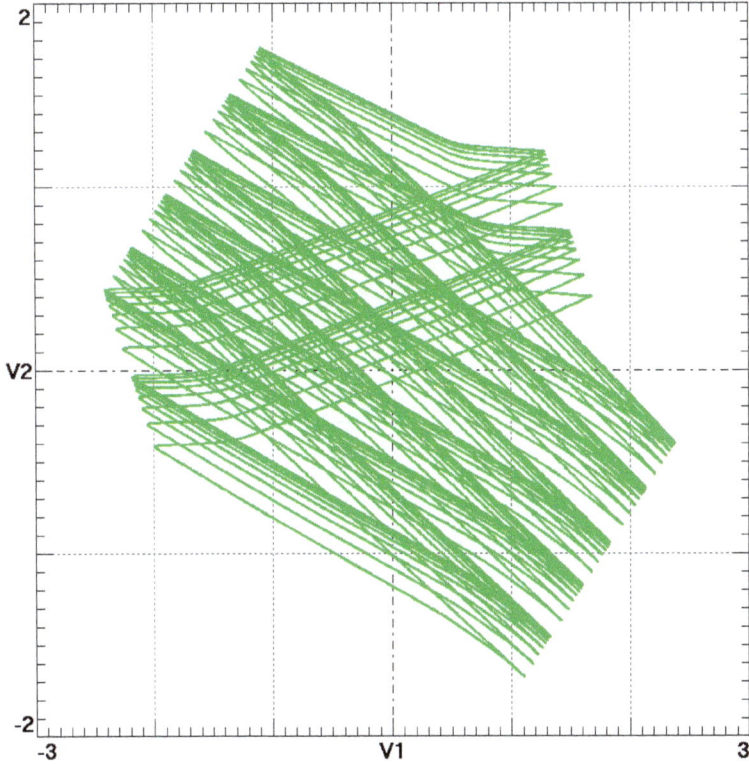

Fig. 5.33 Oscilloscope phase space plot of many coupled oscillators from Fig. 5.30.

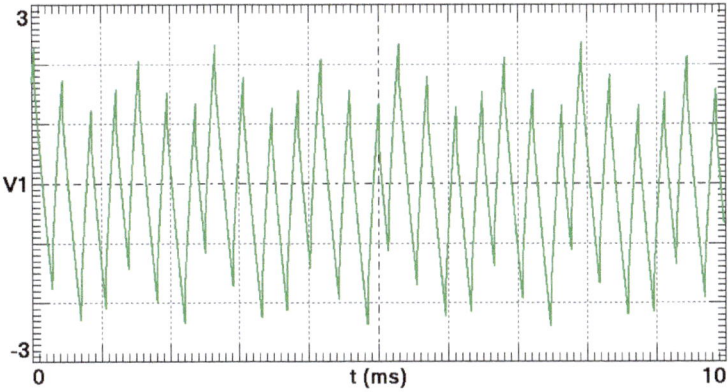

Fig. 5.34 Oscilloscope time plot of many coupled oscillators from Fig. 5.30.

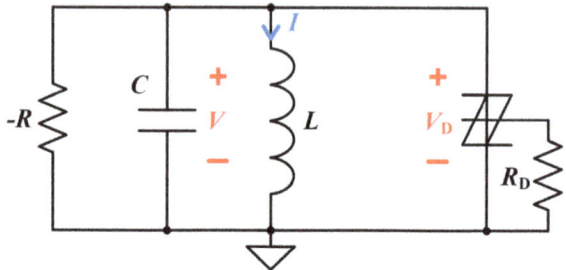

Fig. 5.35 Saito family thyristor circuit.

Table 5.6 Component values for Fig. 5.35.

Component	Numerical	Simulated	Experimental
R	1.5 kΩ	1.5 kΩ	1.5 kΩ (adjust)
R_D	-	-	100 kΩ
C	0.1 μF	0.1 μF	0.1 μF
L	0.01 H	0.01 H	0.01 H
V_{BO}	4 V	4 V	4 V
Thyristor	-	Ideal switch	BS08D-T112

The parameters $a = 10$, $b = 1.5$, and $c = 100$ give the attractor shown in Fig. 5.36 and a time series as shown in Fig. 5.37. Because of the discontinuity in the equations, it is not possible to calculate the Lyapunov exponents, although we confirmed the chaos using the method described in Section 4.2. The attractor has a Class 3 basin of attraction with $P \approx 8/r^2$ and an unstable focus at the origin with eigenvalues $(3.333 \pm 31.4466i, 0)$.

The component values are shown in Table 5.6, where the active negative resistor R can be implemented using a current feedback amplifier and a resistor as shown in Fig. 2.42. The phase space plot of I versus V is shown in Fig. 5.38, and the time plot of $V(t)$ is shown in Fig. 5.39. Both experimental plots differ slightly from the theoretical results since the thyristor switches off when $V(t)$ reaches 0.6 V, while this voltage drop is neglected in the model.

Although the circuit can theoretically work at larger inductance and capacitor values, the silicon bilateral switch tends to become stuck in the conducting state if the frequency is too low. The circuit has one resistor, one capacitor, one inductor, one current feedback amplifier for the active resistor, and one thyristor for a total of five components.

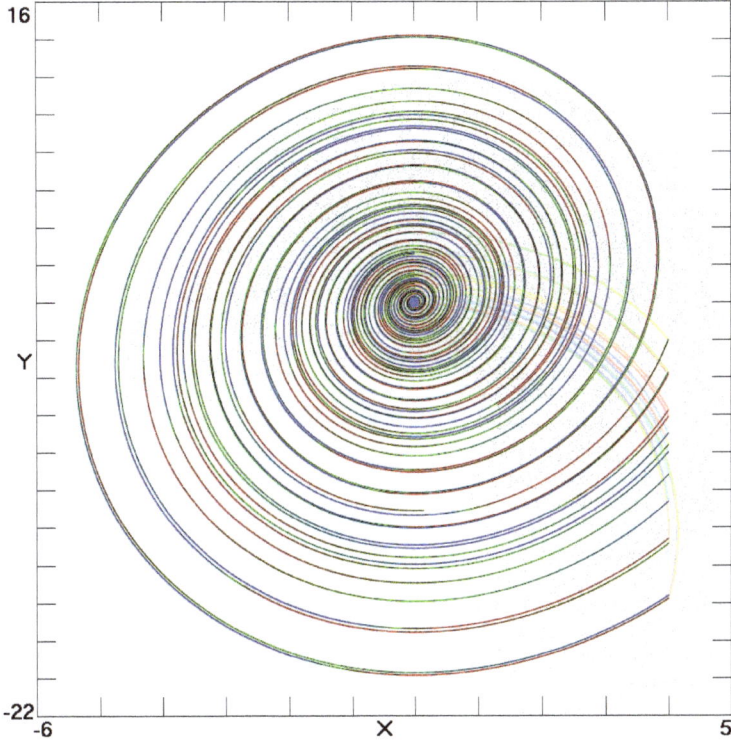

Fig. 5.36 Numerical solution of Eq. (5.13) with $a = 10$, $b = 1.5$, and $c = 100$.

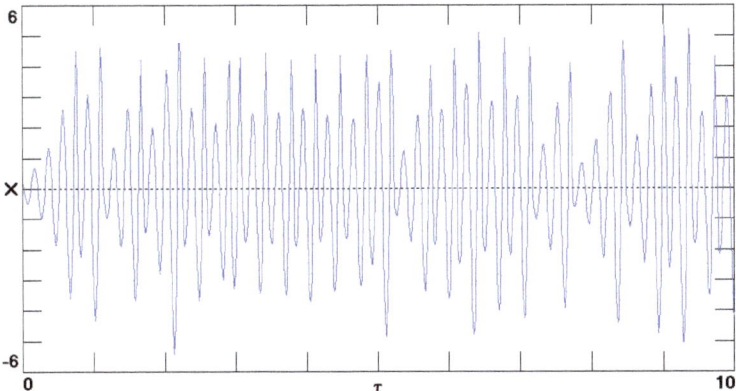

Fig. 5.37 Numerical waveform for $x(\tau)$ from Eq. (5.13) with $a = 10$, $b = 1.5$, and $c = 100$.

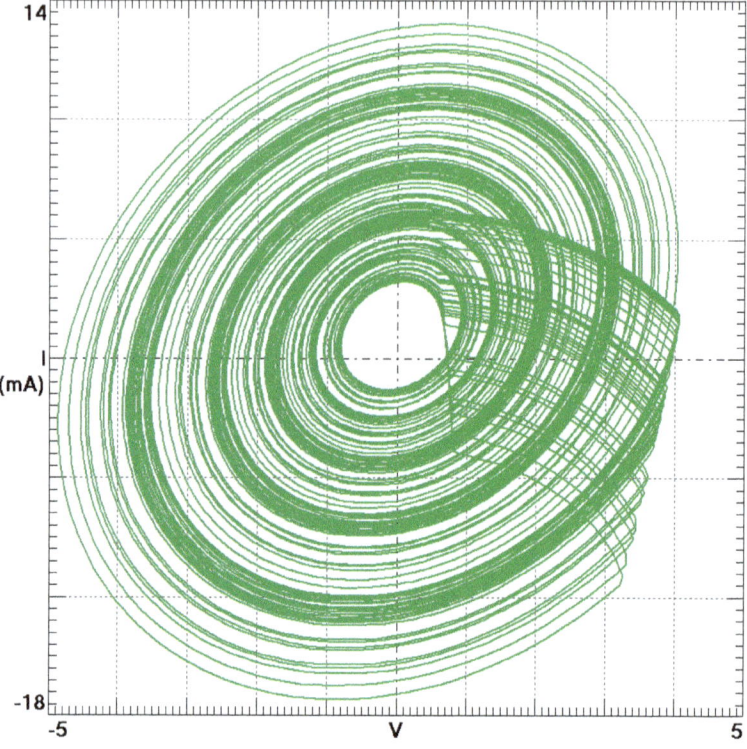

Fig. 5.38 Oscilloscope phase space plot of the Saito family thyristor circuit from Fig. 5.35.

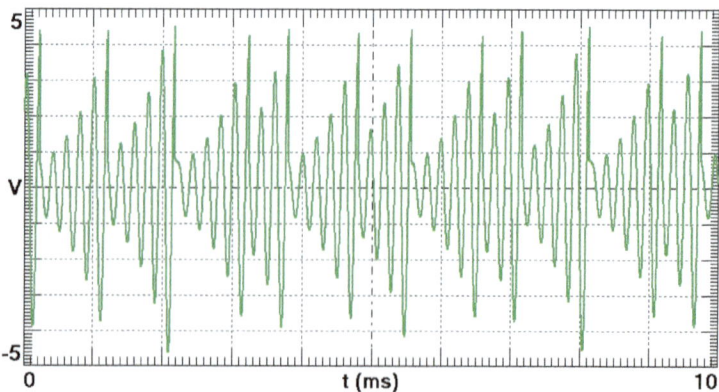

Fig. 5.39 Oscilloscope time plot of the Saito family thyristor circuit from Fig. 5.35.

Chapter 6

Saturating Amplifier Circuits

Previously, we used the operational amplifier in its linear region to amplify, buffer, or integrate a signal. However, this is insufficient to produce chaos unless a nonlinear device such as a diode is added or if the operational amplifier becomes nonlinear when it saturates. Here we briefly review the operational amplifier and show examples of chaotic circuits that rely solely on its saturation.

6.1 Operational Amplifiers

An operational amplifier as shown schematically in Fig. 6.1 has two inputs, called the inverting input $(-)$ and the non-inverting input $(+)$, respectively, and one output, V_{out}. It amplifies the voltage difference between the two inputs according to $V_{out} = A(V_+ - V_-)$, where the voltage gain A is very large and can often be taken as infinity. As such, the operational amplifier is known as a *high-gain amplifier*.

Ideal operational amplifiers with negative feedback obey two rules. The first is that the voltage difference between the input terminals is zero ($V_+ - V_- = 0$). The second is that the inputs draw no current ($I_+ = I_- = 0$). These rules apply in the active region where the operational amplifier behaves in a linear manner.

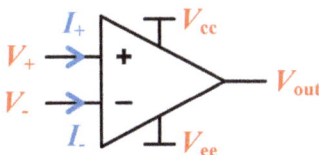

Fig. 6.1 Operational amplifier.

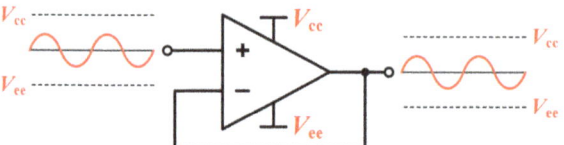

Fig. 6.2 Operational amplifier buffer in the linear region.

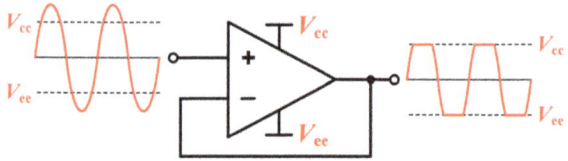

Fig. 6.3 Operational amplifier buffer when driven into saturation.

6.1.1 *Operational amplifier transfer characteristic*

A previous operational amplifier circuit from Section 2.4 is the buffer amplifier which has a voltage gain of $K = 1$. Any input signal entering the buffer is left unchanged at the output as illustrated in Fig. 6.2. A reason you might want such a circuit is so that you can drive a low-impedance load without drawing much current from the source.

However, if the input voltage is increased to the point where the output would exceed the power supply voltage ($> V_{cc}$ or $< V_{ee}$), as shown in Fig. 6.3, the output *saturates* and is prevented from increasing further. This saturation can be modeled by the piecewise-linear equation

$$V_{out} = \begin{cases} V_{ee} & V_{out} < V_{ee} \\ V_+ & V_{ee} \leq V_{out} \leq V_{cc} \\ V_{cc} & V_{out} > V_{cc}, \end{cases} \qquad (6.1)$$

where typically $V_{cc} = -V_{ee} \approx 15$ V. We will refer to this limiting voltage as V_{sat}. Equation (6.1) can be written more compactly as $V_{out} = \min[\max(V_{out}, V_{ee}), V_{cc}]$ or $V_{out} = \min[\max(V_{out}, -V_{sat}), V_{sat}]$.

The linear and saturation regions can be plotted through a *transfer characteristic* which graphs the output V_{out} versus the input $(V_+ - V_-)$ as shown in Fig. 6.4. This behavior can be approximated by the *hyperbolic tangent* given by

$$\tanh x = \frac{e^x - e^{-x}}{e^x + e^{-x}} = \frac{e^{2x} - 1}{e^{2x} + 1}. \qquad (6.2)$$

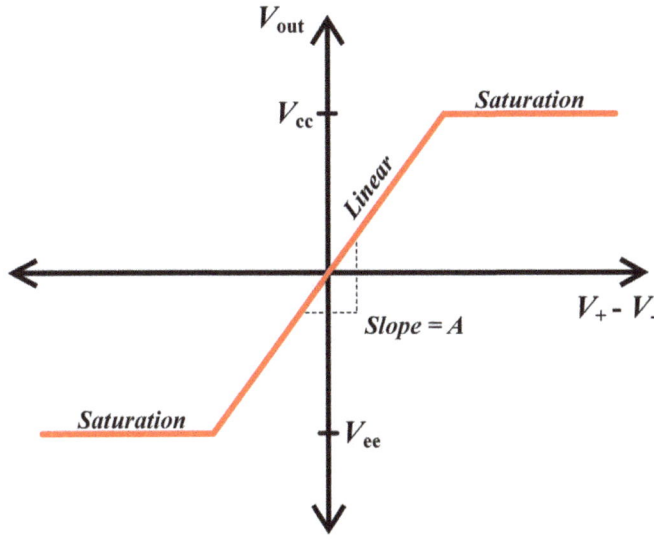

Fig. 6.4 Operational amplifier V_{out} versus $V_+ - V_-$.

6.1.2 Comparators

If the amplifier is driven hard into saturation, ($|x| \gg 1$), it often suffices to further approximate the hyperbolic tangent by the signum function $\operatorname{sgn} x$ which is $+1$ if $x > 0$ and -1 if $x < 0$. In this limit, the operational amplifier behaves like a comparator, producing a constant positive voltage of approximately V_{cc} if $V_+ > V_-$ and a constant negative voltage of approximately V_{ee} if $V_+ < V_-$ so that $V_{out} = V_{sat} \operatorname{sgn}(V_+ - V_-)$. In fact, special digital comparator devices are available that are designed specifically for this application and that have very fast switching and other desirable properties.

In the saturation region, there is no guarantee that the rules that apply in the linear region hold. In particular, the voltages V_+ and V_- are no longer equal. However, the input resistance usually remains sufficiently large that the currents I_+ and I_- can be neglected. We will refer to the operational amplifiers used in simulation as 'nonideal' if the operational amplifier no longer behaves as an ideal linear amplifier.

In conclusion, both the saturation and ideal operational amplifier rules allow a wide variety of piecewise-linear functions to be generated, which will be useful in the circuits that follow.

6.2 Saturating Wien Bridge Oscillator

To give an example of how saturation leads to chaos, we revisit the Wien bridge oscillator. This oscillator was modified to be chaotic by adding a diode as described in Section 2.6, or by using a junction field effect transistor as described in Section 3.9. In those cases, the non-inverting operational amplifier operated in the linear region with a gain of $K = 2$ to sustain the oscillation.

González-López (1998) proposed an alternate way to make the circuit chaotic by allowing the operational amplifier to saturate. His circuit shown in Fig. 6.5 is modeled by the equations

$$\dot{V}_1 = (IR - V_1)/RC_1$$
$$\dot{V}_2 = I/C_2 \qquad (6.3)$$
$$\dot{I} = (V_{out} - IR_L - V_1 - V_2)/L,$$

where $V_{out} = \min[\max(KV_1, -V_{sat}), V_{sat}]$ from Eq. (6.1) and $K = 1 + R_1/R_2$.

Elegant circuit values that give chaos are $R = 20$ kΩ, $R_1 = R_2 = 1$ kΩ, $R_L = 0.2$ kΩ, $C_1 = C_2 = 1$ μF, $L = 1$ H, and $V_{sat} = 15$ V. The dimensionless equations are

$$\dot{x} = z - ax$$
$$\dot{y} = z \qquad (6.4)$$
$$\dot{z} = \min[\max(kx, -c), c] - bz - x - y.$$

The parameters $a = 0.05$, $b = 0.2$, $c = 15$, and $k = 2$ give the attractor shown in Fig. 6.6 and a time series as shown in Fig. 6.7. Note that c is an amplitude parameter that does not affect the dynamics. The Lyapunov exponents are $(0.0392, 0, -0.2892)$, the Kaplan–Yorke dimension is $D_{ky} = 2.1357$, and the attractor has a Class 2 basin of attraction that attracts half of the space. The single equilibrium point at the origin is an unstable saddle focus with eigenvalues $(0.1062 \pm 0.3113i, -0.4623)$.

Note that this system is invariant under the transformation $(x, y, z) \to (-x, -y - z)$. Thus any attractor must either have that symmetry or there must be a symmetric pair of attractors. In fact, with $k = \infty$ where V_{out} can be approximated by a signum function, there is a symmetric two-lobe attractor. For the chosen parameters, there is a second attractor displaced slightly from the one shown and that links it with intricate basins of attraction.

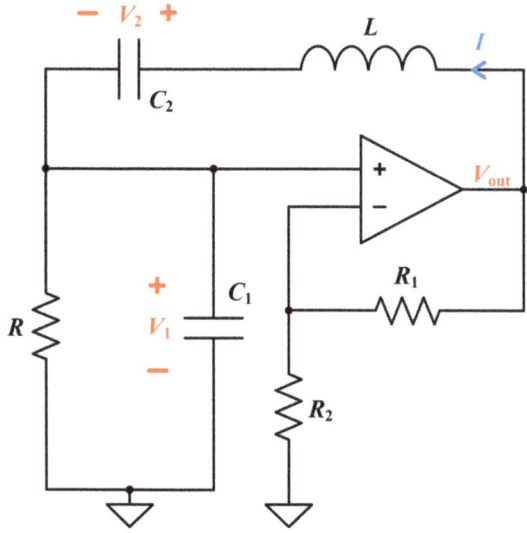

Fig. 6.5 Saturating Wien bridge oscillator.

Table 6.1 Component values for Fig. 6.5.

Component	Numerical	Simulated	Experimental
R	20 kΩ	20 kΩ	20 kΩ
R_1	1 kΩ	1 kΩ	1 kΩ
R_2	1 kΩ	1 kΩ	1 kΩ
R_L	0.2 kΩ	0.2 kΩ	0.2 kΩ
C_1	1 μF	0.1 μF	0.1 μF
C_2	1 μF	0.1 μF	0.1 μF
L	1 H	0.1 H	0.1 H
V_{sat}	15 V	15 V	15 V
Op amp	-	Nonideal	LM741

This is an example of *spontaneous symmetry breaking*, and it is common in symmetric systems of differential equations. The origin lies on the basin boundary, and the attractors nearly touch their basin boundaries in places. Thus small changes in initial conditions or other perturbations will determine which attractor is observed.

You can observe this behavior for yourself by shorting out C_1 and C_2 with wires and quickly removing them while the circuit is operating so that the initial conditions of V_1 and V_2 are zero. You should see that the two attractors are visited equally often when you do this many times. However,

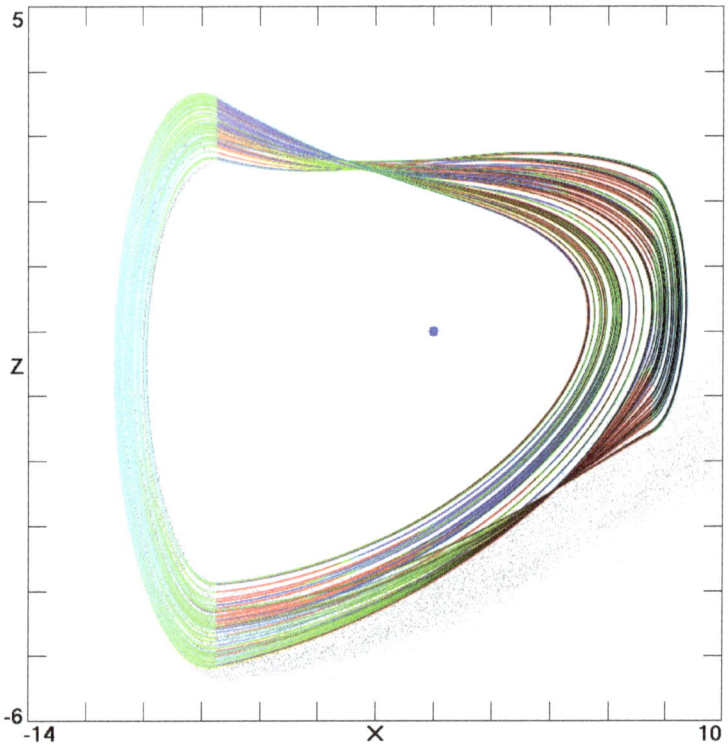

Fig. 6.6 Numerical solution of Eq. (6.4) with $a = 0.05$, $b = 0.2$, $c = 15$, and $k = 2$.

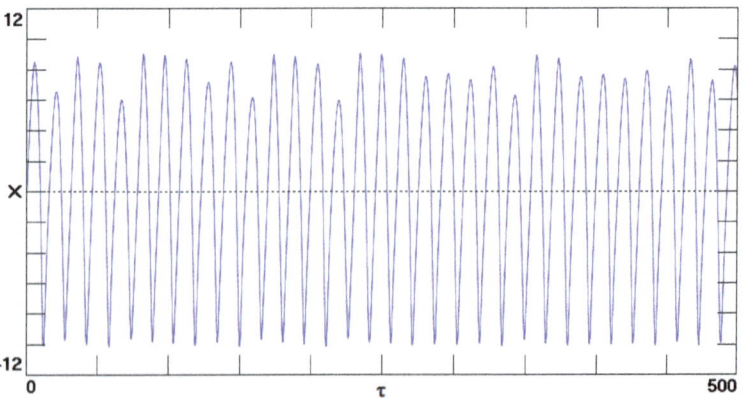

Fig. 6.7 Numerical waveform for $x(\tau)$ from Eq. (6.4) with $a = 0.05$, $b = 0.2$, $c = 15$, and $k = 2$.

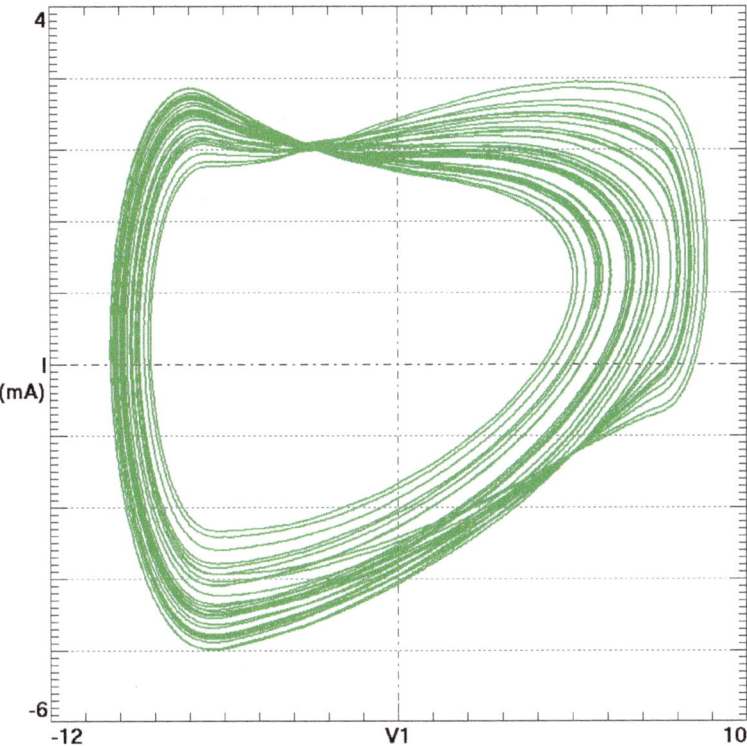

Fig. 6.8 Oscilloscope phase space plot of the saturating Wien bridge oscillator from Fig. 6.5.

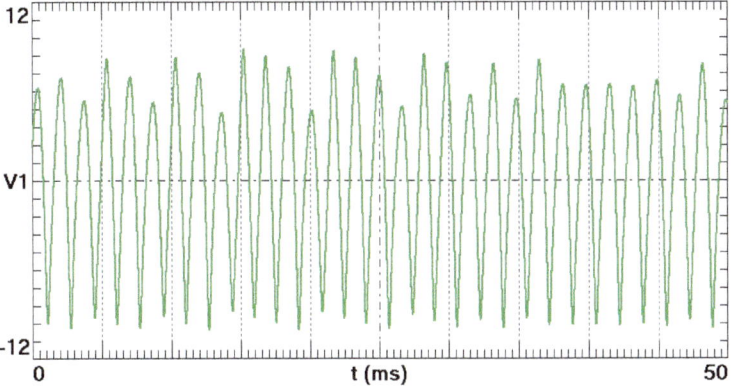

Fig. 6.9 Oscilloscope time plot of the saturating Wien bridge oscillator from Fig. 6.5.

it is necessary that the saturation voltages be equal and opposite within a few millivolts; otherwise only one attractor will appear.

None of the parameters in Table 6.1 are critical, and the circuit should be chaotic when it is powered on without additional adjustment. If you measure V_{out}, you can observe the operational amplifier saturating when KV_1 reaches the limit of V_{ee} and V_{cc} and cannot increase further. Since V_1 and V_2 are close in phase due to the large value of R, we opted to plot I versus V_1 instead as shown in Fig. 6.8. We also obtained better agreement when the frequency was increased to 10 kHz, although the circuit is still chaotic at lower frequencies.

Figure 6.9 shows the time plot of $V_1(t)$, which you can find by probing the top of C_1. You can then find I by placing R_L between C_1 and C_2 and measuring the voltage at the other end, which can be used to calculate the current through the resistor. This circuit has three resistors, two capacitors, one inductor, and one operational amplifier for a total of seven components.

6.3 Murali–Lakshmanan–Chua Circuit

Among the vast research on Chua's circuit are studies that have attempted to improve its elegance such as the one by Barboza and Chua (2008) who reduced the original circuit by two resistors. Generally, the main change is to modify Chua's diode, which uses operational amplifier saturation to produce a piecewise-linear I-V characteristic.

Another elegant circuit that uses Chua's diode was introduced by Murali et al. (1994) who simply replaced the nonlinear device in the nonautonomous circuit of Ueda and Akamatsu (1981). Later, Lacy (1996) found that Chua's diode in this circuit could be implemented with a single operational amplifier and three resistors. We have further simplified their circuit as shown in Fig. 6.10 by using the saturation of the current feedback amplifier as shown in Fig. 2.42. The current feedback amplifier was originally meant to emulate a linear active resistor, but it becomes nonlinear when the current exiting the TZ node of the AD844 saturates at about 10 mA when the power supply voltage is ± 15 V. The equations for the circuit are

$$\dot{V} = (I + I_R)/C \\ \dot{I} = (V_S - V - IR_L)/L, \qquad (6.5)$$

where $V_S = V_0 \sin \omega t$ and $I_R = V_R/R$. The saturation current at the TZ node is 10 mA, which means that $V_R = \min[\max(V, -V_{sat}), V_{sat}]$ saturates at $V_{sat} = 10$ V with $R = 1$ kΩ.

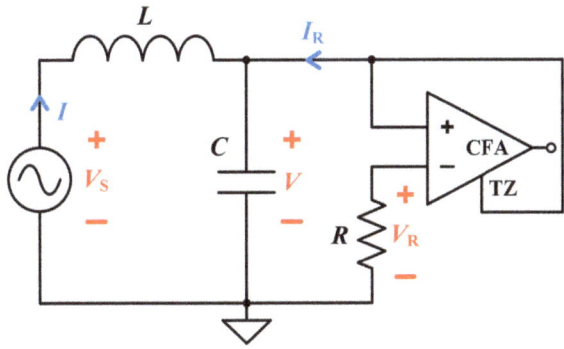

Fig. 6.10   Murali–Lakshmanan–Chua circuit.

Table 6.2   Component values for Fig. 6.10.

Component	Numerical	Simulated	Experimental
R	1 kΩ	1 kΩ	1 kΩ
R_L	1 kΩ	1 kΩ	1 kΩ
C	1 μF	1 μF	1 μF
L	1.5 H	1.5 H	1.5 H
V_0	1 V	1 V	1.5 V (adjust)
V_{sat}	10 V	7.5 V	+11.8/ − 10.8 V
ω	0.6 rad/ms	0.2π rad/ms	0.2π rad/ms
Op amp	-	AD844	AD844

Elegant circuit values that give chaos are $R = R_L = 1$ kΩ, $C = 1$ μF, $L = 1.5$ H, $V_0 = 1$ V, $V_{sat} = 10$ V, and $\omega = 0.6$ rad/ms. The dimensionless equations are

$$\dot{x} = y + \min[\max(x, -a), a] \\ \dot{y} = (\sin c\tau - x - y)/b.$$
(6.6)

The parameters $a = 10$, $b = 1.5$, and $c = 0.6$ give the attractor shown in Fig. 6.11 and a time series as shown in Fig. 6.12. The Lyapunov exponents are $(0.0255, 0, -0.1665)$, the Kaplan–Yorke dimension is $D_{ky} = 2.1530$, and the attractor has a global Class 1a basin of attraction.

It may be useful for you to measure the saturation current of your current feedback amplifier by applying a large voltage V in Fig. 2.42 and measuring the current through R with an ammeter. Our AD844 saturates at 11.8 mA for $V = 15$ V and at -10.8 mA for $V = -15$ V. With $R = 1$ kΩ, the voltage V_R then saturates at 11.8 V and at -10.8 V. The varying

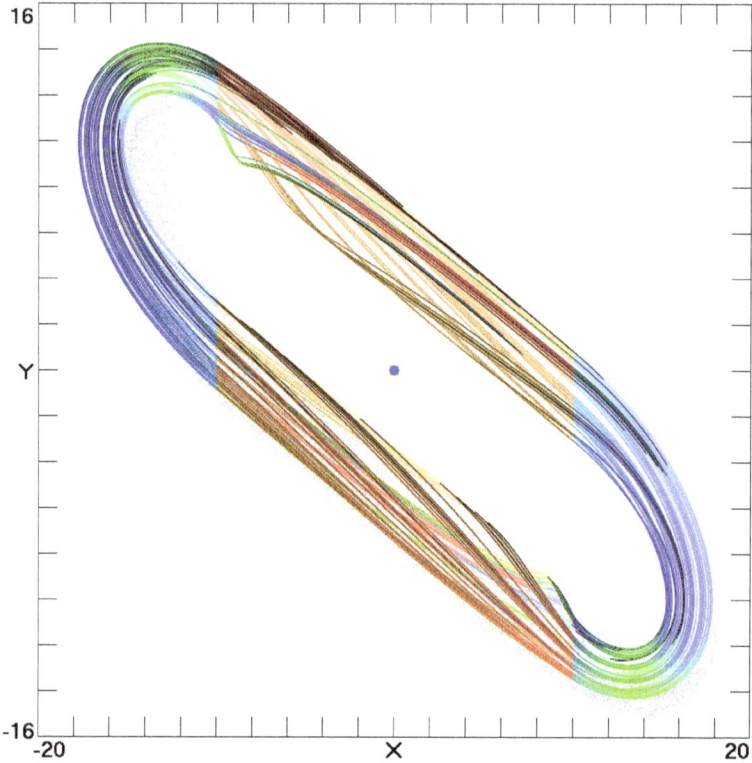

Fig. 6.11 Numerical solution of Eq. (6.6) with $a = 10$, $b = 1.5$, and $c = 0.6$.

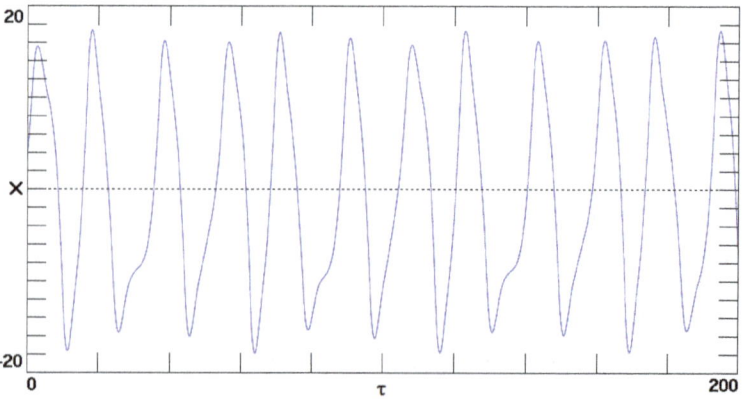

Fig. 6.12 Numerical waveform for $x(\tau)$ from Eq. (6.6) with $a = 10$, $b = 1.5$, and $c = 0.6$.

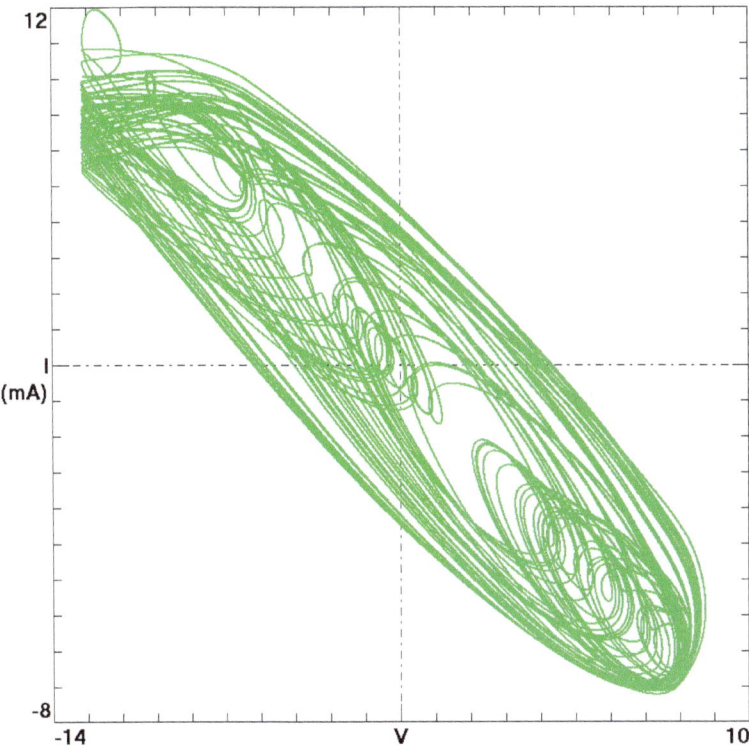

Fig. 6.13 Oscilloscope phase space plot of the Murali–Lakshmanan–Chua circuit from Fig. 6.10.

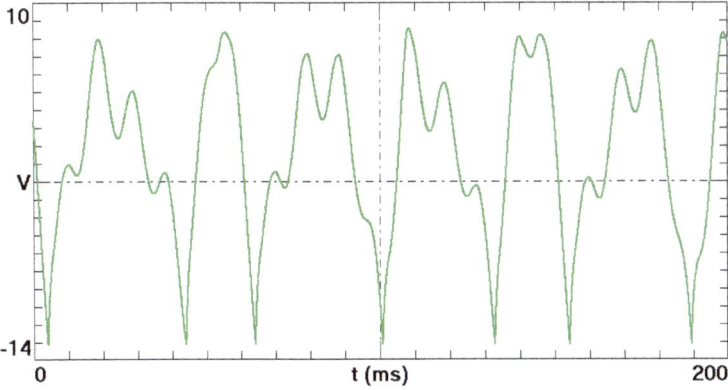

Fig. 6.14 Oscilloscope time plot of the Murali–Lakshmanan–Chua circuit from Fig. 6.10.

saturation voltages in the numerical, simulated, and experimental cases are likely the reason for the other parameter differences in Table 6.2. You should be able to obtain a similar attractor by adjusting the parameter V_0.

The phase space plot of I versus V is shown in Fig. 6.13, and a time plot of $V(t)$ is shown in Fig. 6.14. You can find I by placing R_L in series with the inductor and using Ohm's law to calculate the current from the voltage across R_L. This circuit has one resistor, one capacitor, one inductor, and one current feedback amplifier for a total of four components.

6.4 Wang–Zhang–Bao Circuit

Wang et al. (2020) designed an autonomous circuit shown in Fig. 6.15 using a saturating operational amplifier with an unusual piecewise-linear characteristic. The output of this operational amplifier is $V_{out} = V_2 - V_1 + V_-$, where $V_- = 0$ in the linear region. When the operational amplifier saturates at $V_{out} = \pm V_{sat}$, it no longer behaves ideally, and V_- becomes $V_- = V_1 - V_2 \pm V_{sat}$. This is basically a current-to-voltage converter in which $V_{out} = V_2 - V_1 = I_1 R$, where I_1 is the current through the capacitor C_1, except that V_{out} saturates at $\pm V_{sat}$. The equations for the circuit are

$$\begin{aligned}
\dot{V}_1 &= (V_2 - V_1)/RC_1 \\
\dot{V}_2 &= [I - (V_2 - V_1)/R]/C_2 \\
\dot{I} &= (V_{out} - V_2)/L,
\end{aligned} \quad (6.7)$$

where $V_{out} = \min[\max(V_2 - V_1, -V_{sat}), V_{sat}]$ from Eq. (6.1). The operational amplifier is powered by ± 15 V, so that $V_{sat} \approx 15$ V.

Elegant circuit values that give chaos are $R = 10$ kΩ, $C_1 = 10$ μF, $C_2 = 1$ μF, $L = 1$ H, and $V_{sat} = 15$ V. The dimensionless equations are

$$\begin{aligned}
\dot{x} &= a(y - x) \\
\dot{y} &= z - b(y - x) \\
\dot{z} &= \min[\max(y - x, -c), c] - y.
\end{aligned} \quad (6.8)$$

The parameters $a = 0.01$, $b = 0.1$, and $c = 15$ give the attractor shown in Fig. 6.16 and a time series as shown in Fig. 6.17. The Lyapunov exponents are $(0.0206, 0, -0.1306)$, the Kaplan–Yorke dimension is $D_{ky} = 2.1574$, and the attractor has a Class 2 basin of attraction that fills half the space. The single equilibrium at the origin is a saddle focus with eigenvalues $(0.0745 \pm 0.1818i, -0.2590)$.

This is another example of a system that is symmetric under the transformation $(x, y, z) \to (-x, -y, -z)$. Thus the attractor shown in the plots has a symmetric mate with the opposite sign of the variables. The basin

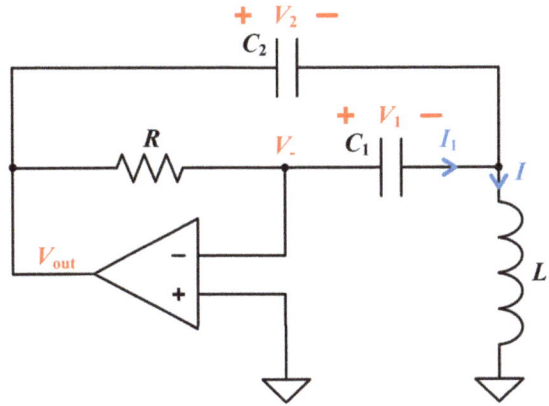

Fig. 6.15 Wang–Zhang–Bao circuit.

Table 6.3 Component values for Fig. 6.15.

Component	Numerical	Simulated	Experimental
R	10 kΩ	10 kΩ	10 kΩ (adjust)
C_1	10 μF	10 μF	10 μF
C_2	1 μF	1 μF	1 μF
L	1 H	1 H	1 H
V_{sat}	15 V	15 V	15 V
Op amp	-	Nonideal	LM741

boundary intersects the origin so that small differences in the initial capacitor voltages or slight asymmetries in the operational amplifier determines which attractor will be observed.

The component values for this circuit are given in Table 6.3, with R being the main parameter to adjust. A phase space plot of V_2 versus V_1 is shown in Fig. 6.18, and a time plot of $V_1(t)$ is in Fig. 6.19. You can view V_1 and V_2 by connecting the oscilloscope probe ground to the common point of C_1 and C_2 and measuring the voltages across both capacitors. The circuit has one resistor, two capacitors, one inductor, and one operational amplifier for a total of five components.

6.5 Coupled RC Circuits

Masuda *et al.* (2009) discovered an elegant chaotic circuit by coupling two RC circuits using comparators, or two RL circuits as Doike *et al.* (2013)

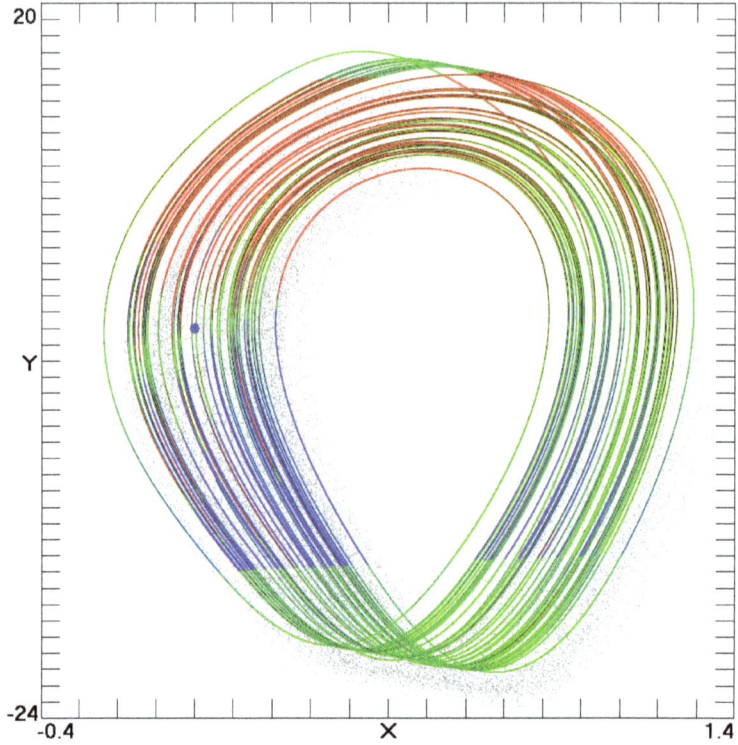

Fig. 6.16 Numerical solution of Eq. (6.8) with $a = 0.01$, $b = 0.1$, and $c = 15$.

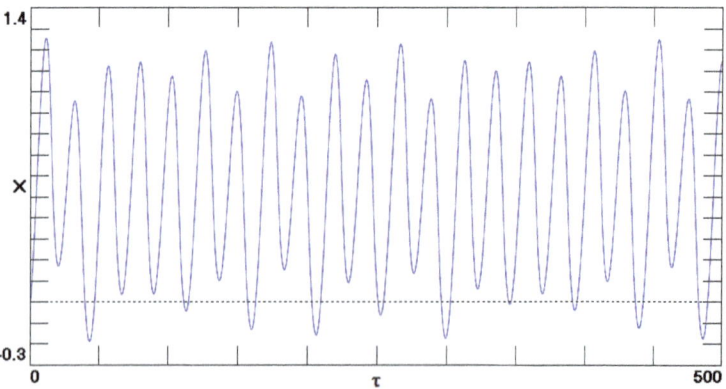

Fig. 6.17 Numerical waveform for $x(\tau)$ from Eq. (6.8) with $a = 0.01$, $b = 0.1$, and $c = 15$.

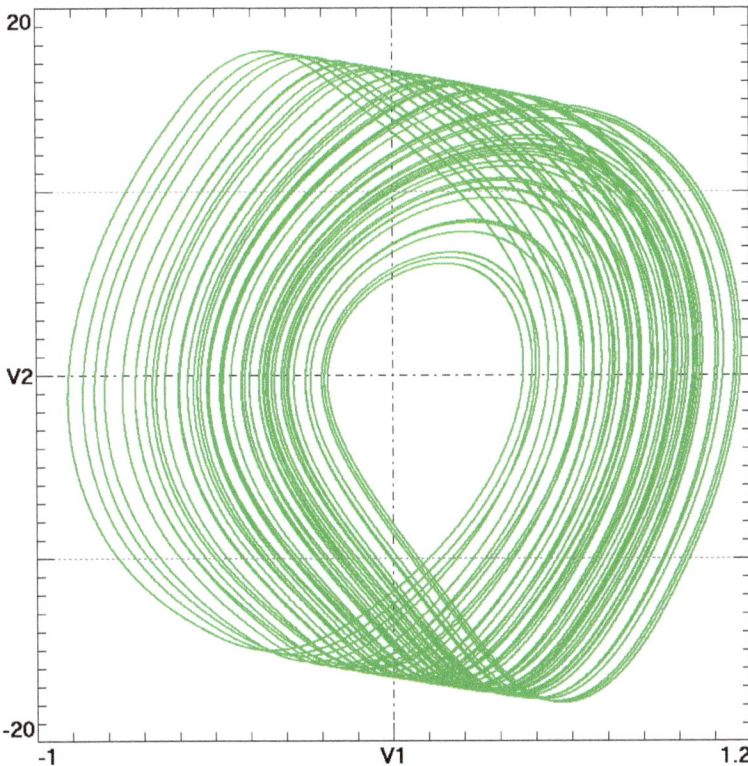

Fig. 6.18 Oscilloscope phase space plot of the Wang–Zhang–Bao circuit from Fig. 6.15.

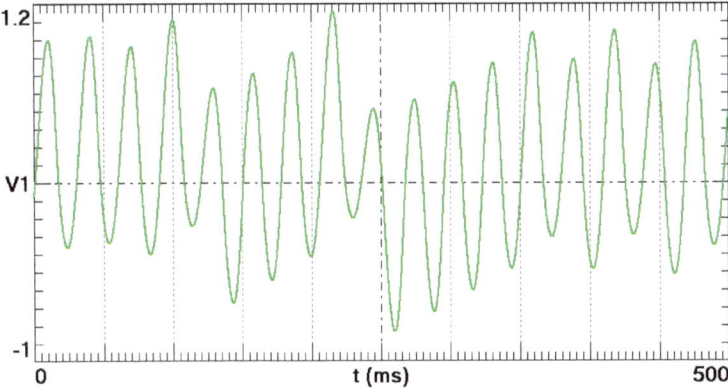

Fig. 6.19 Oscilloscope time plot of the Wang–Zhang–Bao circuit from Fig. 6.15.

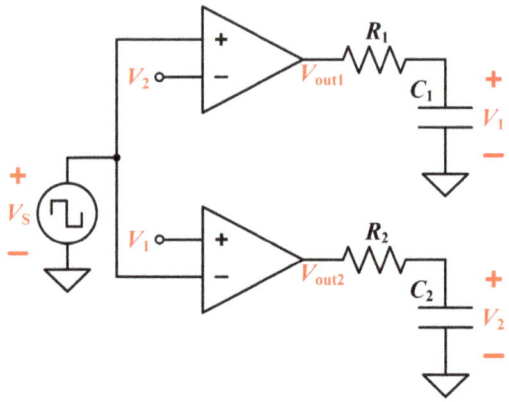

Fig. 6.20 Coupled RC circuits.

Table 6.4 Component values for Fig. 6.20.

Component	Numerical	Simulated	Experimental
R_1	3 kΩ	3 kΩ	3 kΩ
R_2	3 kΩ	3 kΩ	3 kΩ
C_1	1 μF	1 μF	1 μF
C_2	1 μF	1 μF	1 μF
V_{sat}	15 V	15 V	15 V
ω	1 rad/ms	1 rad/ms	1 rad/ms
Op amp	-	Nonideal	LM741

subsequently described. The equations for the circuit in Fig. 6.20 are

$$\dot{V}_1 = (V_{out1} - V_1)/R_1 C_1 \\ \dot{V}_2 = (V_{out2} - V_2)/R_2 C_2, \qquad (6.9)$$

where $V_{out1} = \text{sgn}(V_S - V_2)$, $V_{out2} = \text{sgn}(V_1 - V_S)$, and V_S is a square wave of angular frequency ω.

Elegant circuit values that give chaos are $R_1 = R_2 = 3$ kΩ, $C_1 = C_2 = 1$ μF, $V_{sat} = 15$ V, and $\omega = 1$ rad/ms. The dimensionless equations are

$$\dot{x} = \{c\,\text{sgn}[\text{sgn}(\sin\tau) - y] - x\}/a \\ \dot{y} = \{c\,\text{sgn}[x - \text{sgn}(\sin\tau)] - y\}/b. \qquad (6.10)$$

The parameters $a = b = 3$ and $c = 15$ give the attractor shown in Fig. 6.21 and a time series as shown in Fig. 6.22. The Lyapunov exponents are $(0.3026, 0, -0.9693)$, the Kaplan–Yorke dimension is $D_{ky} = 2.3122$, and the attractor has a global Class 1a basin of attraction.

You should easily obtain chaos in this circuit with little or no adjustment of the parameters in Table 6.4. The phase space plot of V_2 versus V_1 is shown in Fig. 6.23, and $V_1(t)$ is in Fig. 6.24. This circuit has two resistors, two capacitors, and two comparators for a total of six components.

6.6 Ketthong–Banlue Circuit

Ketthong et al. (2017) proposed a circuit that uses the inherent nonlinearity of the operational amplifier approximated by the hyperbolic tangent. This circuit uses the same method as the Banlue–Rattikarn circuit in Section 2.4 to initiate and maintain oscillations. The equations for the circuit in Fig. 6.25 are

$$\dot{V}_1 = I/C_1$$
$$\dot{V}_2 = [I + (V_{out} - V_2)/R - I_0]/C_2 \qquad (6.11)$$
$$\dot{I} = -(V_1 + V_2 + IR_L)/L,$$

where $V_{out} = V_{sat} \tanh(AV_1)$ and $V_{sat} = 15$ V. The constant A is a large positive number representing the gain of the amplifier in the absence of feedback and whose value is not critical. In fact, A can be taken as infinity, in which case $\tanh(AV_1) = \operatorname{sgn} V_1$.

Elegant circuit values that give chaos are $R = 12.4$ kΩ, $R_L = 1$ kΩ, $C_1 = C_2 = 1$ μF, $L = 1$ H, $V_{sat} = 15$ V, $I_0 = 1$ mA, and $A = \infty$. The dimensionless equations are

$$\dot{x} = z$$
$$\dot{y} = z + (a \operatorname{sgn} x - y)/b - c \qquad (6.12)$$
$$\dot{z} = -x - y - z.$$

Chaos occurs over a very narrow range of parameters (a 1.6% change in parameters is likely to destroy the chaos).

The parameters $a = 15$, $b = 12.4$, and $c = 1$ give the attractor shown in Fig. 6.26 and a time series as shown in Fig. 6.27. The Lyapunov exponents are $(0.0430, 0, -1.1237)$, the Kaplan–Yorke dimension is $D_{ky} = 2.0383$, and the attractor has a global Class 1a basin of attraction. To calculate the Lyapunov exponents, $\operatorname{sgn} x$ was replaced by $\tanh(100x)$ to avoid the problematic discontinuity in the equations.

Table 6.5 lists the components, with R and R_L being the main parameters to adjust. The phase space plot of V_2 versus V_1 is shown in Fig. 6.28, and the time plot of $V_1(t)$ is shown in Fig. 6.29, which you can find by measuring the top of the capacitors ($V_1 + V_2$) and subtracting V_2 from it. You can use the circuit in Fig. 2.16 to implement the current source whose

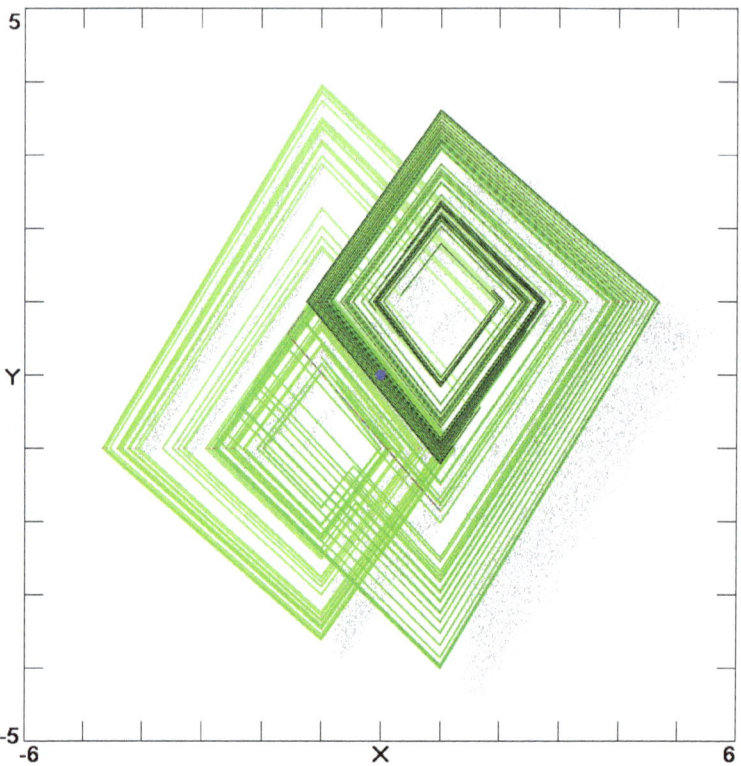

Fig. 6.21 Numerical solution of Eq. (6.10) with $a = b = 3$ and $c = 15$.

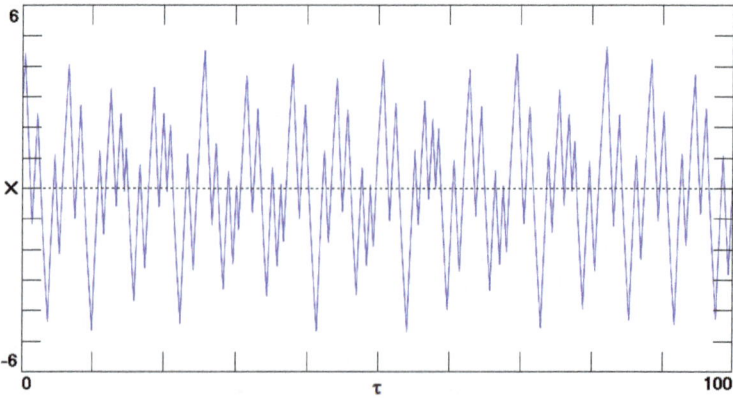

Fig. 6.22 Numerical waveform for $x(\tau)$ from Eq. (6.10) with $a = b = 3$ and $c = 15$.

Saturating Amplifier Circuits 223

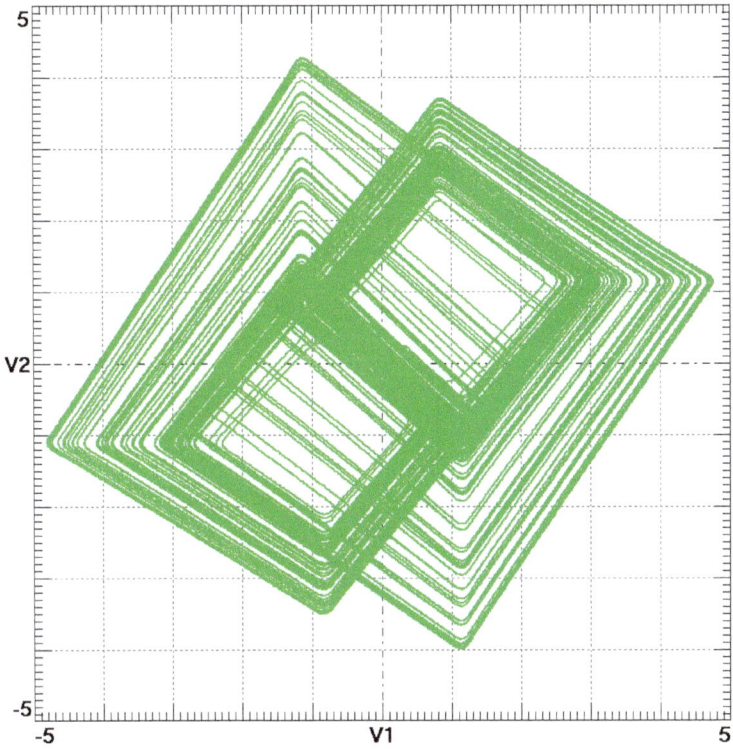

Fig. 6.23 Oscilloscope phase space plot of the coupled RC circuits from Fig. 6.20.

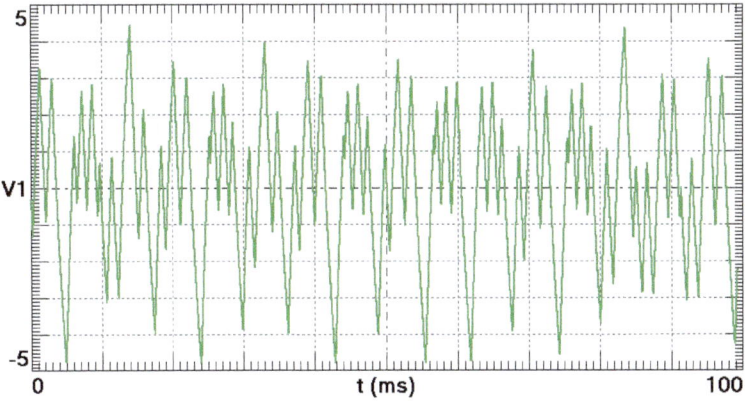

Fig. 6.24 Oscilloscope time plot of the coupled RC circuits from Fig. 6.20.

224 Elegant Circuits

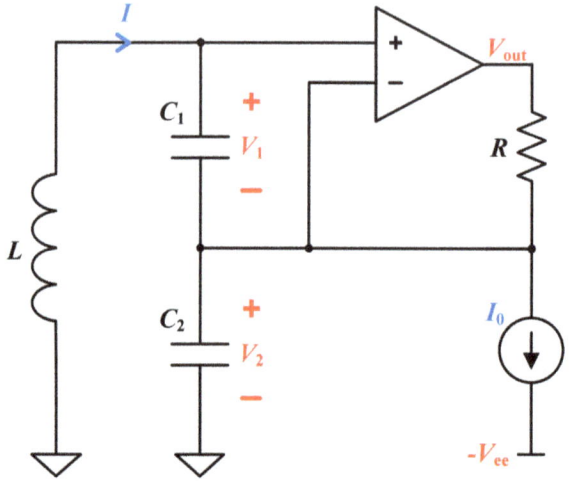

Fig. 6.25 Ketthong-Banlue circuit.

Table 6.5 Component values for Fig. 6.25.

Component	Numerical	Simulated	Experimental
R	12.4 kΩ	12.4 kΩ	12.4 kΩ (adjust)
R_L	1 kΩ	1 kΩ	0.5 kΩ (adjust)
C_1	1 μF	1 μF	1 μF
C_2	1 μF	1 μF	1 μF
L	1 H	1 H	1 H
V_{sat}	15 V	15 V	15 V
I_0	1 mA	1 mA	1 mA
Op amp	-	Nonideal	LM741

direction is not critical unlike in the Banlue–Rattikarn circuit. This circuit has one resistor, two capacitors, one inductor, and one operational amplifier for a total of five components.

6.7 Saito Family Hysteresis Circuit

In addition to discovering a family of chaotic circuits with diodes and thyristors, Saito (1985) investigated circuits with operational amplifier saturation as the nonlinearity. The right part of the circuit in Fig. 6.30 formed by the operational amplifier and by R_1, R_2, and R_3 is called a *Schmitt trigger*, which is a comparator with hysteresis. A Schmitt trigger will switch at two threshold limits when the input voltage V reaches $\pm V_T$ rather than switching at zero to avoid unwanted oscillations triggered by noise when V is small.

The equations for the circuit in Fig. 6.30 are

$$\dot{V} = [V/R - I + (V_{out} - V)/R_1]/C$$
$$\dot{I} = V/L.$$
(6.13)

The threshold voltages are $V_T = (V_0 R_2 \pm V_{sat} R_3)/(R_2 + R_3)$. When V reaches $(V_0 R_2 + V_{sat} R_3)/(R_2 + R_3)$, V_{out} switches from V_{sat} to $-V_{sat}$. When V reaches $(V_0 R_2 - V_{sat} R_3)/(R_2 + R_3)$, V_{out} switches from $-V_{sat}$ to V_{sat}.

Elegant circuit values that give chaos are $R = 1$ kΩ, $R_1 = 2$ kΩ, $R_2 = 4$ kΩ, $R_3 = 1$ kΩ, $C = 1$ μF, $L = 1$ H, $V_0 = 4$ V, and $V_{sat} = 15$ V. The dimensionless equations are

$$\dot{x} = x - y + a(z - x)$$
$$\dot{y} = x$$
$$z = \begin{cases} -c & x > (b+c)/5 \text{ (latch)} \\ c & x < (b-c)/5 \text{ (latch)}, \end{cases}$$
(6.14)

where 'latch' means that z retains its value until it is reset.

The parameters $a = 0.5$, $b = 16$, and $c = 15$ give the attractor shown in Fig. 6.31 and a time series as shown in Fig. 6.32. The Lyapunov exponents cannot be easily calculated because of the discontinuities in the equation. The attractor has a Class 3 basin of attraction with $P \approx 4/r^2$, and there is an unstable focus at the origin with eigenvalues $(0.2500 \pm 0.9682i, 0)$.

The main parameter to adjust is V_0, which controls V_T and can be constructed using the circuit in Fig. 1.29. If you are having trouble getting this circuit to work, you can check your Schmitt trigger by plotting V_{out} versus V and checking that the operational amplifier switches between 6.2 V and 0.2 V for the component values in Table 6.6.

The phase space plot of I versus V is shown in Fig. 6.33, and the time plot of $V(t)$ is shown in Fig. 6.34. You can use a grounded 50 Ω resistor in series with L to measure I. The active resistor $-R$ is constructed from the circuit in Fig. 2.42. Finally, you must maintain the ratio of V_0/V_{sat} if you change V_{sat}, which controls the amplitude of all the voltages and currents. This circuit has four resistors, one capacitor, one inductor, one current feedback amplifier for the active resistor, and one operational amplifier for a total of eight components.

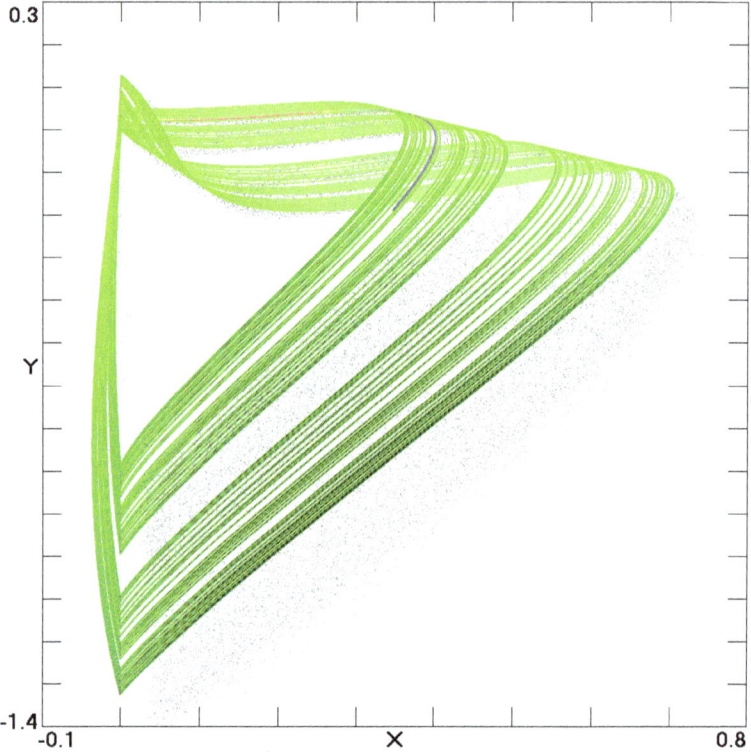

Fig. 6.26 Numerical solution of Eq. (6.12) with $a = 15$, $b = 12.4$, and $c = 1$.

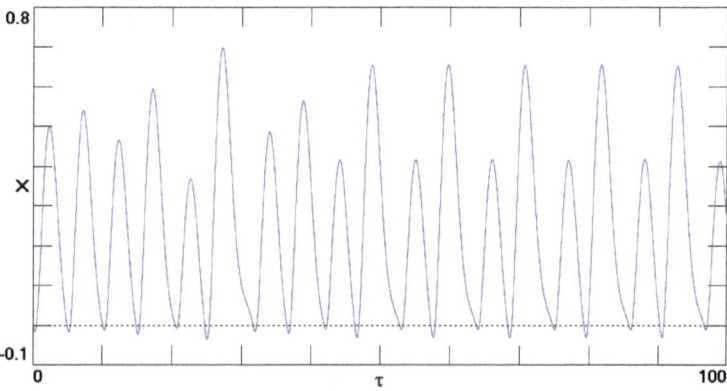

Fig. 6.27 Numerical waveform for $x(\tau)$ from Eq. (6.12) with $a = 15$, $b = 12.4$, and $c = 1$.

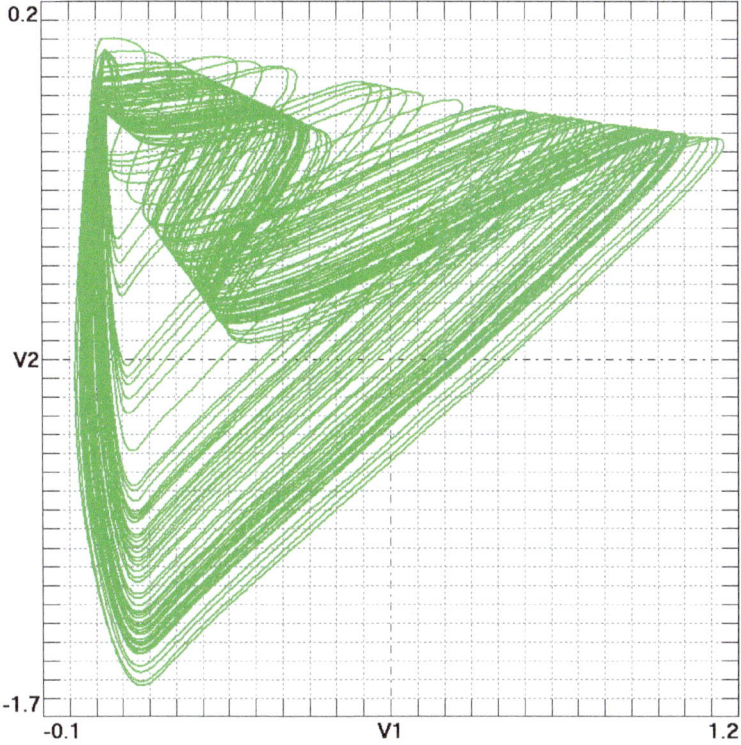

Fig. 6.28   Oscilloscope phase space plot of the Ketthong–Banlue circuit from Fig. 6.25.

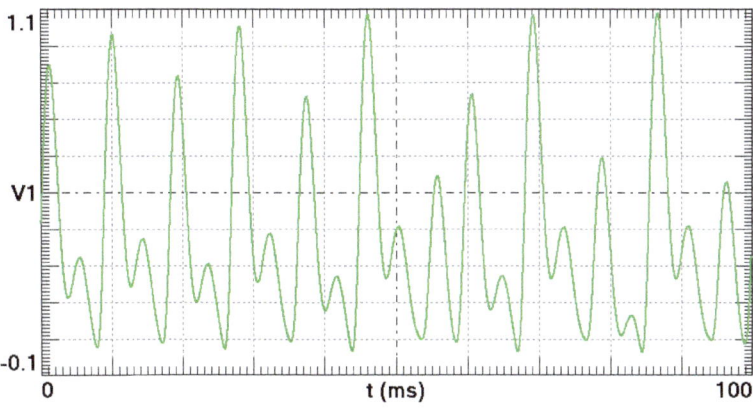

Fig. 6.29   Oscilloscope time plot of the Ketthong–Banlue circuit from Fig. 6.25.

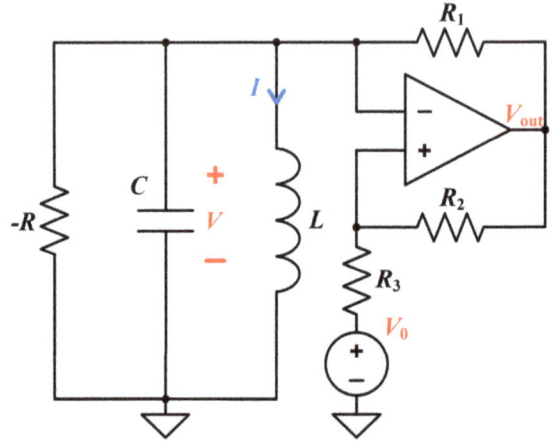

Fig. 6.30 Saito family hysteresis circuit.

Table 6.6 Component values for Fig. 6.30.

Component	Numerical	Simulated	Experimental
R	1 kΩ	1 kΩ	1 kΩ
R_1	2 kΩ	2 kΩ	2 kΩ
R_2	4 kΩ	4 kΩ	4 kΩ
R_3	1 kΩ	1 kΩ	1 kΩ
C	1 μF	1 μF	1 μF
L	1 H	1 H	1 H
V_0	4 V	4 V	4 V (adjust)
V_{sat}	15 V	15 V	15 V
Op amp	-	Nonideal	LM741

6.8 Saito Family Switch Circuit

Saito and Fujita (1981) found another unusual circuit family that uses a comparator to control a switch that momentarily closes when $V \geq V_T$. This switch connects the capacitor to a DC voltage source V_0 less than V_T that restarts the oscillations. Several variants of this circuit exist such as the one described by Elwakil and Ozoguz (2003), which uses the comparator to control a forcing square wave instead of a voltage source. We will use the circuit of Nakano and Saito (2002) in Fig. 6.35 whose equations are

$$\begin{aligned} \dot{V} &= (V/R - I)/C \\ \dot{I} &= (V - IR_L)/L. \end{aligned} \tag{6.15}$$

When $V \geq V_T$, the switch closes which causes $V = V_0$, where $V_0 < V_T$. We assume no current enters the noninverting terminal of the operational amplifier.

Elegant circuit values that give chaos are $R = 1$ kΩ, $R_L = 0.8$ kΩ, $C = 1$ μF, $L = 1$ H, $V_0 = -1$ V, and $V_T = 1$ V. The dimensionless equations are

$$\dot{x} = x - y$$
$$\dot{y} = x - ay \qquad (6.16)$$
$$x = -b \text{ when } x \text{ reaches } c$$

The parameters $a = 0.8$ and $b = c = 1$ give the attractor shown in Fig. 6.37 and a time series as shown in Fig. 6.38. The Lyapunov exponents cannot be easily calculated because of the discontinuities in the equation. The attractor has a global Class 1a basin of attraction, and there is an unstable focus at the origin with eigenvalues $(0.1000 \pm 0.4359i)$.

Details of the switch and comparator are shown in Fig. 6.36, where pins 5 and 6 of the comparator can be left disconnected. The output of the comparator will switch between 5 V and 0 V to control the analog switch, where 5 V closes the switch and 0 V opens the switch. It would be more convenient if you use ± 5 V as the power supply for all the components in this circuit, since you can use V_{cc} to set the logic threshold of the switch V_L rather than the circuit from Fig. 1.29. The comparator also requires a 'pull up' resistor R_0 that we consider to be part of the comparator.

You might also observe that the comparator switches rapidly between states, which is due to the comparator being triggered by noise around 0 V. This can be fixed by either using the Schmitt trigger from the previous section to set the threshold limits above zero, or you can add a 0.1 μF ceramic capacitor between V_{cc} and ground to reduce the noise.

The component values are given in Table 6.7, and you can refer to Fig. 2.42 for the active resistor. We use the single-pole, single-throw *analog switch* MAX318CPA, but you can use an analog switch of your choice that has a series resistance of less than a few ohms. You will need to use the circuit from Fig. 1.29 to set V_0 and V_T, where adjusting V_T below V_0 will cause the circuit to stop oscillating.

The phase space plot of I versus V is shown in Fig. 6.39, where I can be observed by measuring the voltage across a small R_L in series with the grounded end of the inductor L. The time plot of $V(t)$ is shown in Fig. 6.40, which shows a rapid drop when V reaches V_T, although never reaching V_0. The circuit has one resistor, one capacitor, one inductor, one current feedback amplifier for the active resistor, one analog switch, and one comparator for a total of six components.

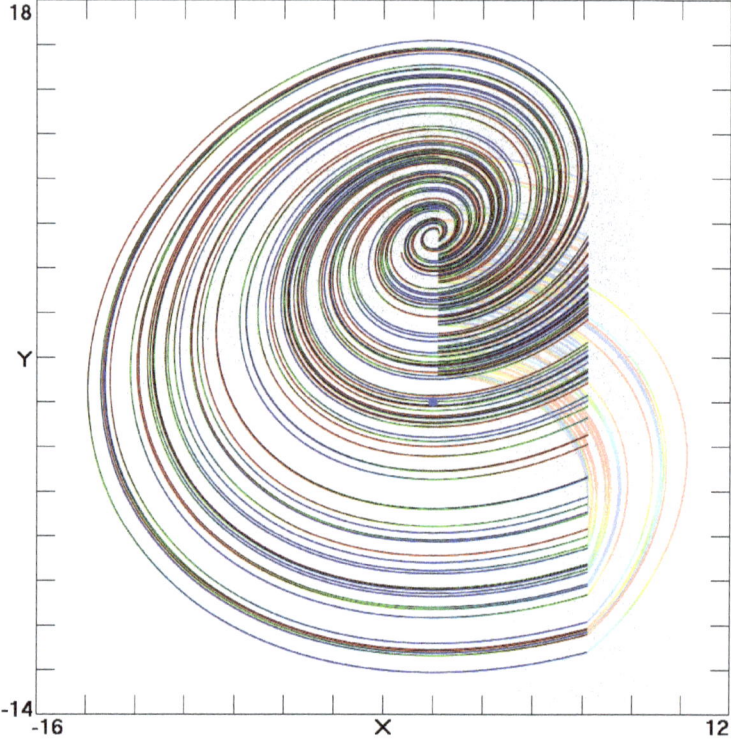

Fig. 6.31 Numerical solution of Eq. (6.14) with $a = 0.5$, $b = 16$, and $c = 15$.

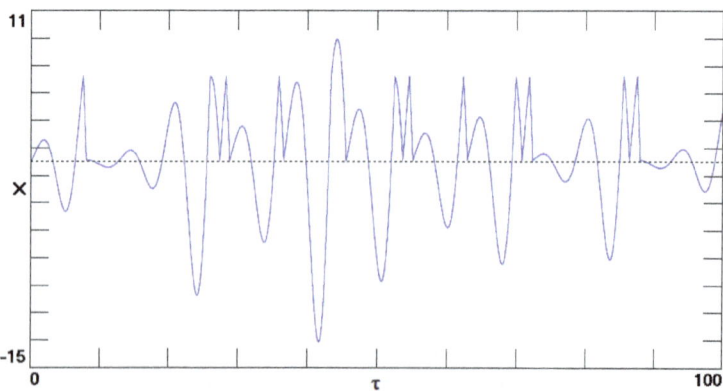

Fig. 6.32 Numerical waveform for $x(\tau)$ from Eq. (6.14) with $a = 0.5$, $b = 16$, and $c = 15$.

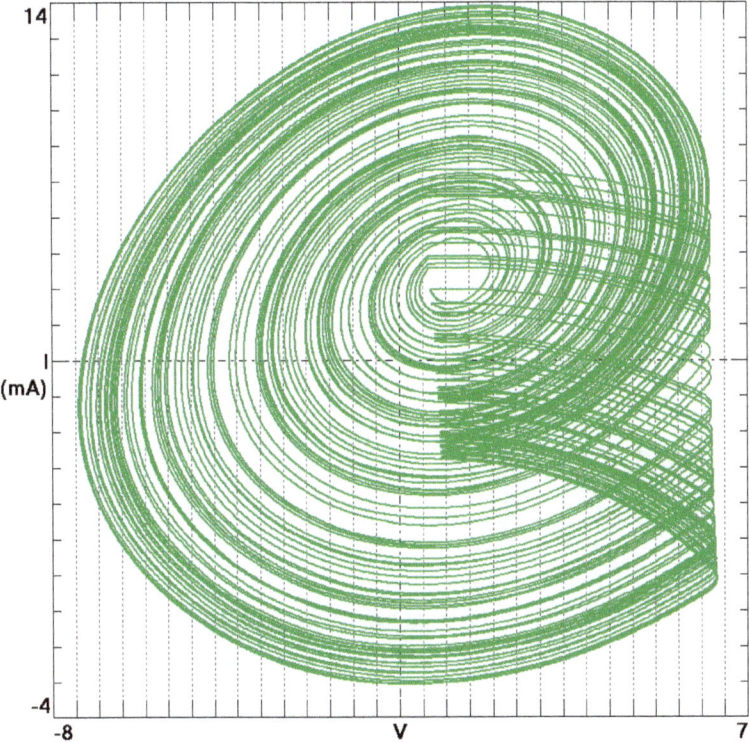

Fig. 6.33 Oscilloscope phase space plot of the Saito family hysteresis circuit from Fig. 6.30.

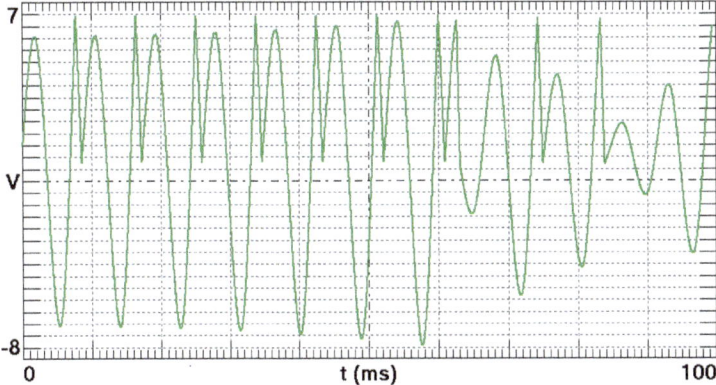

Fig. 6.34 Oscilloscope time plot of the Saito family hysteresis circuit from Fig. 6.30.

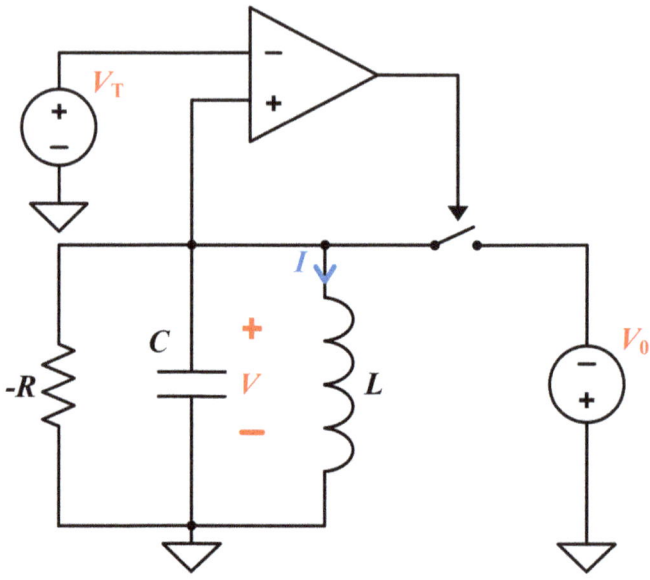

Fig. 6.35 Saito family switch circuit.

Table 6.7 Component values for Fig. 6.35.

Component	Numerical	Simulated	Experimental
R	1 kΩ	1 kΩ	1 kΩ
R_L	0.8 kΩ	0.8 kΩ	0.8 kΩ
R_0	-	-	220 Ω
C	1 µF	1 µF	1 µF
L	1 H	1 H	1 H
V_0	-1 V	-1 V	-1 V
V_T	1 V	1 V	1 V (adjust)
Switch	-	ADG1612	MAX318CPA
Comparator	-	Nonideal	LM311

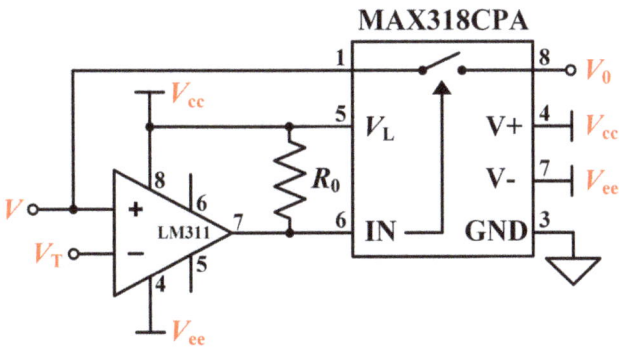

Fig. 6.36 Comparator and switch connections.

6.9 Simplified Piper–Sprott Circuit

Piper and Sprott (2010) described a simple chaotic circuit that uses a saturating operational amplifier (a comparator) along with an operational amplifier integrator and an LC low-pass filter whose operation is modeled by a simple jerk equation. A version of that circuit in which the saturating amplifier is replaced by a diode was described in Section 2.5.

We discovered that the comparator can be eliminated in that circuit as shown in Fig. 6.41 if the nonlinearities in the integrator are exploited. In particular, an operational amplifier will saturate if the output current exceeds some value I_{sat} just as it does if the output voltage exceeds V_{sat}. When that occurs, the operational amplifier will no longer keep the inverting terminal voltage V_- at zero, and the voltage at that terminal will be determined by the other components in the circuit.

The circuit can be modeled by assuming the operational amplifier operates in the linear region when $-I_{sat} < I_{out} < I_{sat}$, where I_{sat} is 25 mA for the LM741, called the 'short circuit current' or 'output current' in the data sheet. For I_{out} outside this range, $I_{out} = \pm I_{sat}$ and $V_- = V_1 - (I - I_{out})R$. The voltage saturation of V_{out} at the power supply can be excluded from this model by making resistor R small enough so that the output current saturates before the output voltage does.

Although the circuit was made more elegant by eliminating a comparator and a resistor from the original Piper and Sprott (2010) circuit, it is not robust since I_{sat} varies for different operational amplifiers, and the

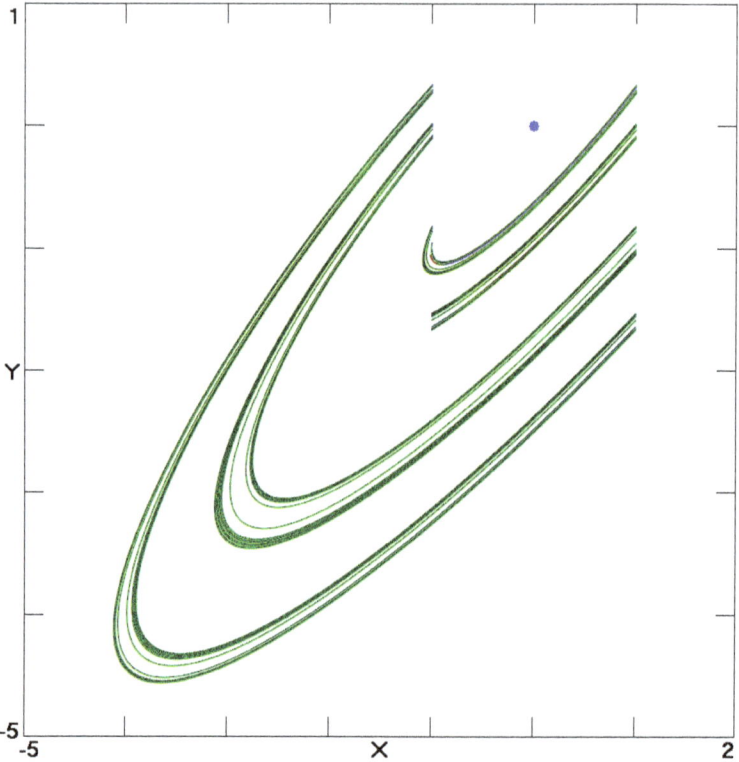

Fig. 6.37 Numerical solution of Eq. (6.16) with $a = 0.8$ and $b = c = 1$.

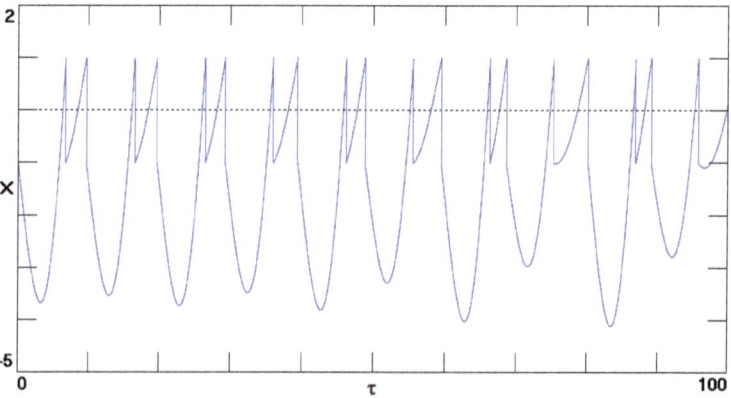

Fig. 6.38 Numerical waveform for $x(\tau)$ from Eq. (6.16) with $a = 0.8$ and $b = c = 1$.

Saturating Amplifier Circuits 235

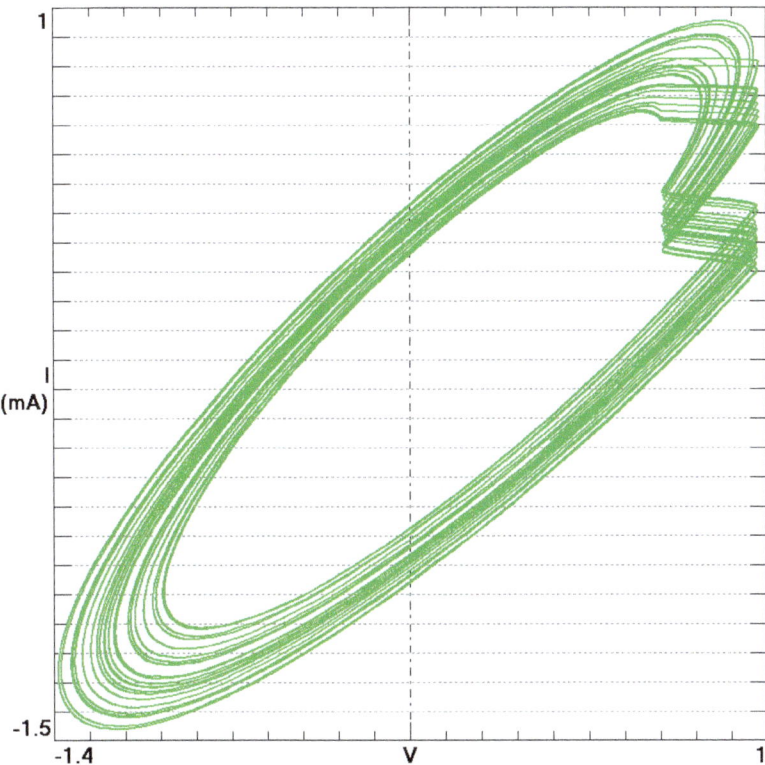

Fig. 6.39 Oscilloscope phase space plot of the Saito family switch circuit from Fig. 6.35.

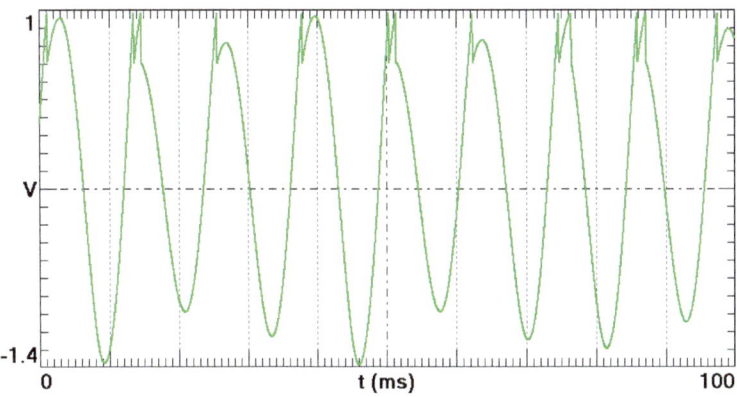

Fig. 6.40 Oscilloscope time plot of the Saito family switch circuit from Fig. 6.35.

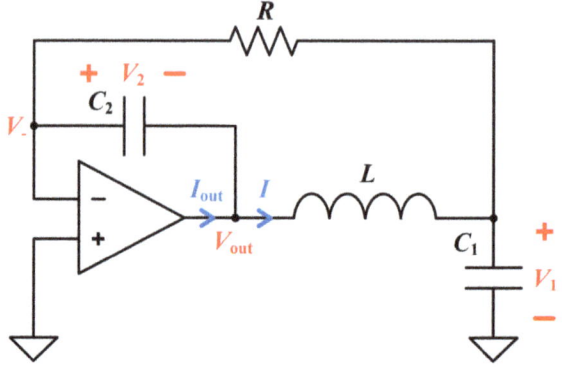

Fig. 6.41 Simplified Piper–Sprott circuit.

Table 6.8 Component values for Fig. 6.41.

Component	Numerical	Simulated	Experimental
R	1 kΩ	1 kΩ	1.2 kΩ (adjust)
C_1	10 μF	10 μF	10 μF
C_2	0.01 μF	0.01 μF	0.01 μF
L	0.05 H	0.05 H	0.05 H
V_{sat}	-	20 V	20 V
I_{sat}	25 mA	25 mA	+26/ − 20 mA
Op amp	-	Nonideal	LM741

resulting equations are less elegant as given by

$$\dot{V}_1 = [I + (V_- - V_1)/R]/C_1$$
$$\dot{V}_2 = (I - I_{out})/C_2 \qquad (6.17)$$
$$\dot{I} = (V_{out} - V_1)/L,$$

where $V_{out} = V_- - V_2$, $V_- = V_1 - (I - I_{out})R$, and $I_{out} = \min[\max(I - V_1/R, -I_{sat}), I_{sat}]$.

Elegant circuit values that give chaos are $R = 1$ kΩ, $C_1 = 10$ μF, $C_2 = 0.01$ μF, $L = 0.05$ H, and $I_{sat} = 25$ mA. The circuit requires an inductor with a relatively small parasitic resistance. For the given parameters, the chaos in the model is suppressed if R_L is greater than about 100 Ω. The dimensionless equations are

$$\dot{x} = a\iota$$
$$\dot{y} = b(z - \iota) \qquad (6.18)$$
$$\dot{z} = c(-z + \iota - y),$$

where $\iota = \min[\max(z - x, -d), d]$.

The parameters $a = 0.1$, $b = 100, c = 20$, and $d = 25$ give the attractor shown in Fig. 6.42 and a time series as shown in Fig. 6.43. The Lyapunov exponents are $(1.1051, 0, -4.0320)$, and the Kaplan–Yorke dimension is $D_{ky} = 2.2741$. The attractor has a global Class 1a basin of attraction, and there is an unstable saddle focus at the origin with eigenvalues $(2.8335 \pm 5.1626i, -5.7669)$.

The most critical parameter in Table 6.8 is I_{sat}, but it is intrinsic to the operational amplifier and thus hard to control. We measured this saturation current by connecting an ammeter between the output of an operational amplifier buffer (Fig. 6.2) and ground. If the input to this circuit is sufficiently large, the current will saturate at I_{sat}.

For large positive voltages, our LM741 can output ('source') $I_{sat} \approx 26$ mA, while for large negative voltages the output can take in ('sink') $-I_{sat} \approx -20$ mA. This difference in saturation currents is likely the cause of the asymmetry in the experimental phase space plot of V_2 versus V_1 in Fig. 6.44 compared to the numerical plot. If you try this test, be sure that your operational amplifier has protection to handle the large output current.

Also be aware that the larger the I_{sat} in your operational amplifier, the larger will be the required V_{sat} so that a plot similar to Fig. 6.42 can be obtained without voltage saturation. Otherwise, you might only see half of the attractor. We used $V_{sat} = 20$ V because the circuit otherwise saturates and does not resemble the model.

For operational amplifiers with particularly large I_{sat} such as 40 mA, the required V_{sat} might be impractical. One way around this at the cost of elegance is to add a resistor in series at the output of operational amplifier to lower I_{sat}. However, if this resistor is too large, additional terms must be added to the model equations. Alternately, you can lower R while keeping $R^2 C/L$ constant.

The time plot for $V_1(t)$ is shown in Fig. 6.45. You should measure the voltage V_2 with an instrumentation amplifier from Fig. 1.23 since the inverting terminal is no longer a virtual ground. There are several other circuits of similar elegance such as described by Tchitnga et al. (2017) and by Yim et al. (2004), but they rely on parasitic properties of the operational amplifier that are difficult to model, and so they will not be described here. This circuit has a resistor, two capacitors, one inductor, and an operational amplifier for a total of five components and is shown on the back cover of the book.

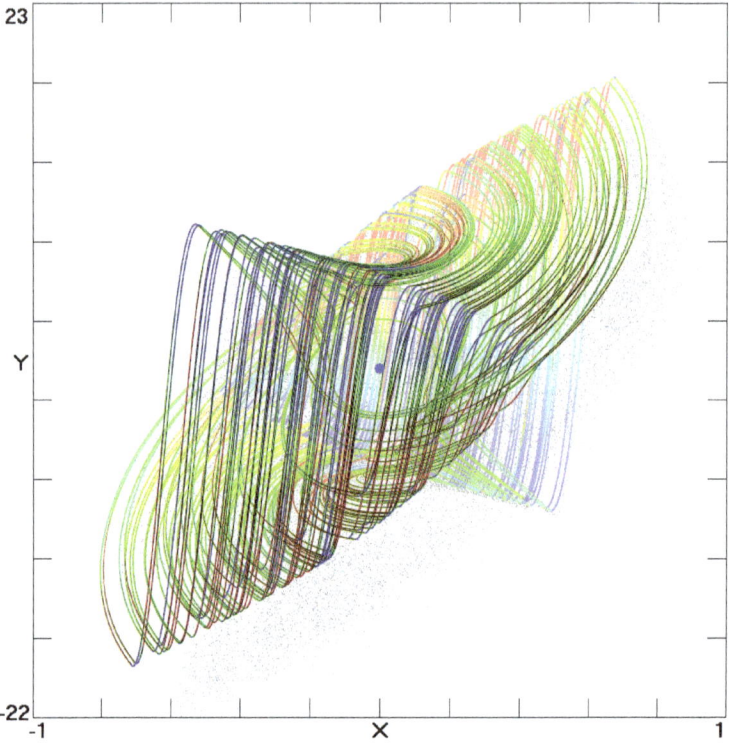

Fig. 6.42 Numerical solution of Eq. (6.18) with $a = 0.1$, $b = 100$, $c = 20$, and $d = 25$.

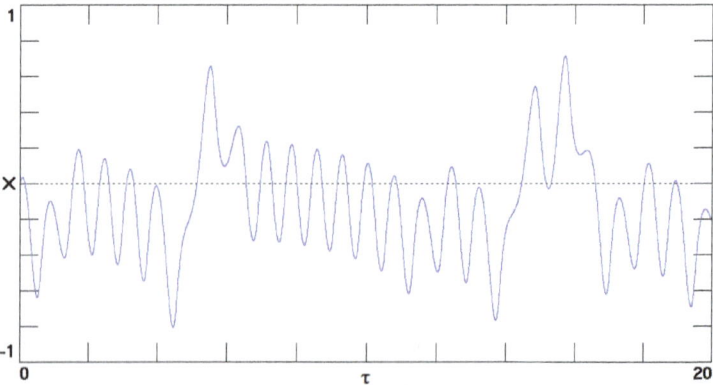

Fig. 6.43 Numerical waveform for $x(\tau)$ from Eq. (6.18) with $a = 0.1$, $b = 100$, $c = 20$, and $d = 25$.

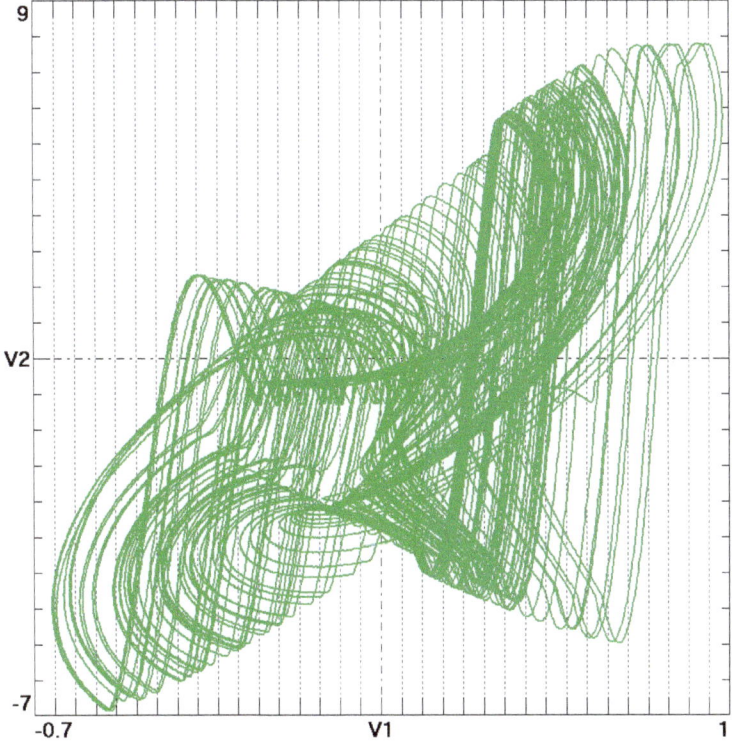

Fig. 6.44 Oscilloscope phase space plot of the simplified Piper–Sprott circuit from Fig. 6.41.

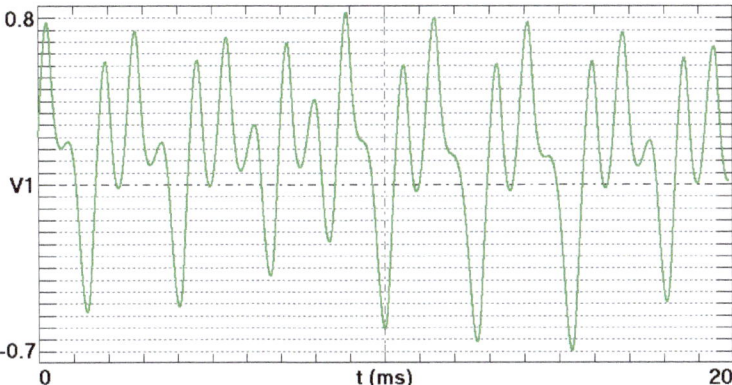

Fig. 6.45 Oscilloscope time plot of the simplified Piper–Sprott circuit from Fig. 6.41.

Chapter 7

Analog Multiplier Circuits

Perhaps the most versatile nonlinear device is the analog multiplier. Many standard chaotic systems involve polynomial nonlinearities and thus lend themselves to electronic implementation using such devices. This chapter gives a small selection of some of the most elegant examples of such systems and their corresponding circuits.

7.1 Analog Multipliers

Analog multipliers are a common electronic component in chaotic circuits used to implement equations involving the multiplication of two variables or the squaring of a single variable. They are widely used for research purposes because of their great versatility.

7.1.1 *Analog computers*

The discovery and study of chaos was made possible mainly because of the computer, which automates repetitive calculations that would be tedious for a human to do. We have used a *digital computer* to find elegant component values for the circuits in this book, calculate Lyapunov exponents and Kaplan–Yorke dimensions, and to simulate the circuits.

Before the digital computer, you would have to build specialized circuits to perform mathematical operations such as addition, subtraction, and integration, and connect them together based on the equations you were solving. Such *analog computers* were quite complicated and were limited in what they could do and have now become obsolete for most purposes.

One property of analog computers is that they use physical quantities (voltage and current), whereas a digital computer uses symbolic values of zero and one. This allows the digital computer to simulate systems without energy dissipation (conservative systems) such as an oscillating LC circuit with no resistance. However, since real-world circuits always have some internal resistance, constructing an analog computer that precisely conserves energy is impossible.

Fortunately, most chaotic systems of practical interest are dissipative, and such systems lend themselves to analysis using analog computers. For that reason, many papers describing new chaotic systems include results from an analog computer to verify that the chaos is real and not a numerical artifact from the inevitable approximations required to solve the equations with a digital computer. Many such papers include only SPICE simulations of the circuit, which amounts to solving the equations with a digital computer and thus does not provide a confirmation of chaos.

Today, analog computers are usually constructed using a *field programmable gate array (FPGA)* or a *field programmable analog array (FPAA)* which allows you to simply program the circuit. You can find an introduction to building chaotic circuits using these tools in references such as Buscarino *et al.* (2014) and Muthuswamy and Banerjee (2015), and you can use them to construct circuits for many chaotic systems.

The circuits described in this chapter use a small number of discrete components since most them do not require any linear amplifiers and instead use passive components to perform the integrations and active analog multipliers to offset the energy loss through dissipation. Thus these circuits are especially elegant compared with those typically found in the research literature.

7.1.2 AD633 multiplier

The analog multipliers frequently used in analog computers, as shown schematically in Fig. 7.1, take two signals V_1 and V_2 and multiplies them to get $V_1 V_2 / u$. They can also be configured to do division, take square roots, or approximate trigonometric functions using their Taylor series.

Since the output signal from multiplication is often large, analog multipliers include a scale factor $u > 1$ to prevent saturating the circuit. Typical voltages reach the order of 10 V so that $u = 10$ is commonly used so that V^2 is also of order 10 V. In principle, the saturation could be exploited to provide an additional nonlinearity, but in practice this is rarely done except by accident, and none of the circuits in the chapter use that mechanism.

In this chapter we will use the AD633 analog multiplier whose pins are shown in Fig. 7.2. This multiplier performs the operation $(x_1 - x_2)(y_1 -$

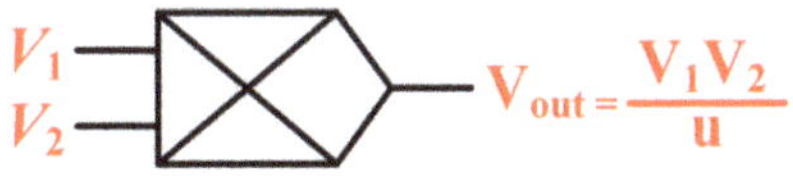

V_1 ⎯⎯⎤
         ⎬⊗⎯ $V_{out} = \dfrac{V_1 V_2}{u}$
V_2 ⎯⎯⎦

Fig. 7.1 Analog multiplier.

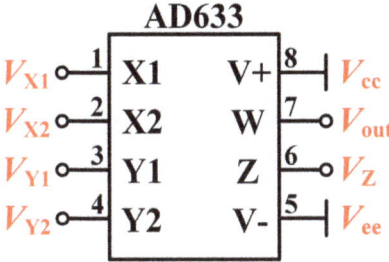

Fig. 7.2 Pin-out of AD633 where $V_{out} = (V_{X1} - V_{X2})(V_{Y1} - V_{Y2})/10 + V_Z$.

$y_2)/u + z$, where the scaling factor is $u = 10$ and z is an offset that is exploited in some of the circuits. The inputs to this integrated circuit have a high impedance and draw negligible current, while the output has a low impedance and hence acts as a voltage source, much like an operational amplifier.

This type of multiplier is also called a *four-quadrant multiplier* since it allows positive and negative signals at both inputs. For the circuits that follow, we will use Fig. 7.1 to represent the multipliers and separately provide details about the connections to the pins in Fig. 7.2.

7.2 Lorenz System

The oldest and most widely studied chaotic system is due to Lorenz (1963). With two multiplicative nonlinearities, this system lends itself to electronic implementation using analog multipliers, and a number of such circuits have been designed and constructed. Most of these circuits have a large number of components and are thus rather inelegant. For example, Cuomo and Oppenheim (1993) designed a system with over thirty components in addition to the two analog multipliers.

Blakely *et al.* (2007) described a much more elegant example that uses RC circuits rather than the operational amplifiers used in most analog computers. Their circuit is shown in Fig. 7.3 and its multiplier connections

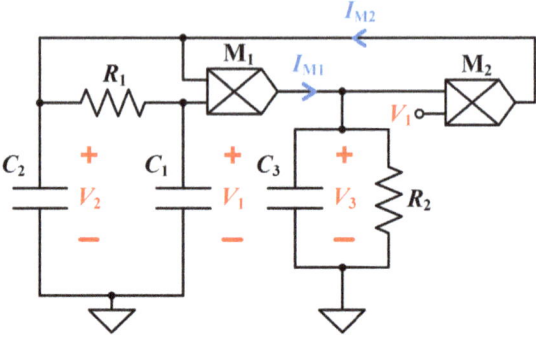

Fig. 7.3 Lorenz system described by Eq. (7.1).

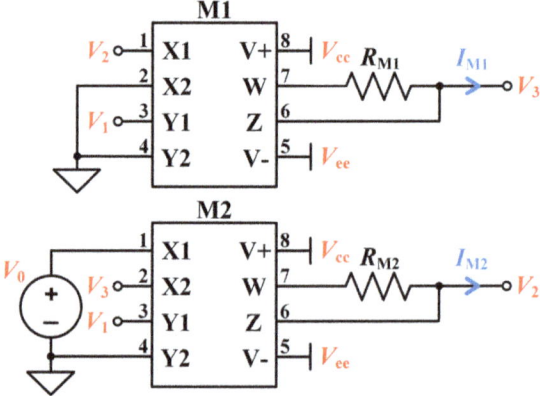

Fig. 7.4 Analog multiplier connections for Fig. 7.3.

Table 7.1 Component values for Fig. 7.3.

Component	Numerical	Simulated	Experimental
R_1	1 kΩ	1 kΩ	1 kΩ
R_2	1 kΩ	1 kΩ	1 kΩ
R_{M1}	0.025 kΩ	0.025 kΩ	0.025 kΩ
R_{M2}	0.025 kΩ	0.025 kΩ	0.025 kΩ
C_1	0.25 μF	0.25 μF	0.25 μF
C_2	1 μF	1 μF	1 μF
C_3	1 μF	1 μF	1 μF
V_0	5.5 V	5.5 V	5.5 V (adjust)
Multiplier	-	Ideal	AD633

in Fig. 7.4. The equations for the circuit are

$$\dot{V}_1 = (V_2 - V_1)/R_1C_1$$
$$\dot{V}_2 = (I_{M2}R_1 + V_1 - V_2)/R_1C_2 \qquad (7.1)$$
$$\dot{V}_3 = (I_{M1}R_2 - V_3)/R_2C_3,$$

where $I_{M1} = V_1V_2/10R_{M1}$ and $I_{M2} = V_1(V_0 - V_3)/10R_{M2}$.

Elegant circuit values that give chaos are $R_1 = R_2 = 1$ kΩ, $R_{M1} = R_{M2} = 0.025$ kΩ, $C_1 = 0.25$ μF, $C_2 = C_3 = 1$ μF, and $V_0 = 5.5$ V. The dimensionless equations are

$$\dot{x} = \sigma(y - x)$$
$$\dot{y} = -xz + rx - y \qquad (7.2)$$
$$\dot{z} = xy - bz.$$

The parameters $\sigma = 4$, $r = 23$, and $b = 1$ give the attractor shown in Fig. 7.5 and a time series as shown in Fig. 7.6. The Lyapunov exponents are (0.4843, 0, −6.483), the Kaplan–Yorke dimension is $D_{ky} = 2.0747$, and the attractor has a global Class 1a basin of attraction. There is an unstable saddle node at the origin with eigenvalues (7.2082, −1, −12.2082) and a symmetric pair of unstable saddle foci at (1.1726, 1.1726, 5.5) and (−1.1726, −1.1726, 5.5) with eigenvalues (0.1067 ± 5.3211i, −6.2134).

There is a subcritical Hopf bifurcation at $r = 16$ where the off-axis equilibria lose their stability. For slightly lower values ($r < 16$), these stable equilibria coexist with the strange attractor. The behavior is similar to the Lorenz system for the standard parameters of $\sigma = 10$, $r = 28$, and $b = 8/3$, but with slightly more elegant parameters.

Both multipliers are configured using the offset Z so that the outputs are currents I_{M1} and I_{M2}. The only parameter you need to adjust in Table 7.1 is V_0 using the circuit in Fig. 1.29, but you can reduce R_{M1} and R_{M2} if you have difficulty with saturation of the multipliers.

The phase space plot of V_2 versus V_1 is shown in Fig. 7.7, and the time plot of $V_1(t)$ is shown in Fig. 7.8. This circuit has four resistors, three capacitors, and two analog multipliers for a total of nine components.

7.3 Rössler Prototype-4 System

In the years following the publication of the Lorenz system, a number of other people found additional examples of low-dimensional chaotic systems. In particular, Otto Rössler, a non-practicing medical doctor, found several systems there were in some sense simpler than the Lorenz system. Perhaps the simplest such system is the Rössler prototype-4 system [Rössler

246   *Elegant Circuits*

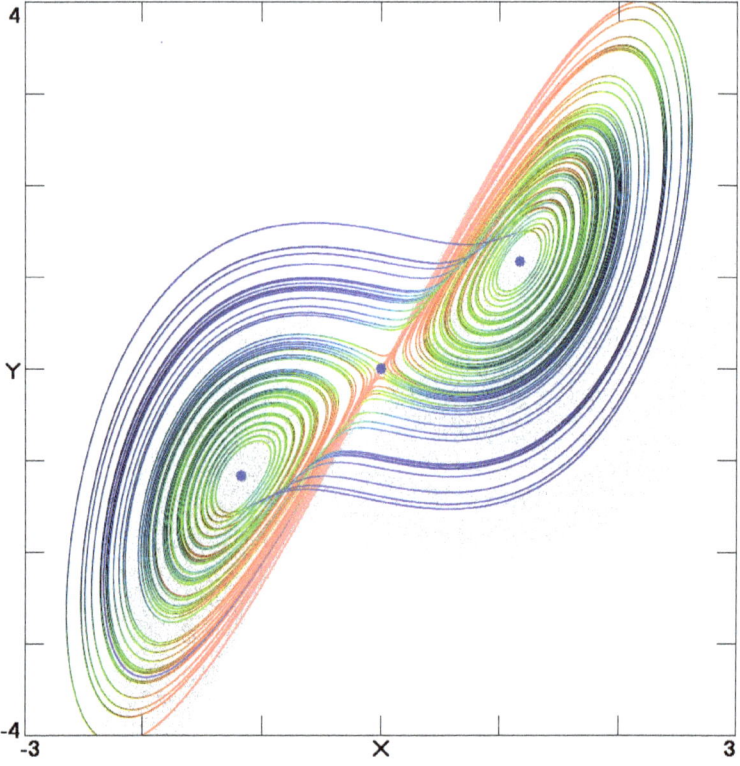

Fig. 7.5   Numerical solution of Eq. (7.2) with $\sigma = 4$, $r = 16$, and $b = 1$.

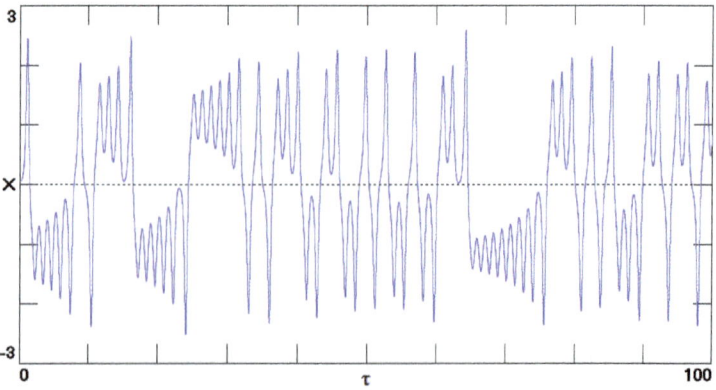

Fig. 7.6   Numerical waveform for $x(\tau)$ from Eq. (7.2) with $\sigma = 4$, $r = 16$, and $b = 1$.

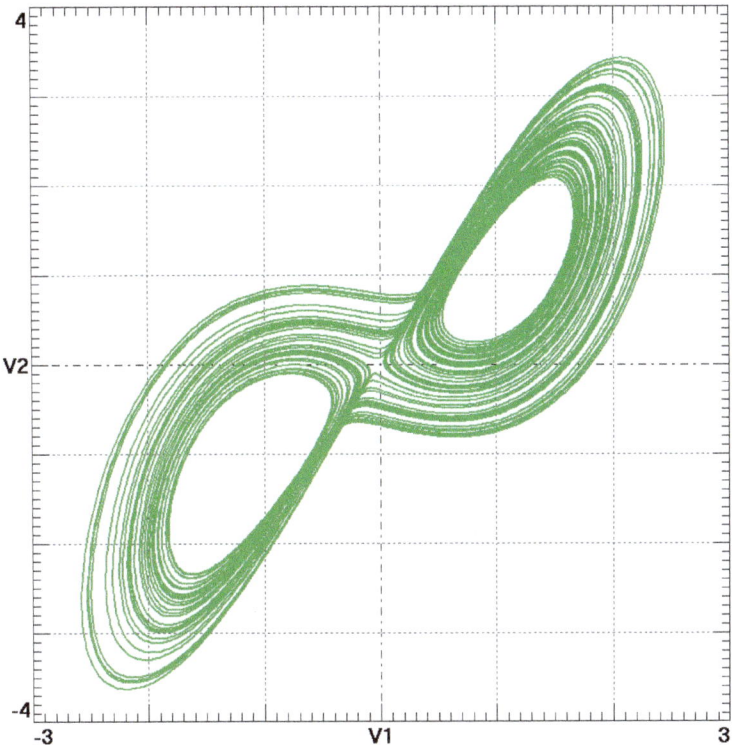

Fig. 7.7 Oscilloscope phase space plot of the Lorenz system from Fig. 7.3.

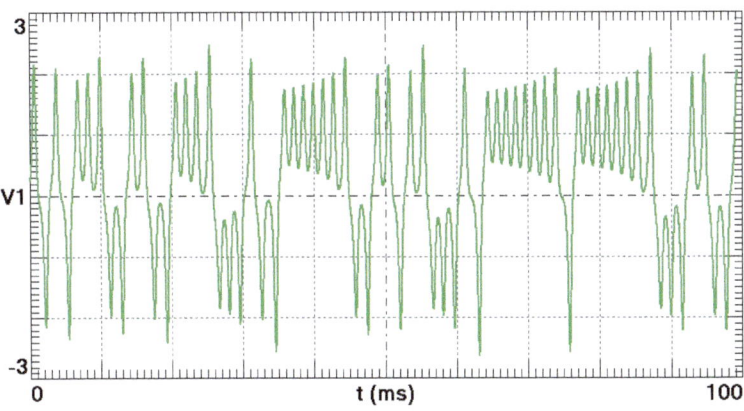

Fig. 7.8 Oscilloscope time plot of the Lorenz system from Fig. 7.3.

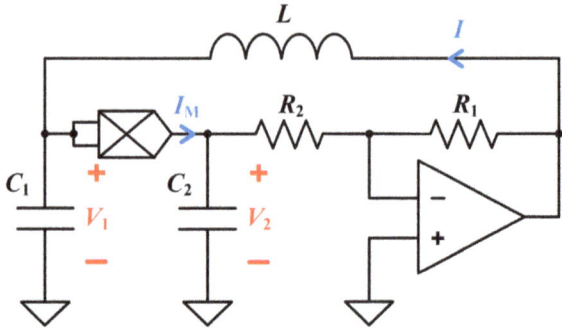

Fig. 7.9 Rössler prototype-4 system described by Eq. (7.3).

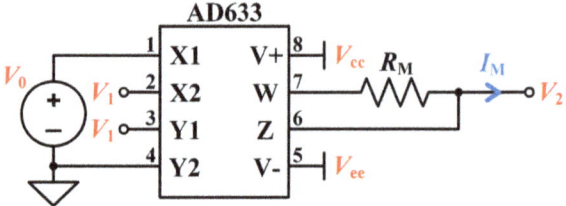

Fig. 7.10 Analog multiplier connections for Fig. 7.9.

Table 7.2 Component values for Fig. 7.9.

Component	Numerical	Simulated	Experimental
R_1	1 kΩ	1 kΩ	1.3 kΩ (adjust)
R_2	1 kΩ	1 kΩ	1 kΩ
R_M	0.1 kΩ	0.1 kΩ	0.1 kΩ
C_1	1 μF	1 μF	1 μF
C_2	2 μF	2 μF	2 μF
L	1 H	1 H	1 H
V_0	1 V	1 V	1 V
Op amp	-	Ideal	LM741
Multiplier	-	Ideal	AD633

(1979)] which has six terms (one less than Lorenz) and a single quadratic nonlinearity (one less than Lorenz).

This system lends itself nicely to electronic implementation using the circuit in Fig. 7.9 with the analog multiplier configured as shown in Fig. 7.10.

The equations that govern the circuit are

$$\dot{I} = (-R_1 V_2/R_2 - V_1)/L$$
$$\dot{V_1} = I/C_1 \qquad (7.3)$$
$$\dot{V_2} = (I_M - V_2/R_2)/C_2,$$

where $I_M = V_1(V_0 - V_1)/10R_M$.

Elegant circuit values that give chaos are $R_1 = R_2 = 1$ kΩ, $R_M = 0.1$ kΩ, $C_1 = 1$ μF, $C_2 = 2$ μF, $L = 1$ H, and $V_0 = 1$ V. The dimensionless equations are

$$\dot{x} = -y - z$$
$$\dot{y} = x \qquad (7.4)$$
$$\dot{z} = a(y - y^2) - bz.$$

The parameters $a = b = 0.5$ give the attractor shown in Fig. 7.11 and a time series as shown in Fig. 7.12. The Lyapunov exponents are (0.0938, 0, −0.5938), the Kaplan–Yorke dimension is $D_{ky} = 2.1580$, and the attractor has a Class 3 basin of attraction with $P \approx 5/r^{2.7}$. The system has unstable saddle foci at (0, 0, 0) and at (0, 2, −2) with eigenvalues ($0.1519 \pm 1.1050i$, −0.8038) and ($0.6015, -0.5507 \pm 1.1659i$), respectively.

The components for this circuit are given in Table 7.2 where V_0 is constructed from the circuit in Fig. 1.29. Make sure V_0 is close to 1 V, and then adjust R_1 slightly from 1 kΩ until you observe chaos. The phase space plot of V_1 versus I is shown in Fig. 7.13, where $I(t)$ in Fig. 7.14 can be measured by adding a 50 Ω resistor between C_1 and L and determining its current using Ohm's law. This circuit has two resistors, two capacitors, one inductor, one operational amplifier, and one analog multiplier for a total of seven components.

7.4 Original Ueda System

Around the time that Lorenz made his discovery, a graduate student in Japan, Yoshisuke Ueda, discovered another chaotic system using an analog computer, but he was prevented from publishing it by his supervisor. The system he was studying [Ueda et al. (1973)] as described by Ueda and Akamatsu (1981) is shown in Fig. 7.15 using the analog multipliers in Fig. 7.16 and governed by the equations

$$\dot{V} = (I + I_M)/C$$
$$\dot{I} = (V_S - V - IR_L)/L, \qquad (7.5)$$

where $I_M = (V_0 - V^2/10)V/10R_M$ and $V_S = V_m \sin \omega t$.

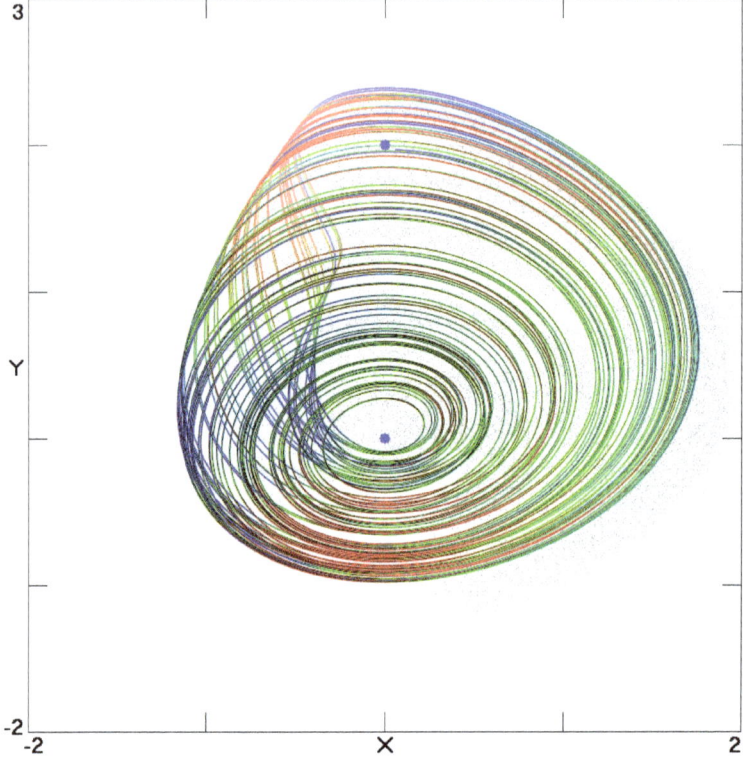

Fig. 7.11 Numerical solution of Eq. (7.4) with $a = b = 0.5$.

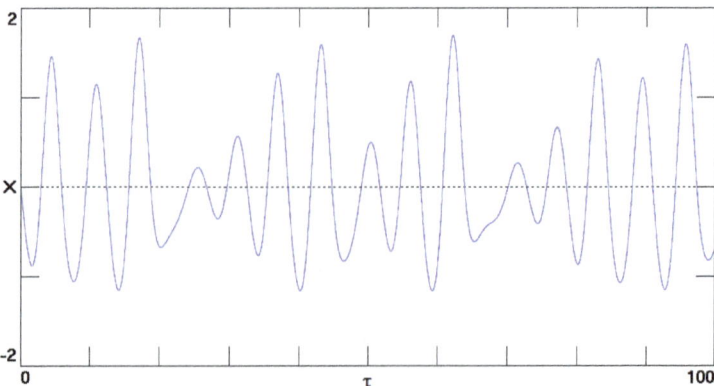

Fig. 7.12 Numerical waveform for $x(\tau)$ from Eq. (7.4) with $a = b = 0.5$.

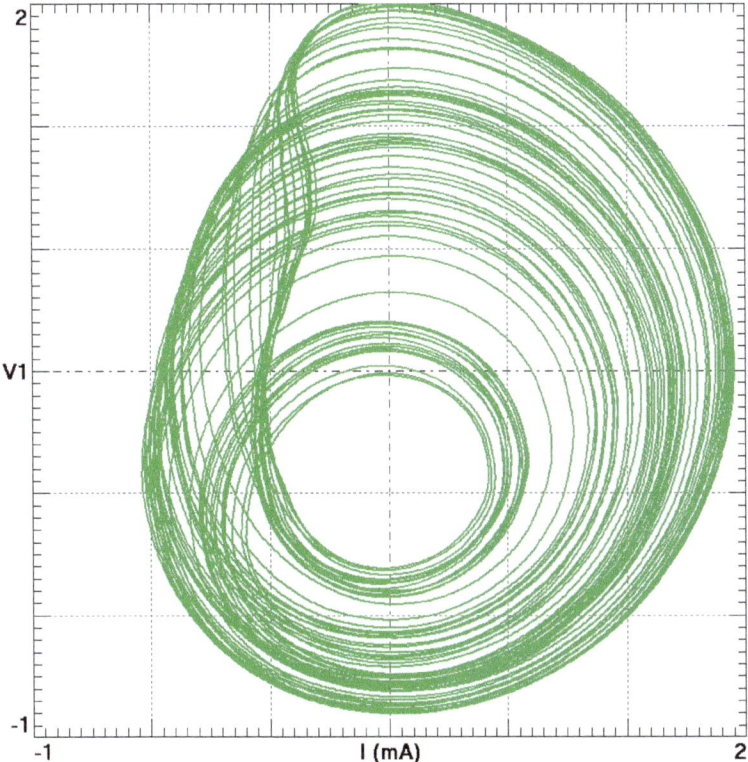

Fig. 7.13   Oscilloscope phase space plot of the Rössler prototype-4 system from Fig. 7.9.

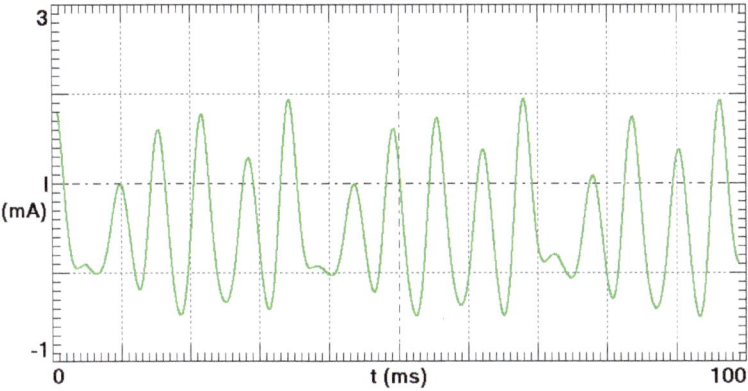

Fig. 7.14   Oscilloscope time plot of the Rössler prototype-4 system from Fig. 7.9.

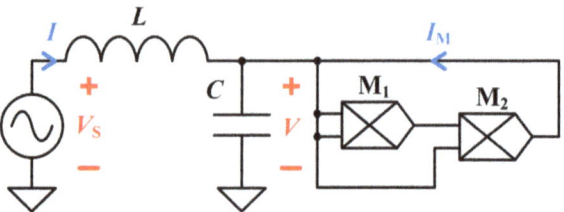

Fig. 7.15 Original Ueda system described by Eq. (7.5).

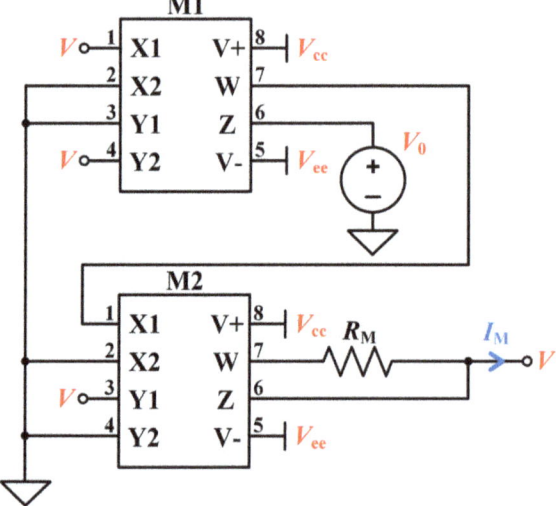

Fig. 7.16 Analog multiplier connections for Fig. 7.15.

Table 7.3 Component values for Fig. 7.15.

Component	Numerical	Simulated	Experimental
R_L	0.7 kΩ	0.7 kΩ	0.7 kΩ
R_M	0.07 kΩ	0.07 kΩ	0.07 kΩ
C	1 µF	1 µF	1 µF
L	1 H	1 H	1 H
V_0	1 V	1 V	0.78 V (adjust)
V_M	1 V	1 V	1 V
ω	1 rad/ms	1 rad/ms	1 rad/ms
Multiplier	-	Ideal	AD633

Elegant circuit values that give chaos are $R_L = 0.7$ kΩ, $R_M = 0.07$ kΩ, $C = 1$ μF, $L = 1$ H, $V_0 = V_m = 1$ V, and $\omega = 1$ rad/ms. Choosing the parameters such that $V_0 R_L = 10/R_M$ eliminates a linear term in the equation. The dimensionless equations are

$$\dot{x} = y + (1 - x^2/10)x/a$$
$$\dot{y} = \sin \tau - x - by. \quad (7.6)$$

For the special case where $a = b = 0.7$, Eq. (7.6) can be written as $\ddot{x} - (5.1 - 3x^2)\dot{x}/7 + 0.1x^3 = \sin \tau$, which is a periodically-forced hybrid van der Pol–Duffing oscillator. Ueda subsequently found that the chaos also occurs when the damping is linear, and the resulting equation $\ddot{x} + k\dot{x} + x^3 = B \sin \tau$ is now known as the *Ueda system* [Thompson and Stewart (1986)].

The parameters $a = b = 0.7$ give the attractor shown in Fig. 7.17 and a time series as shown in Fig. 7.18. The Lyapunov exponents are (0.0481, 0, −0.5831), the Kaplan–Yorke dimension is $D_{ky} = 2.0825$, and the attractor has a global Class 1a basin of attraction.

This circuit is straightforward to construct, and it requires only minimal adjustment of the components in Table 7.3. We adjusted V_0 using the circuit from Fig. 1.29 to obtain better agreement with the numerical plot. The phase space plot of I versus V is shown in Fig. 7.19, where I can be obtained through Ohm's law by putting R_L between the inductor and V. The time plot of $V(t)$ is shown in Fig. 7.20. This circuit has one resistor, one capacitor, one inductor, and two analog multipliers for a total of five components.

7.5 Simple Jerk System

The search for ever simpler systems of equations with chaotic solutions ultimately led to what is the simplest such system [Sprott (1997a)]. Zhang and Heidel (1997, 1999) rigorously proved that all simpler systems of differential equations with quadratic nonlinearities cannot be chaotic.

Unfortunately, its circuit implementation is considerably less elegant than most of the other circuits in this book. Furthermore, it is very delicate in the sense that the chaos exits over only a small range of the parameters, and the basin of attraction is small, especially in the vicinity of the origin.

However, it is only the first in a hierarchy of similar systems [Eichhorn et al. (1998)], the next one of which is more robust (a 0.86% variation

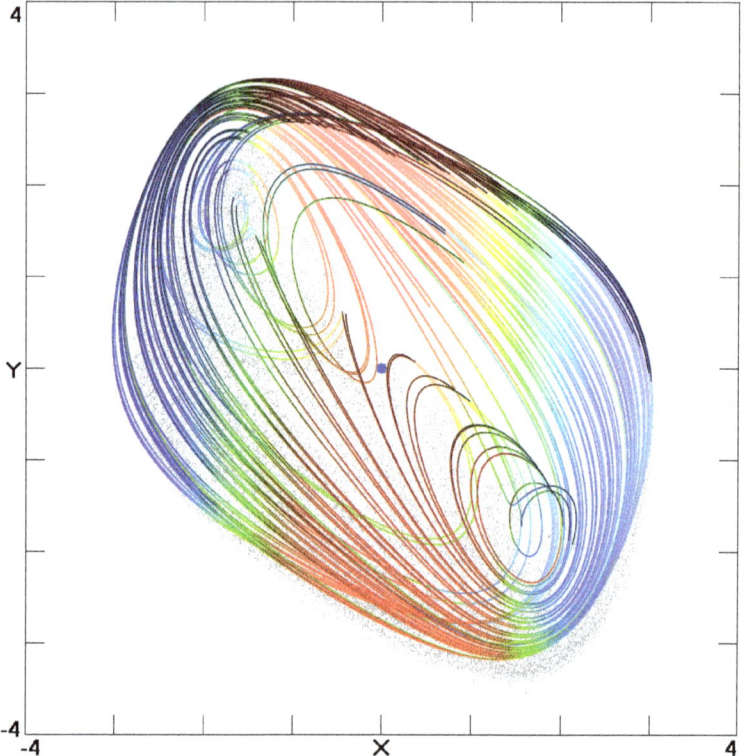

Fig. 7.17   Numerical solution of Eq. (7.6) with $a = b = 0.7$.

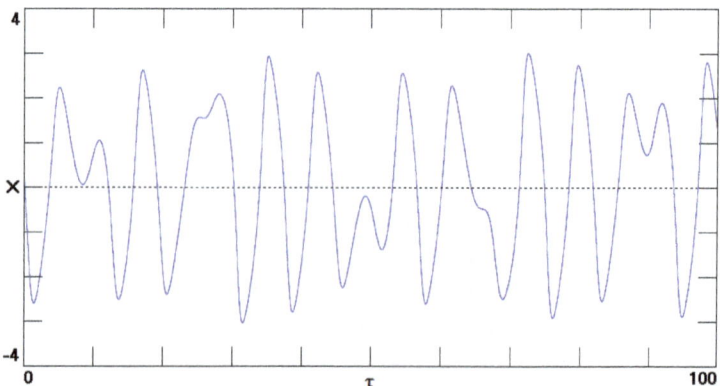

Fig. 7.18   Numerical waveform for $x(\tau)$ from Eq. (7.6) with $a = b = 0.7$.

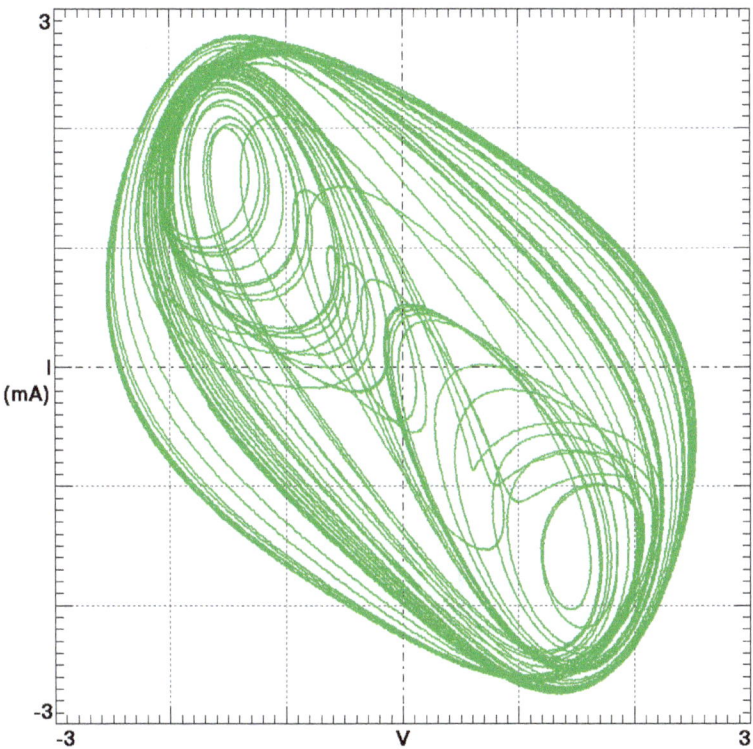

Fig. 7.19 Oscilloscope phase space plot of the original Ueda system from Fig. 7.15.

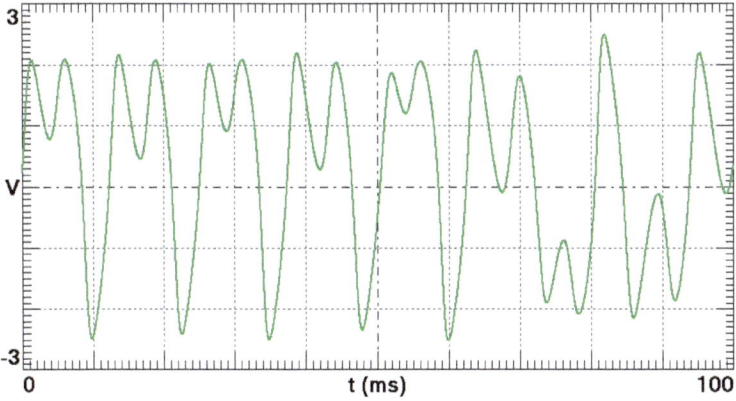

Fig. 7.20 Oscilloscope time plot of the original Ueda system from Fig. 7.15.

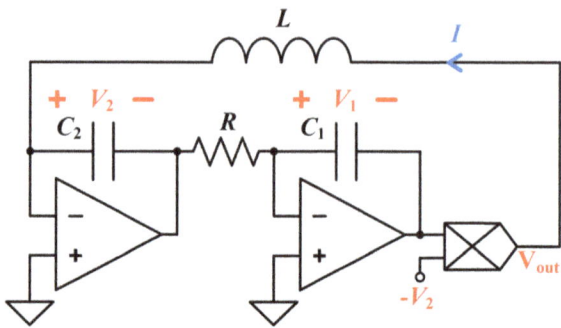

Fig. 7.21 Simple jerk system described by Eq. (7.7).

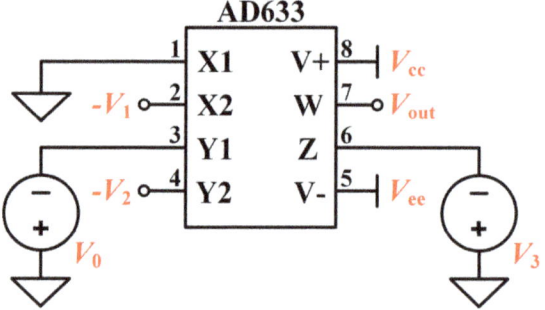

Fig. 7.22 Analog multiplier connections for Fig. 7.21.

in parameters is likely to destroy the chaos) and gives an elegant circuit as shown in Fig. 7.21 with the analog multiplier connected as shown in Fig. 7.22. The equations that describe the circuit are

$$\begin{aligned}\dot{V}_1 &= -V_2/RC_1 \\ \dot{V}_2 &= I/C_2 \\ \dot{I} &= (V_{out} - IR_L)/L,\end{aligned} \quad (7.7)$$

where $V_{out} = V_1(V_0 + V_2)/10 + V_3$.

Elegant circuit values that give chaos are $R = 1$ kΩ, $R_L = 0.18$ kΩ, $C_1 = C_2 = 1$ μF, $L = 0.1$ H, $V_0 = 2$ V, and $V_3 = 0.1$ V. The dimensionless equations are

$$\begin{aligned}\dot{x} &= -y \\ \dot{y} &= z \\ \dot{z} &= ax + xy + 1 - bz,\end{aligned} \quad (7.8)$$

which can be written more compactly as $\dddot{x} = -ax + x\dot{x} - 1 - b\ddot{x}$.

Table 7.4 Component values for Fig. 7.21.

Component	Numerical	Simulated	Experimental
R	1 kΩ	1 kΩ	1 kΩ
R_L	0.18 kΩ	0.18 kΩ	0.18 kΩ
C_1	1 μF	1 μF	1 μF
C_2	1 μF	1 μF	1 μF
L	0.1 H	0.1 H	0.1 H
V_0	2 V	2 V	2.4 V (adjust)
V_3	0.1 V	0.1 V	0.1 V
Op amp	-	Ideal	LM348
Multiplier	-	Ideal	AD633

Equation (7.8) is called a *jerk system* since $\dot{x}$ can be considered as a velocity and $\ddot{x}$ as an acceleration whose derivative $\dddot{x}$ is a jerk [Schot (1978)]. It turns out that any explicit three-dimensional system of equations with a single nonlinearity (and some with multiple nonlinearities) can be written in jerk form, and Eq. (7.8) is one of the simplest such system with chaotic solutions. Eichhorn *et al.* (2002) studied this system in some detail.

The parameters $a = 2$ and $b = 1.8$ give the attractor shown in Fig. 7.23 and a time series as shown in Fig. 7.24. The Lyapunov exponents are (0.0647, 0, −1.8647), the Kaplan–Yorke dimension is $D_{ky} = 2.0347$, and the attractor has a Class 2 basin of attraction that fills half the space of initial conditions. The system has an unstable saddle focus at (−0.5, 0, 0) with eigenvalues (0.1183 ± 0.9839i, −2.0367).

The output of both operational amplifiers will start at either V_{cc} or V_{ee} when the circuit is energized, where $V_{cc} = -V_{ee} = 15$ V for our circuit. If the system starts outside the basin of attraction, the circuit will not oscillate. In particular, the basin of attraction does not include the case of $V_1 = V_{ee}$ and $V_2 = V_{cc}$, while other combinations do. To fix this, you should make sure that all the components are close to those listed in Table 7.4 where V_Z and V_0 are implemented using the circuit from Fig. 1.29. Since inputs to the multiplier have a high impedance, the buffer from Fig. 1.29 is unnecessary.

You can then use a wire to directly connect the outputs $-V_1$ and $-V_2$ to a power supply rail V_{cc} or V_{ee}. Any combination that is not $V_1 = V_{ee}$ and $V_2 = V_{cc}$ will do. When these wires are simultaneously disconnected, the circuit starts at an initial condition within the basin of attraction and should oscillate. You may have to carefully adjust V_0 from 2 to 2.4 V until you observe chaos.

258 Elegant Circuits

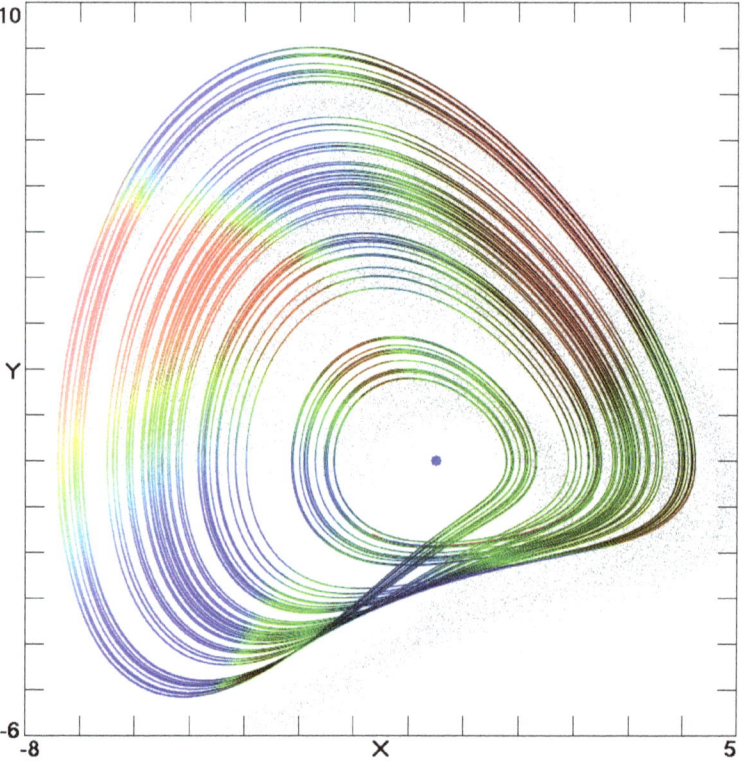

Fig. 7.23 Numerical solution of Eq. (7.8) with $a = 2$ and $b = 1.8$.

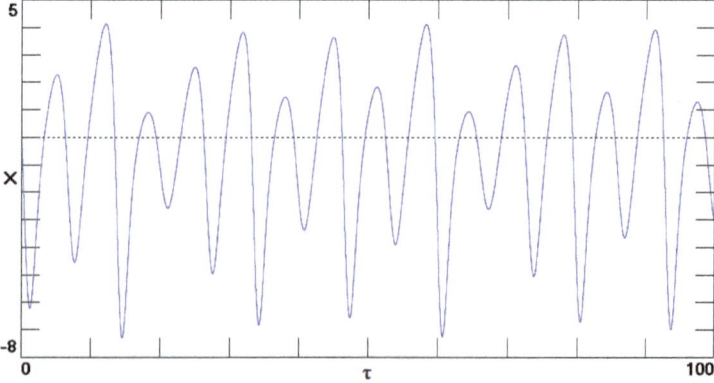

Fig. 7.24 Numerical waveform for $x(\tau)$ from Eq. (7.8) with $a = 2$ and $b = 1.8$.

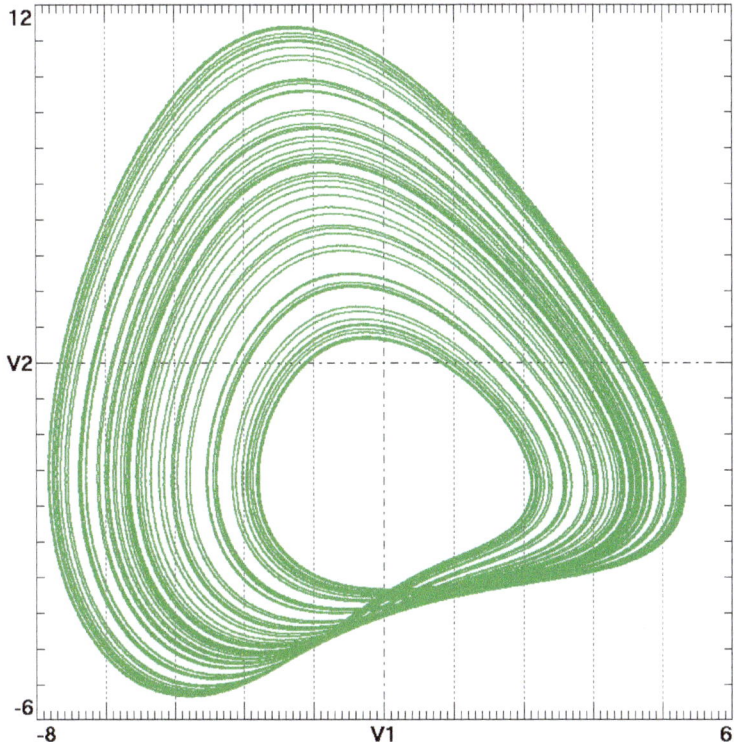

Fig. 7.25 Oscilloscope phase space plot of the simple jerk system from Fig. 7.21.

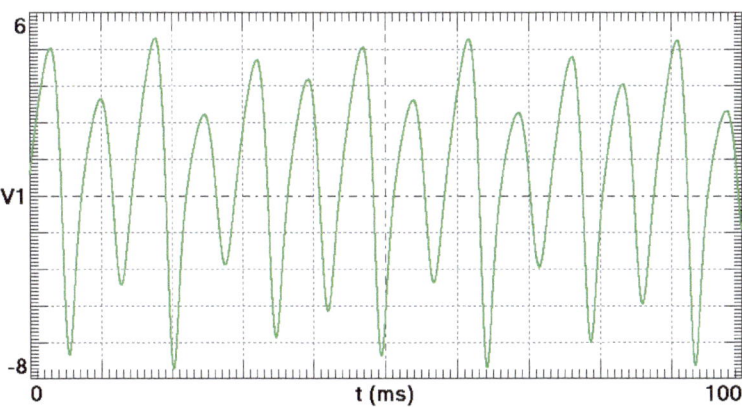

Fig. 7.26 Oscilloscope time plot of the simple jerk system from Fig. 7.21.

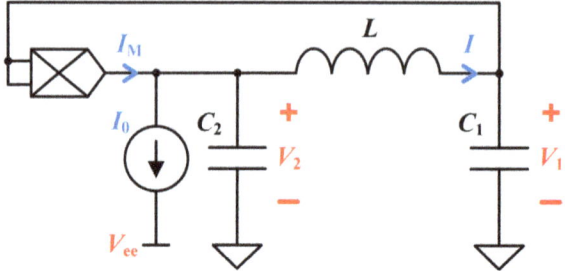

Fig. 7.27 Petrzela–Polak circuit described by Eq. (7.9).

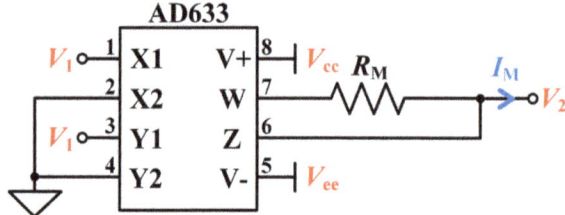

Fig. 7.28 Analog multiplier connections for Fig. 7.27.

Table 7.5 Component values for Fig. 7.27.

Component	Numerical	Simulated	Experimental
R_L	0.5 kΩ	0.5 kΩ	0.5 kΩ
R_M	0.08 kΩ	0.08 kΩ	0.08 kΩ
C_1	1 μF	1 μF	1 μF
C_2	1 μF	1 μF	1 μF
L	1 H	1 H	1 H
I_0	1 mA	1 mA	5 mA (adjust)
Multiplier	-	Ideal	AD633

The phase space plot of V_2 versus V_1 is shown in Fig. 7.25, and the time plot of $V_1(t)$ is shown in Fig. 7.26. This circuit has one resistor, two capacitors, one inductor, two operational amplifiers, and one analog multiplier for a total of seven components.

7.6 Petrzela–Polak Circuit

Petrzela and Polak (2019) found that a variety of chaotic circuits can be created using passive RC and LC filters and one analog multiplier. The most elegant of their circuits is shown in Fig. 7.27 with the multiplier connections in Fig. 7.28. The equations for the circuit are

$$\begin{aligned} \dot{V}_1 &= I/C_1 \\ \dot{V}_2 &= (I_M - I - I_0)/C_2 \\ \dot{I} &= (V_2 - V_1 - IR_L)/L, \end{aligned} \qquad (7.9)$$

where $I_M = V_1^2/10R_M$.

Elegant circuit values that give chaos are $R_L = 0.5$ kΩ, $R_M = 0.08$ kΩ, $C_1 = C_2 = 1$ μF, $L = 1$ H, and $I_0 = 1$ mA. The dimensionless equations are

$$\begin{aligned} \dot{x} &= z \\ \dot{y} &= ax^2 - z - 1 \\ \dot{z} &= y - x - bz. \end{aligned} \qquad (7.10)$$

The parameters $a = 1.25$ and $b = 0.5$ give the attractor shown in Fig. 7.29 and a time series as shown in Fig. 7.30. The Lyapunov exponents are $(0.1269, 0, -0.6269)$, the Kaplan–Yorke dimension is $D_{ky} = 2.2025$, and the attractor has a Class 3 basin of attraction with $P \approx 6/r^{2.8}$. The system has a symmetric pair of unstable saddle foci at $(0.8944, 0.8944, 0)$ and $(-0.8944, -0.8944, 0)$ with eigenvalues $(0.2156 \pm 1.5346i, -0.9311)$.

The components are shown in Table 7.5 where the current source can be constructed using the circuit described in Fig. 2.16. This circuit is sensitive to the value of I_0 (a 14% variation in I_0 is likely to destroy the chaos), which you must carefully adjust to obtain chaos. You also may need to restart the power supply if I_0 is adjusted too high and the multiplier saturates since the basin of attraction is relatively small. The plot of V_2 versus V_1 is shown in Fig. 7.31, and the time plot of $V_1(t)$ is shown in Fig. 7.32. This circuit has one resistor, two capacitors, one inductor, and one multiplier for a total of five components.

7.7 Dissipative Nosé–Hoover System

This circuit arose out of our failed attempt to construct a circuit for the Nosé–Hoover system [Nosé (1984); Hoover (1985)]. Unlike most of the other systems in this book that are dissipative, the Nosé–Hoover system is conservative but only by virtue of nonlinear feedback that acts as a *thermostat* to keep the average energy of the linear oscillator constant when averaged over a sufficiently long time. Thus it has intervals of dissipation that alternate with intervals of anti-dissipation and that exactly cancel.

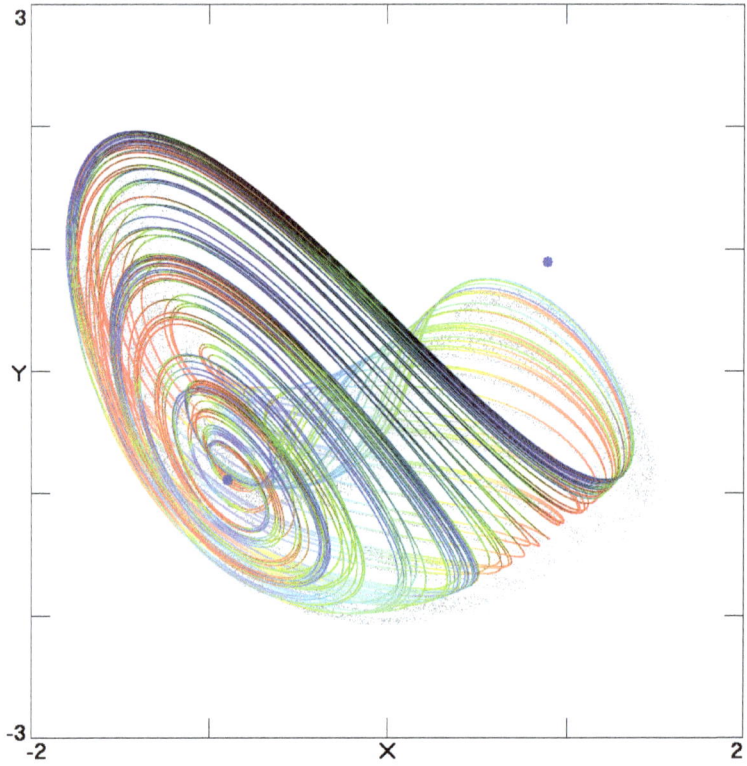

Fig. 7.29   Numerical solution of Eq. (7.10) with $a = 1.25$ and $b = 0.5$.

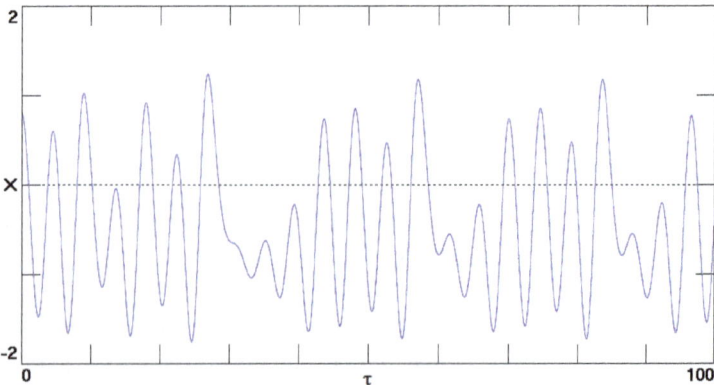

Fig. 7.30   Numerical waveform for $x(\tau)$ from Eq. (7.10) with $a = 1.25$ and $b = 0.5$.

Analog Multiplier Circuits 263

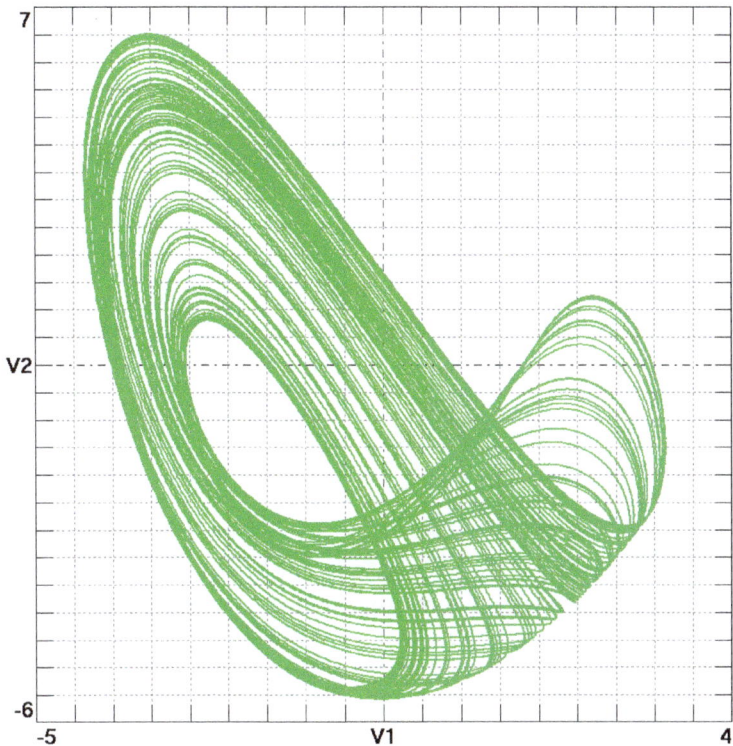

Fig. 7.31 Oscilloscope phase space plot of the Petrzela–Polak circuit from Fig. 7.27.

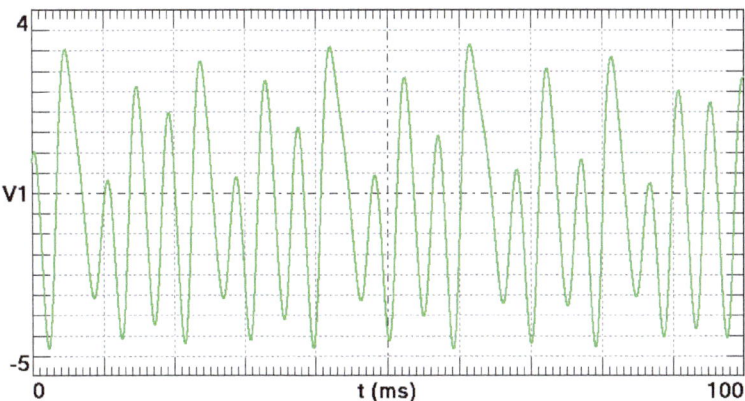

Fig. 7.32 Oscilloscope time plot of the Petrzela–Polak circuit from Fig. 7.27.

Such a system is said to be *nonuniformly conservative* [Heidel and Zhang (1999)] and is *isothermal* rather than *isoenergetic*.

We discovered that despite the thermostat that controls the energy of the oscillation, it is apparently impossible to construct a circuit in which the dissipation is precisely zero. Any small dissipation destroys the time-reversibility and causes the system to attract to a limit cycle, giving a circuit that oscillates periodically rather than chaotically.

However, if the dissipation is intentionally increased sufficiently, the limit cycle undergoes a period-doubling route to chaos with a strange attractor that resembles the Lorenz attractor. The circuit is shown in Fig. 7.33 with the analog multipliers connected as shown in Fig. 7.34. The equations for the circuit are

$$\dot{I} = (V_1 - IR_L)/L$$
$$\dot{V}_1 = (I_{M1} - I)/C_1 \qquad (7.11)$$
$$\dot{V}_2 = (I_{M2} - I_0)/C_2,$$

where $I_{M1} = -V_1 V_2 / 10 R_{M1}$ and $I_{M2} = V_1^2 / 10 R_{M2}$.

If the inductor were ideal with no parasitic resistance ($R_L = 0$), these equations would give exactly the Nosé–Hoover system with a chaotic sea that surrounds an intricate set of nested invariant tori, but we exploit the inevitable R_L, enhanced somewhat, to give a dissipative variant of the system with a strange attractor.

Elegant circuit values that give chaos are $R_L = 1$ kΩ, $R_{M1} = R_{M2} = 0.1$ kΩ, $C_1 = C_2 = 1$ μF, $L = 1$ H, and $I_0 = 1$ mA. The dimensionless equations are

$$\dot{x} = y - ax$$
$$\dot{y} = -x - yz \qquad (7.12)$$
$$\dot{z} = y^2 - b.$$

For $a = 0$, this is exactly the conservative Nosé–Hoover system with temperature b. For a small and positive, the system is dissipative with a limit cycle that loses its stability at $a \approx 0.658$ with $b = 1$, giving way to a strange attractor.

The parameters $a = b = 1$ give the attractor shown in Fig. 7.35 and a time series as shown in Fig. 7.36. The Lyapunov exponents are (0.1370, 0, −0.5599), the Kaplan–Yorke dimension is $D_{ky} = 2.2447$, and the attractor has a global Class 1a basin of attraction. The system has a symmetric pair of unstable saddle foci at (1, 1, −1) and (−1, −1, −1) with eigenvalues (0.3855 ± 1.5639i, −0.7709).

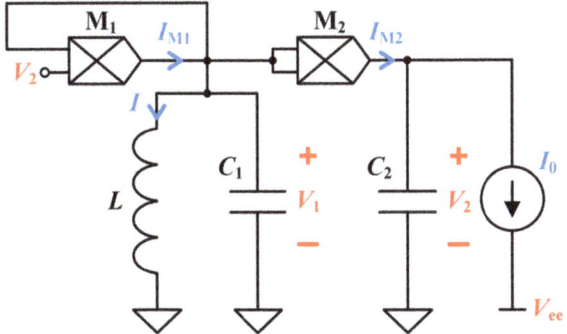

Fig. 7.33 Dissipative Nosé–Hoover system described by Eq. (7.11).

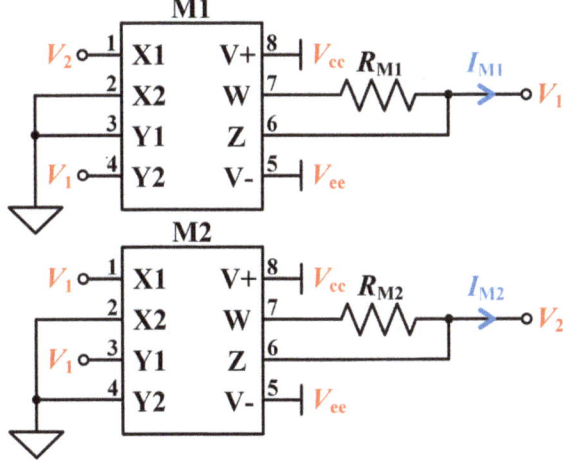

Fig. 7.34 Analog multiplier connections for Fig. 7.33.

Table 7.6 Component values for Fig. 7.33.

Component	Numerical	Simulated	Experimental
R_L	1 kΩ	1 kΩ	1 kΩ (adjust)
R_{M1}	0.1 kΩ	0.1 kΩ	0.1 kΩ
R_{M2}	0.1 kΩ	0.1 kΩ	0.1 kΩ
C_1	1 μF	1 μF	1 μF
C_2	1 μF	1 μF	1 μF
L	1 H	1 H	1 H
I_0	1 mA	1 mA	1 mA
Multiplier	-	Ideal	AD633

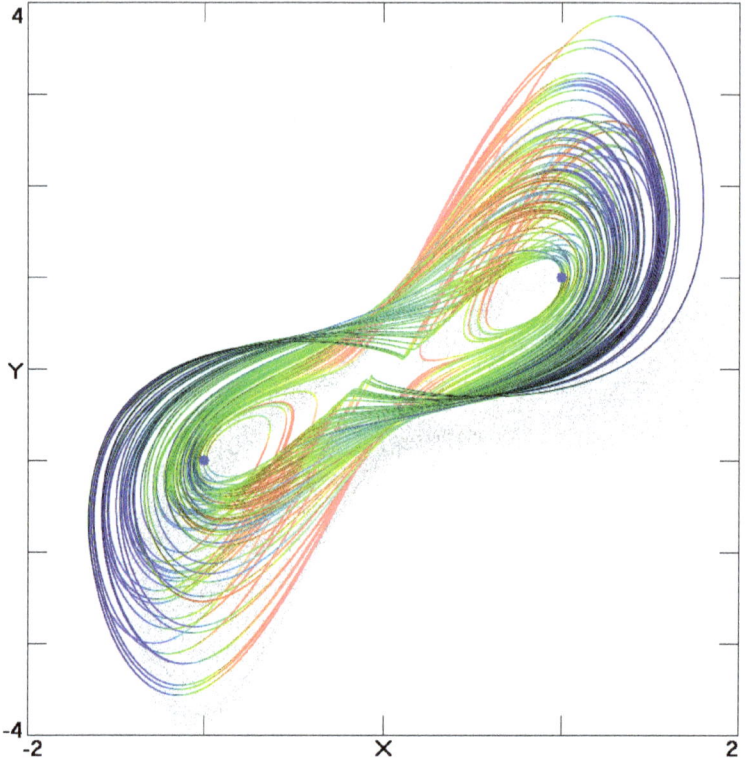

Fig. 7.35 Numerical solution of Eq. (7.12) with $a = b = 1$.

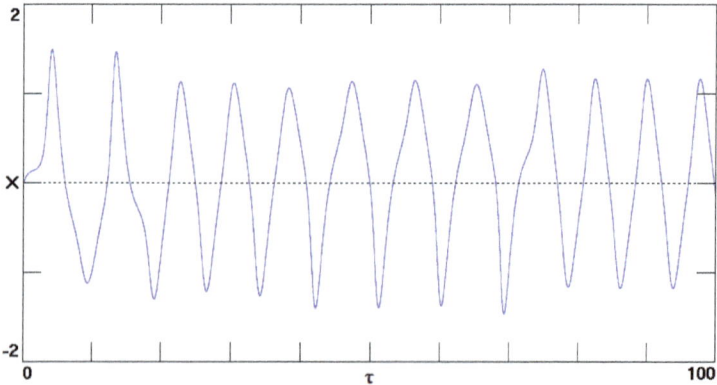

Fig. 7.36 Numerical waveform for $x(\tau)$ from Eq. (7.12) with $a = b = 1$.

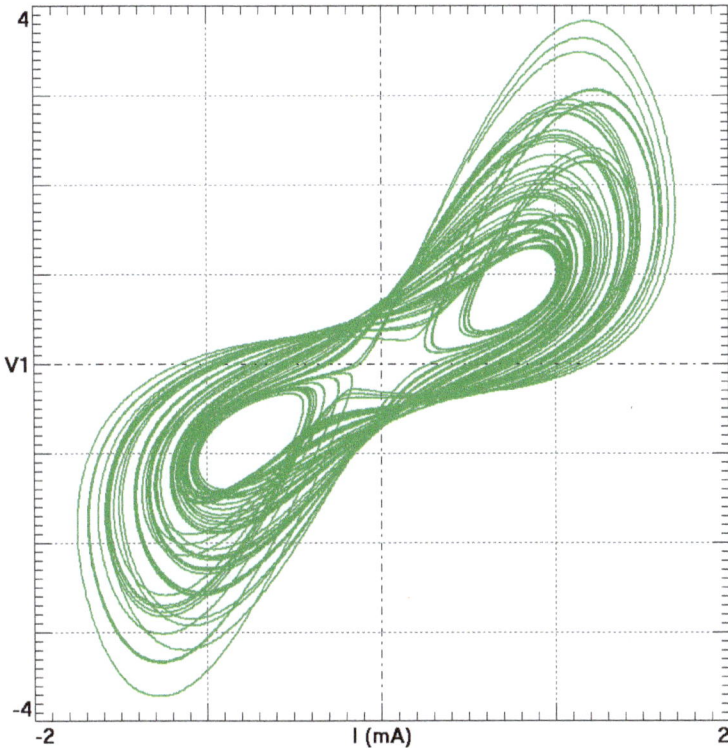

Fig. 7.37 Oscilloscope phase space plot of the dissipative Nosé–Hoover system from Fig. 7.33.

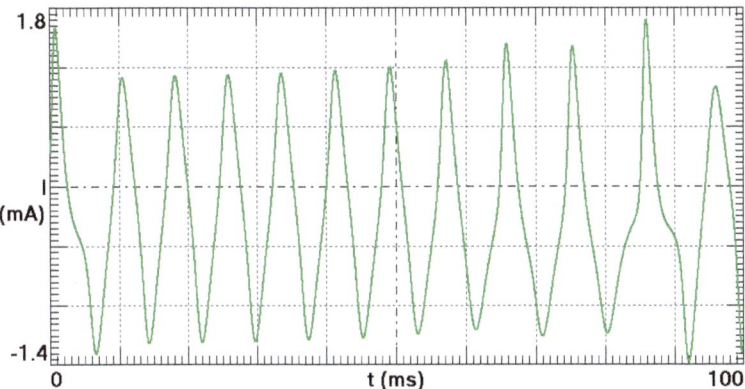

Fig. 7.38 Oscilloscope time plot of the dissipative Nosé–Hoover system from Fig. 7.33.

Table 7.7 Component values for Fig. 7.39.

Component	Numerical	Simulated	Experimental
R_{M1}	0.5 kΩ	0.5 kΩ	0.5 kΩ
R_{M2}	0.1 kΩ	0.1 kΩ	0.1 kΩ
C_1	1 μF	1 μF	1 μF
C_2	1 μF	1 μF	1 μF
L	1 H	1 H	1 H (gyrator)
I_0	1 mA	1 mA	1 mA
V_{sat}	10 V	10 V	10 V
Op amp	-	Ideal	LM348
Multiplier	-	Ideal	AD633

It is straightforward to obtain chaos in this circuit using the component values given in Table 7.6. You can change the dissipation by making R_L an adjustable parameter. When R_L is small, the circuit is closer to the Nosé–Hoover system but gives a limit cycle rather than chaos as previously explained. Increasing R_L to 1 kΩ should lead to chaos through period doubling. Note that using a gyrator for the inductor will not remove the damping, and it is unnecessary for this intentionally dissipative system.

The phase space plot of V_1 versus I is shown in Fig. 7.37, where we observed $I(t)$ in Fig. 7.38 by measuring the voltage across a grounded R_L using Ohm's law. This circuit has two resistors, two capacitors, one inductor, and two analog multipliers for a total of seven components.

7.8 Signum Thermostat

Whereas the circuit in Fig. 7.33 fails to produce the Nosé–Hoover chaotic sea because of parasitic resistance in the inductor, a slight modification of the circuit in which a comparator is inserted in the feedback loop gives a more robust circuit that is chaotic for even small values of the resistance. This circuit is shown in Fig. 7.39 with the multiplier connections in Fig. 7.40.

The circuit is a simplification of one described by Sprott and Thio (2020) that has 21 components. The circuit equations are

$$\dot{I} = V_1/L$$
$$\dot{V}_1 = (I_{M1} - I)/C_1 \quad\quad (7.13)$$
$$\dot{V}_2 = (I_{M2} - I_0)/C_2,$$

where $I_{M1} = -V_1 V_{out}/10 R_{M1}$, $I_{M2} = V_1^2/10 R_{M2}$, and $V_{out} = V_{sat} \operatorname{sgn} V_2$.

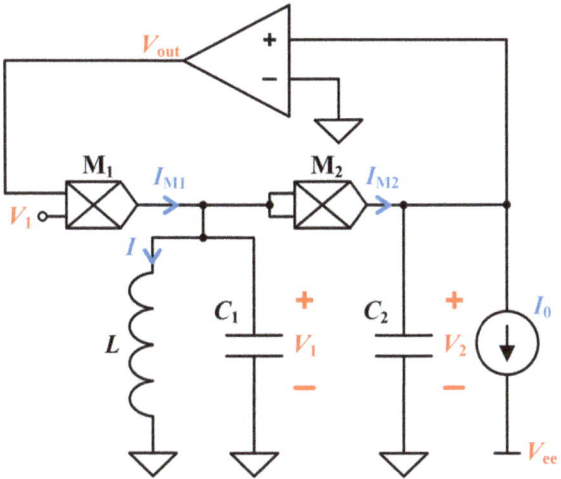

Fig. 7.39 Signum thermostat described by Eq. (7.13).

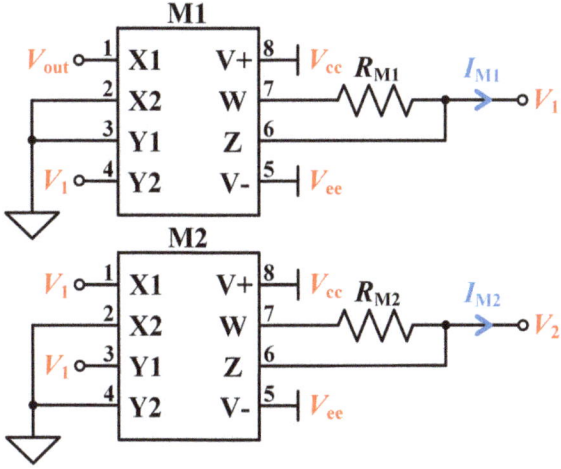

Fig. 7.40 Analog multiplier connections for Fig. 7.39.

Note that this circuit has three nonlinear components, a comparator to produce the sgn V_2 function and two analog multipliers to perform the V_1 sgn V_2 and V_1^2 operations.

Elegant circuit values that give chaos are $R_{M1} = 0.5$ kΩ, $R_{M2} = 0.1$ kΩ, $C_1 = C_2 = 1$ μF, $L = 1$ H, $I_0 = 1$ mA, and $V_{sat} = 10$ V. The dimensionless

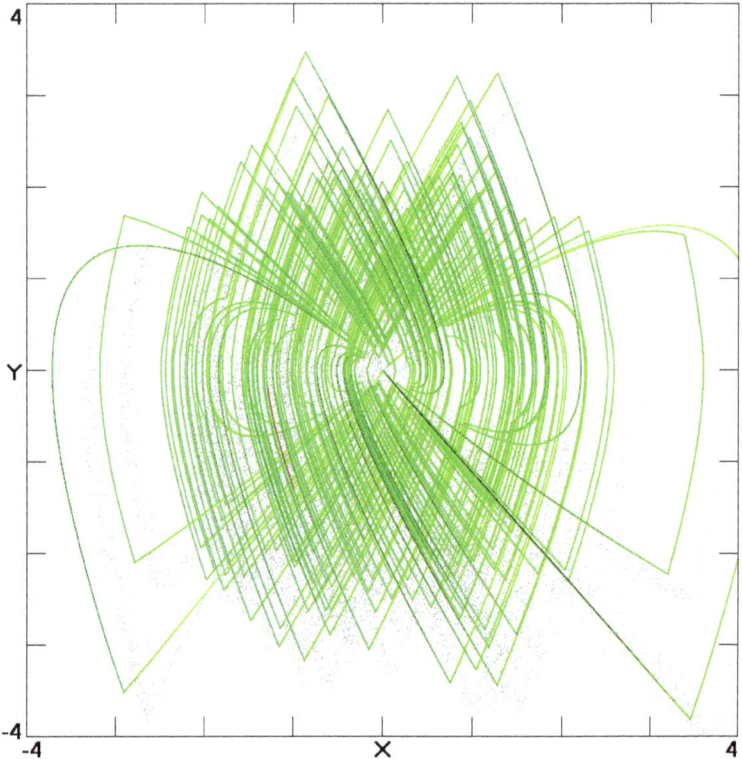

Fig. 7.41 Numerical solution of Eq. (7.14) with $a = 2$ and $b = 1$.

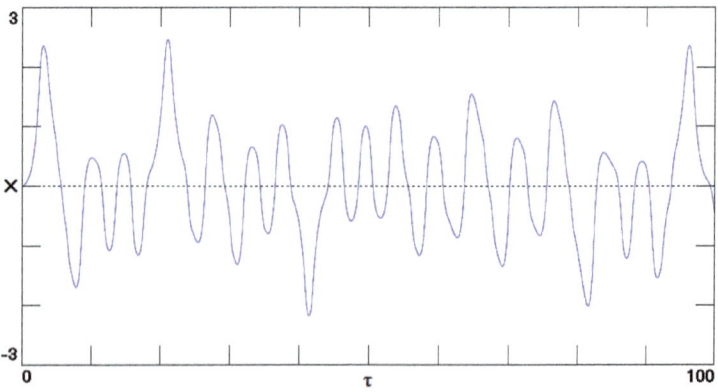

Fig. 7.42 Numerical waveform for $x(\tau)$ from Eq. (7.14) with $a = 2$ and $b = 1$.

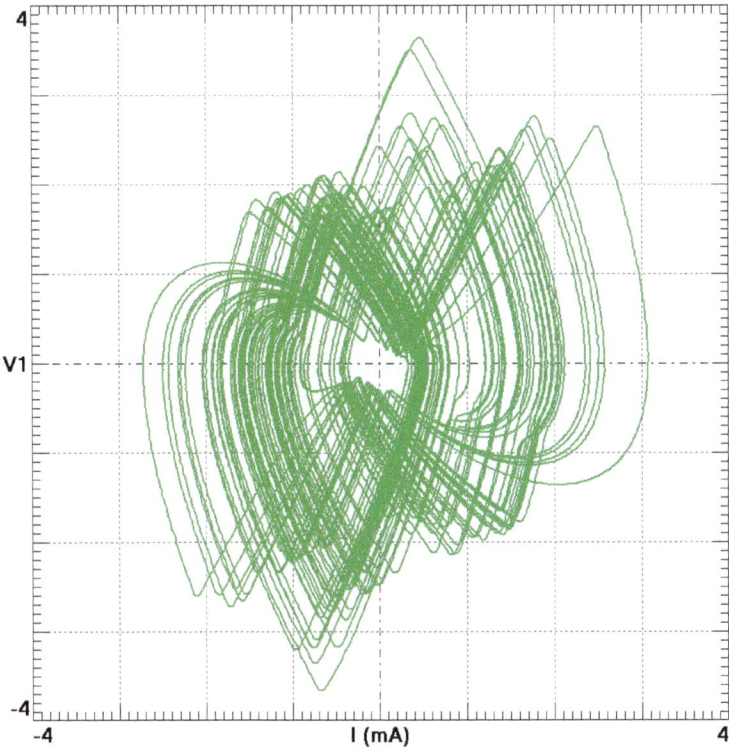

Fig. 7.43   Oscilloscope phase space plot of the signum thermostat from Fig. 7.39.

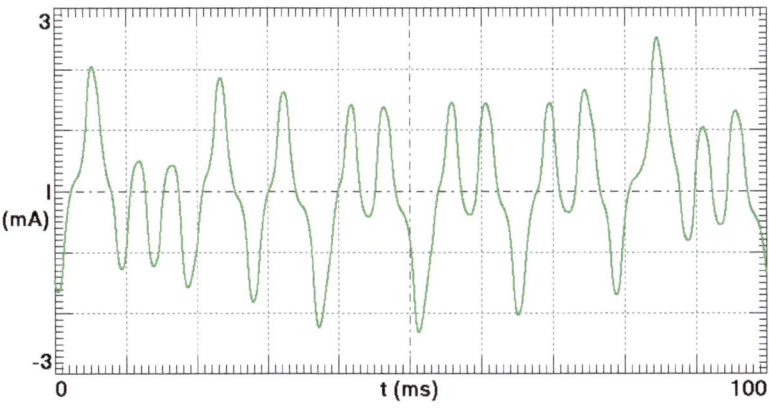

Fig. 7.44   Oscilloscope time plot of the signum thermostat from Fig. 7.39.

equations are

$$\dot{x} = y$$
$$\dot{y} = -x - ay\,\text{sgn}\,z \qquad (7.14)$$
$$\dot{z} = y^2 - b.$$

This system was proposed by Sprott (2018) as an *ergodic* variant of the Nosé–Hoover system. It is ergodic (for $a > 1.8$) in the sense that for any initial conditions (except along the line $x = y = 0$), it will eventually come arbitrarily close to every point in (x, y, z) space. Furthermore, it is nonuniformly conservative and time-reversible. Hence it does not have an attractor, but rather it has a *chaotic sea*.

The parameters $a = 2$ and $b = 1$ give the chaotic sea in Fig. 7.41 and a time series as shown in Fig. 7.42. The Lyapunov exponents are (0.3032, 0, −0.3032), and the Kaplan–Yorke dimension is $D_{ky} = 3.0$ as expected for a conservative system. The system is unusual in that it does not have any equilibrium points, and hence it cannot fail to oscillate, although the oscillation may not be chaotic if a is too small.

This system can be viewed as a model of a harmonic oscillator in equilibrium with a heat bath at temperature b, with the variable z playing the role of a thermostat, turning up the heat when the oscillator energy falls too low and turning it down when the energy gets too high. Because the sgn z factor abruptly switches between $+1$ and -1 when z crosses zero, it is called a *bang-bang controller*. Unlike the Nosé–Hoover system, the temperature b is an amplitude parameter that does not affect the dynamics.

The system is also special because the probability distribution function of the x and y variables are exactly Gaussian, and hence it can be used to generate Gaussian random numbers as described by Sprott and Thio (2020). It is only necessary to sample the instantaneous value of x or y whenever a new value is desired. If the interval between samples is sufficiently large (much larger than the Lyapunov time of $1/0.3032$), successive samples are uncorrelated through Lyapunov instability, although that may not be necessary for some applications. The probability distribution function for z is exponential given by $P_z = (a/2b)e^{-a|z|/b}$.

The components of this circuit are given in Table 7.7, where L is implemented using the gyrator from Fig. 1.25 to keep the parasitic resistance low. The phase space plot of V_1 versus I is shown in Fig. 7.43, and the time plot of $I(t)$ is shown in Fig. 7.44. Of course the actual circuit has some inevitable dissipation primarily due to parasitic resistance in the inductor, and so it gives a strange attractor rather than a conservative chaotic sea.

However, the attractor seems to be ergodic (filling the whole of the three-dimensional state space) with properties indistinguishable from a chaotic sea. The only signatures of the dissipation are a slight asymmetry of the experimental phase space plot and a small notch in the x probability distribution function in the vicinity of $x = 0$ (not shown here).

The circuit has two resistors, two capacitors, one inductor, one comparator, and two analog multipliers for a total of eight components. The circuit is important and potentially useful, and its governing equations have a particular elegance.

Chapter 8

Nonlinear Inductor Circuits

Although inductors are usually considered to be linear circuit components, they can behave in a very nonlinear manner when they use a ferromagnetic core. The previous circuits have largely avoided such nonlinearities, but they can be exploited to produce chaos, much like the circuits in Chapter 6 used the saturation of operational amplifiers for that purpose. This short chapter describes nonlinear inductors and gives three examples of circuits that employ them to produce chaos.

8.1 Ferromagnetism

To achieve a high value of inductance, it is customary to wrap the windings around a ferromagnetic core (either iron or ferrite). The symbol we will use for such an inductor is shown in Fig. 8.1. The resulting nonlinearity arises from the remarkable but somewhat complicated magnetic properties of ferromagnetic materials.

8.1.1 *Magnetic properties of ferromagnets*

An iron core will enhance the externally applied magnetic field H by allowing magnetic domains in the iron to partially orient in the direction of the applied field so that the magnetic flux density in the iron B is proportional to H with a proportionality factor μ, called the *permeability* so that $B = \mu H$. The permeability relative to free space is typically of order 10^3 and can be as high as 10^4.

However, there is a limit on the order of $B \sim 1$ Tesla ($= 10^4$ Gauss) where the magnetic domains become fully aligned and very little increase in B occurs with increasing H. As a result, the relative permeability decreases to order unity as if the iron were absent, and we say the core is *saturated*

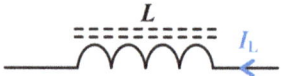

Fig. 8.1 Ferromagnetic-core inductor.

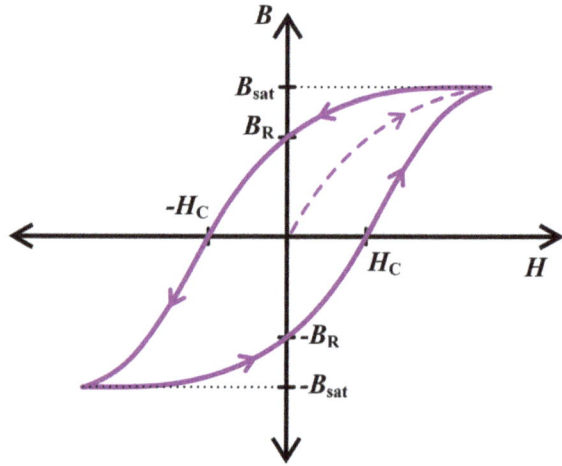

Fig. 8.2 Magnetization (B-H) curve for a typical ferromagnetic material.

at a value of $\pm B_{sat}$ as shown in Fig. 8.2. This *magnetization curve* (also called a *B-H curve* or a *hysteresis curve*) is purely a property of the iron and is independent of the windings that surround it.

Furthermore, iron has a tendency to become permanently magnetized so that B generally does not vanish when $H = 0$ but has a magnitude of B_R, called the *retentivity*. Such *permanent magnets* are useful for many purposes and were used as memory elements in early digital computers since they can be magnetized in either direction.

To demagnetize the iron, an opposing applied field with a strength equal to the *coercivity* or H_C of the material must be applied. As a result the iron exhibits hysteresis, which is a memory of its past that we have encountered in previous circuits. Although the demagnetized iron starts with an initial permeability corresponding to the slope of the dotted line in the figure, this value changes along with the magnetization of the material.

The hysteresis in the B-H curve is another nonlinear effect that can be exploited in chaotic circuits. However, it often suffices just to treat the hysteresis as an additional parasitic resistance in the inductor, but whose value depends on the current and frequency.

Iron is also a moderately good electrical conductor, and the time varying magnetic flux will induce so-called *eddy currents* in the iron causing additional frequency-dependent parasitic resistance. These eddy current losses are usually reduced by *laminating* the iron or by using a *ferrite*, which is a poorly conducting ceramic material made by mixing large proportions of iron oxide with small proportions of one or more additional metallic elements.

Our circuits will use *soft ferrites* (or *transformer ferrites*) which have a small value of coercivity so that the hysteresis losses are small. This makes it easier to change their magnetization compared to *hard ferrites* which are more useful as permanent magnets.

8.1.2 Saturating inductor model

The basic operation of an inductor is that a current I though its windings produces a proportional magnetic field H and a corresponding magnetic flux Φ that is proportional to B. The proportionality constants depend on the size and shape of the core and the number and placement of the windings. The time-varying flux induces a voltage $V = \dot{\Phi}$ in the winding, called a *back electromotive force*. For an ideal inductor, the flux is proportional to the current $\Phi = LI$, where the constant of proportionality L is the inductance. Thus we obtain the defining relation of $V = L\dot{I}$ for an ideal inductor.

In a saturating inductor, the flux is no longer proportional to the current, but rather $\Phi(I)$ takes the same functional form as $B(H)$. Ignoring hysteresis, it can be modeled by a nonlinear function such as $\Phi(I) = L_0 I_{sat} \tanh(I/I_{sat})$, where L_0 is the inductance for $|I| \ll I_{sat}$ and I_{sat} is the current at which in the inductor begins to saturate. From the time derivative of $\dot{\Phi}(I) = \frac{d\Phi}{dI}\dot{I}$, we can obtain a relation between voltage and current given by $V = L_0\dot{I}/\cosh^2(I/I_{sat})$.

For some purposes, it is convenient to define a current-dependent inductance given by $L(I) = L_0/\cosh^2(I/I_{sat})$, whose lowest-order Taylor approximation is

$$L(I) = L_0/(1 + I^2/I_{sat}^2). \tag{8.1}$$

This approximation is exact if $\Phi(I) = L_0 I_{sat} \arctan(I/I_{sat})$ and is actually a better model of $\Phi(I)$ than is the hyperbolic tangent which saturates too strongly. The inductance is large and independent of current ($L = L_0$) when $|I| \ll I_{sat}$, but approaches zero for $|I| \gg I_{sat}$.

8.1.3 Ferrite-core inductor construction

Modern inductors are usually designed to have a large saturation current to maintain a constant value of the inductance and to prevent damage to other circuit components due to the high currents when the inductor saturates. In this section we will describe how you can construct an inductor with a sufficiently low saturation current so that it will be easier to obtain chaos in your circuit with only modest power sources.

It is easier to saturate a toroidal inductor since the magnetic field is mostly confined in its core, whereas a solenoid has more leakage magnetic flux. You can estimate the saturation current of a toroidal inductor (in mA) from $I_{sat} = 5 \times 10^6 B_{sat} r / \mu N$ where B_{sat} is the magnetic field at saturation (in Tesla), r is the major radius of the toroid in mm (approximately the mean between the inner and outer radius), μ is the relative permeability, and N is the number of turns.

The permeability of the toroid μ is the most important parameter and should be large to make I_{sat} small as desired for saturation at modest currents. Our toroid (Amidon™ FT-140A-W) is rated at $B_{sat} = 390$ mT, $r = 6.27$ mm, and $\mu = 10,000 \pm 30\%$. The actual value of μ at saturation is smaller than this nominal value, but it does provide a reasonable estimate of I_{sat}.

The number of turns N required for a desired inductance can be found from $N = \sqrt{L/A_L}$ where A_L is the inductance of a single turn, which is $A_L = 0.0134$ mH for our toroid. Thus an inductor with $L = 100$ mH would require $N = 86$ turns with a corresponding $I_{sat} = 14$ mA. If you use an inductor with a different size and permeability, check that the saturation current is a few milliamperes and that there is sufficient space for the required windings.

We wound the toroid with a bifilar winding made by twisting two strands of 31 AWG magnet wires to create two coupled inductors (a *transformer*) called the *primary* and *secondary*. The primary inductor will be used to find the current I_L, while the secondary will be used to measure the magnetic flux Φ by integrating the induced voltage without having to correct for resistive losses.

In the chaotic circuits, only one winding is used, although they can be connected in parallel to reduce the DC resistance by half or in series to increase the inductance by a factor of four while doubling the resistance and decreasing I_{sat} by half. If you parallel two windings, be sure they have

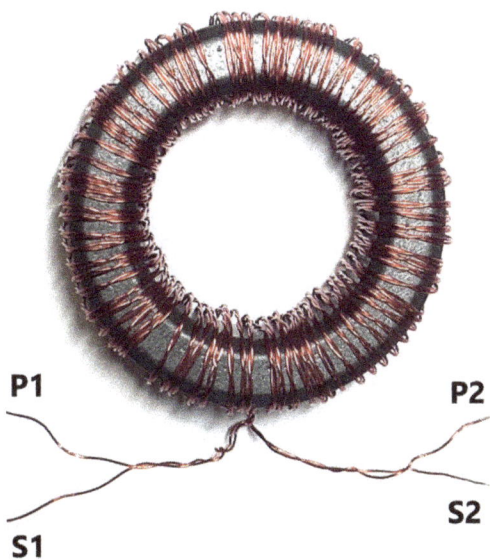

Fig. 8.3 Ferrite-core inductor with primary (P) and secondary terminals (S) marked.

exactly the same number of turns and the same polarity; otherwise, you will end up with a short circuit in parallel with the inductor.

Unfortunately, winding a toroid with sufficiently many turns to achieve a large inductance is a tedious task. One winding method involves breaking the toroid in half and winding each half separately and then gluing it back together, but we advise against doing that since any small remaining gap reduces the effective permeability of the core and the desired nonlinearity.

As you wind, keep the spaces between each turn as even as possible to reduce magnetic flux leakage. When you are finished with the winding, tin the ends with solder for a better connection. Our completed toroid is shown in Fig. 8.3, where the primary and secondary windings are marked and measured to be 100 mH with a DC resistance of 3 Ω.

8.1.4 *Ferrite-core inductor measurement*

You can characterize your inductor using the circuit in Fig. 8.4 which drives the primary winding with a square-wave AC voltage and measures the primary current I_L while obtaining the flux Φ by integrating the voltage

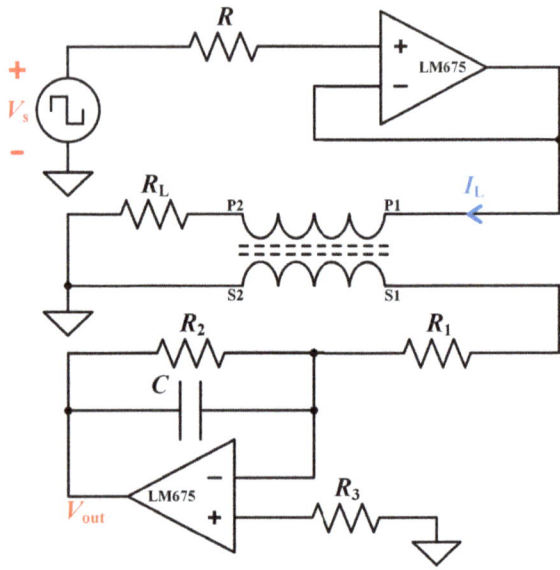

Fig. 8.4 $\Phi(I)$ curve tracer circuit. Component values are $R = 30$ kΩ, $R_L = 5$ Ω (1W), $R_1 = 10$ kΩ, $R_2 = 1$ MΩ, $R_3 = 10$ kΩ, and $C = 0.1$ μF.

across the secondary winding. You can then adjust the inductor parameters L_0 and I_{sat} in Eq. (8.1) to get a good fit for use in the chaotic circuit models.

This circuit requires a dual power supply as shown in Fig. 1.21, but that circuit is limited to 20 mA, which may not adequately saturate the inductor. You can instead use the LM675, a high-voltage operational amplifier, as a split power supply as described in its data sheet. This operational amplifier has a maximum output current of 3 A and will be used when a conventional operational amplifier is insufficient. However, it is more prone to noise, which can be reduced by using 0.1 μF decoupling capacitors placed between the ground and either power supply terminal. Also add a polarized 220-μF capacitor across V_{cc} and V_{ee}; otherwise you will see distortion when the current increases rapidly. Finally, avoid touching these operational amplifiers since they can become very hot.

If your function generator cannot supply the necessary driving current when the inductor saturates, you can use the LM675 as a buffer amplifier and a resistor R to reduce the output noise. The sensing resistor R_L should be only a few ohms to keep the current large, but it needs to be rated for at least 1 W. You can then determine I_L from the voltage across R_L using Ohm's law.

You can obtain Φ by integrating the voltage across the secondary winding using an operational amplifier with a *corner frequency* (the lowest frequency for which the circuit properly integrates) of $f_0 = 1/2\pi R_2 C$. The output of the operational amplifier is then $-\Phi/R_1 C$, which can be scaled to obtain Φ. The feedback resistor R_2 prevents small DC offsets at the input from being integrated and saturating the amplifier. You can remove any additional offset in Φ by AC coupling to the oscilloscope.

To obtain a $\Phi(I)$ plot, start with the amplitude of V_S at zero and increase it while observing $I_L(t)$ and $\Phi(t)$. The current I_L is sensitive to DC offsets which can cause the inductor to saturate asymmetrically and can be corrected by slightly adjusting the DC offset and duty cycle. Increase the amplitude very slowly because the current can reach several hundred milliamperes if the inductor is driven very hard. If the inductor is driven too hard in one direction, it may saturate and get stuck in that direction, which will require restarting the power supply.

The $\Phi(I)$ plot is generally independent of the frequency, but it is easier to saturate the inductor at 1 kHz using a square wave as shown in Fig. 8.5 or by using a lower-frequency sine wave. We then fit the data to a circuit model $\dot{I} = (V_S - IR_L)/L$ using Eq. (8.1) with $L_0 = 0.1$ H, $I_{sat} = 20$ mA, and $R_L = 0.08$ kΩ to obtain the curve in Fig. 8.6.

The main disagreement between the experimental and numerical curves comes from assuming the hysteresis is a result of a series resistance R_L that is independent of current and frequency. In reality, most of the hysteresis loss occurs at small values of I, and the corresponding resistance falls to a low value once the inductor saturates. In fact, the area of the hysteresis loop, which has units of V×ms×mA is exactly the energy lost (in microjoules) per cycle.

Although the equivalent resistance of the inductor R_L, which also includes DC resistance and eddy current losses, depends strongly on current and weakly on frequency, we will take it as a constant but adjustable parameter in the circuits to follow, and that is probably the main reason for the differences between the theoretical models and the experimental results.

8.2 Forced Ferroresonant Circuit

Perhaps the simplest chaotic circuit with a nonlinear inductor is a simple series LC circuit driven by a periodic voltage source as shown in Fig. 8.7. This circuit was first studied by Martienssen (1910) and experimentally

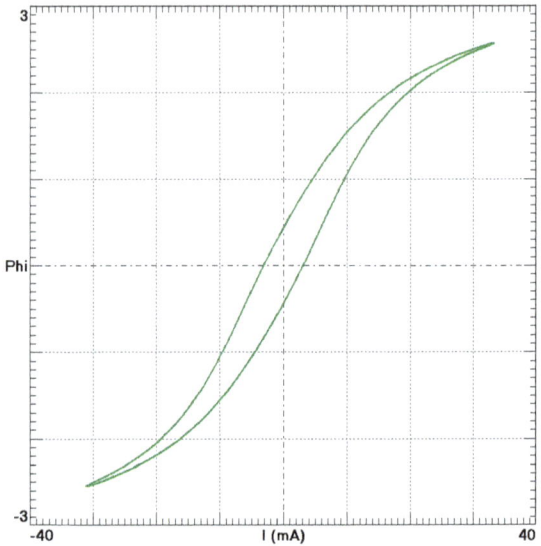

Fig. 8.5 Experimental hysteresis plot for the test inductor driven by a 1 kHz square wave.

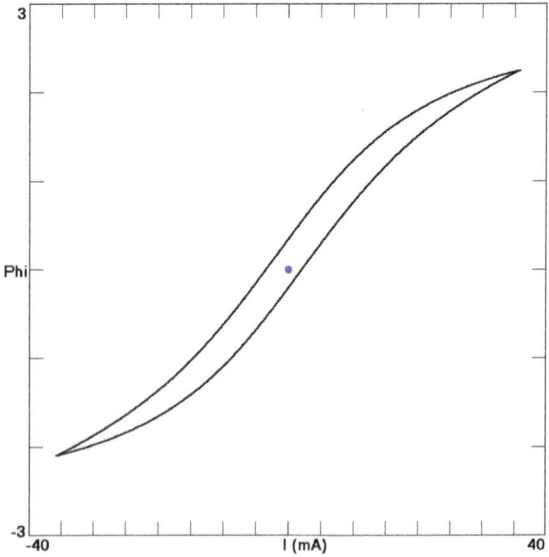

Fig. 8.6 Numerical hysteresis plot for $L(I) = L_0/(1 + I^2/I_{sat}^2)$ with $L_0 = 0.1$ H, $I_{sat} = 20$ mA, and $R_L = 0.08$ kΩ.

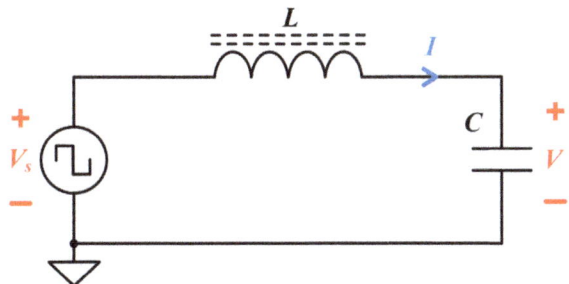

Fig. 8.7 Forced ferroresonant circuit described by Eq. (8.2).

Table 8.1 Component values for Fig. 8.7.

Component	Numerical	Simulated	Experimental
R_L	0.012 kΩ	5 Ω	2 Ω (1W)
C	1 μF	1 μF	1 μF
L_0	0.1 H	0.1 H	0.1 H
V_0	5 V	5 V	4.7 V (adjust)
I_{sat}	20 mA	20 mA	20 mA
ω	1 rad/ms	1 rad/ms	1 rad/ms (adjust)

implemented by Deane and Hamill (1990) who identified it as chaotic although they did not have a numerical model. Subsequently, Deane (1994) provided a somewhat complicated piecewise-linear model that includes hysteresis, which raises the dimension of the system by one.

We were able to obtain reasonable agreement using a simpler model and a system of equations given by

$$\begin{aligned} \dot{V} &= I/C \\ \dot{I} &= (V_S - V - IR_L)/L, \end{aligned} \quad (8.2)$$

where $V_S = V_0 \, \text{sgn}(\sin \omega t)$ and $L = L_0/(1 + I^2/I_{sat}^2)$ from Eq. (8.1). The hysteresis is included in the inductor parasitic resistance R_L and is assumed to be independent of current and frequency.

Elegant circuit values that give chaos are $R_L = 0.012$ kΩ, $C = 1$ μF, $L_0 = 0.1$ H, $V_0 = 5$ V, $I_{sat} = 20$ mA, and $\omega = 1$ rad/ms. The dimensionless equations are

$$\begin{aligned} \dot{x} &= y \\ \dot{y} &= c(1 + dy^2)[a \, \text{sgn}(\sin \tau) - x - by]. \end{aligned} \quad (8.3)$$

The parameters $a = 5$, $b = 0.012$, $c = 10$, and $d = 0.0025$ give the attractor shown in Fig. 8.8 and a time series as shown in Fig. 8.9. The Lyapunov exponents are (0.1708, 0, −0.5957), the Kaplan–Yorke dimension is $D_{ky} = 2.2867$, and the attractor has a global Class 1a basin of attraction.

Deane and Hamill (1990) observed that it was necessary to use a square wave voltage source rather than a sine wave to obtain chaos in this circuit, and we made the same observation. However, numerical modelling shows that two sine waves, for example chosen as the lowest two Fourier components of a square wave with one being a third harmonic of the other works just as well as a square wave. We also found chaotic solutions in our numerical modeling with a single sine wave, but they are relatively delicate and require a higher frequency drive ($\omega \gtrsim 10$ rad/ms).

You are more likely to obtain chaos at lower frequencies and larger amplitudes when the inductor is driven to saturation, but be cautious as you drive it harder since the circuit can draw currents as large as 1 A. If you have constructed an inductor similar to ours, you should be able to observe chaos at a few kilohertz or lower.

If your frequency generator cannot supply the current necessary in this circuit, you will see distortion of the input square wave. You can instead use the operational amplifier buffer and resistor R from Fig. 8.4. Just as with that circuit, use a 220 μF capacitor across V_{cc} and V_{ee} along with 0.1 μF decoupling capacitors, and avoid touching the operational amplifier while it is running since it gets very hot.

If you have additional difficulty obtaining chaos, it might help to adjust the offset and duty cycle slightly, but not too much lest your inductor will saturate in one direction, and you will need to restart the power supply. You might also try increasing $\omega_0 = 1/\sqrt{L_0 C}$ by reducing C from the value in Table 8.1. Also make sure that the secondary winding is not connected to anything, or connect it in parallel with the primary, being careful to observe the polarity of the windings and that the number of turns on each winding are the same.

The phase space plot of I versus V is shown in Fig. 8.10, and the time plot of $V(t)$ is shown in Fig. 8.11. The current I can be measured using an additional resistor R_L which should have a low resistance and be rated for at least 1 W. If you are driving the circuit hard, it might be easier to view the larger signal using ×10 attenuation on your oscilloscope probe. This circuit has one capacitor and one inductor for a total of two components.

8.3 Saito Family Inductor Circuit

It is more difficult to construct an autonomous chaotic circuit in which the inductor is the only nonlinear component. Nishio et al. (1990b) described such a circuit with two inductors, one capacitor, one ordinary resistor, and an active resistor. They did not use an actual saturating inductor but rather simulated the circuit using the piecewise equation from Saito (1988) to model their $\Phi(I)$ curve. The circuit was later implemented in an operational amplifier circuit described by Nishio et al. (1992).

Nishio and Mori (1991) subsequently simplified this circuit to one with a single inductor and capacitor and an active resistor. However, they only simulated their circuit, and their inductors required rather extreme hysteresis. We were unable to find chaos in numerical models of their circuit using our inductor.

These circuits belong to the family of hysteresis circuits described by Saito (1988), and we were able to find chaotic solutions using another one of its members that consists of two inductors, two capacitors, and an active resistor as shown in Fig. 8.12. The equations that describe this circuit are

$$\begin{aligned}
\dot{V}_1 &= (-V_1/R - I_1)/C_1 \\
\dot{V}_2 &= (I_1 - I_2)/C_2 \\
\dot{I}_1 &= (V_1 - V_2 - I_1 R_{L1})/L_1 \\
\dot{I}_2 &= (V_2 - I_2 R_{L2})/L_2,
\end{aligned} \quad (8.4)$$

where $L_2 = L_0/(1 + I_2^2/I_{sat}^2)$ from Eq. (8.1).

Elegant circuit values that give chaos are $R = -1$ kΩ, $R_{L1} = R_{L2} = 0.01$ kΩ, $C_1 = 2$ μF, $C_2 = 8$ μF, $L_0 = L_1 = 0.1$ H, and $I_{sat} = 20$ mA. The dimensionless equations are

$$\begin{aligned}
\dot{x} &= a(x - z) \\
\dot{y} &= b(z - u) \\
\dot{z} &= c(x - y - dz) \\
\dot{u} &= c(1 + eu^2)(y - du).
\end{aligned} \quad (8.5)$$

The parameters $a = 0.5$, $b = 0.125$, $c = 10$, $d = 0.01$, and $e = 0.0025$ give the attractor shown in Fig. 8.14 and a time series as shown in Fig. 8.15. The Lyapunov exponents are $(0.0670, 0.0013, 0, -0.7752)$, the Kaplan–Yorke dimension is $D_{ky} = 3.0869$. The system with the chosen parameters is unusual because it is hyperchaotic (two positive Lyapunov exponents). It is also inversion symmetric under the transformation $(x, y, z, u) \to (-x, -y, -z, -u)$ with an unstable focus at the origin with

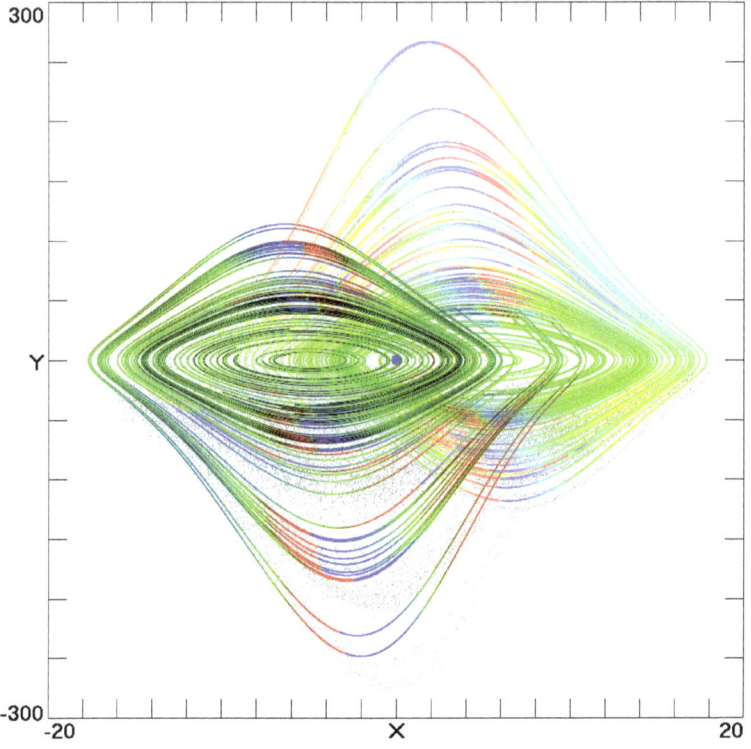

Fig. 8.8 Numerical solution of Eq. (8.3) with $a = 5$, $b = 0.012$, $c = 10$, and $d = 0.0025$.

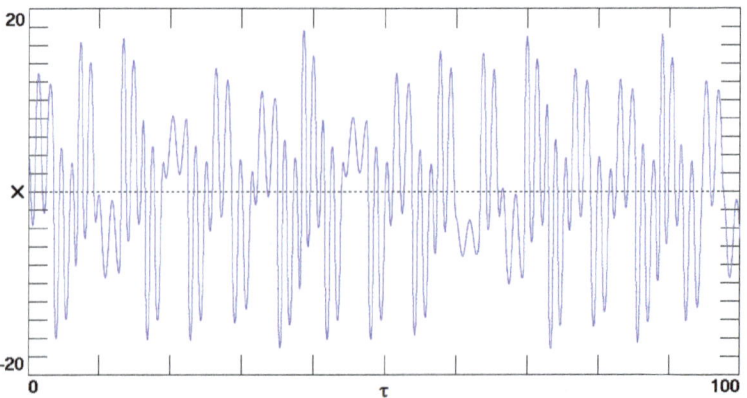

Fig. 8.9 Numerical waveform for $x(\tau)$ from Eq. (8.3) with $a = 5$, $b = 0.012$, $c = 10$, and $d = 0.0025$.

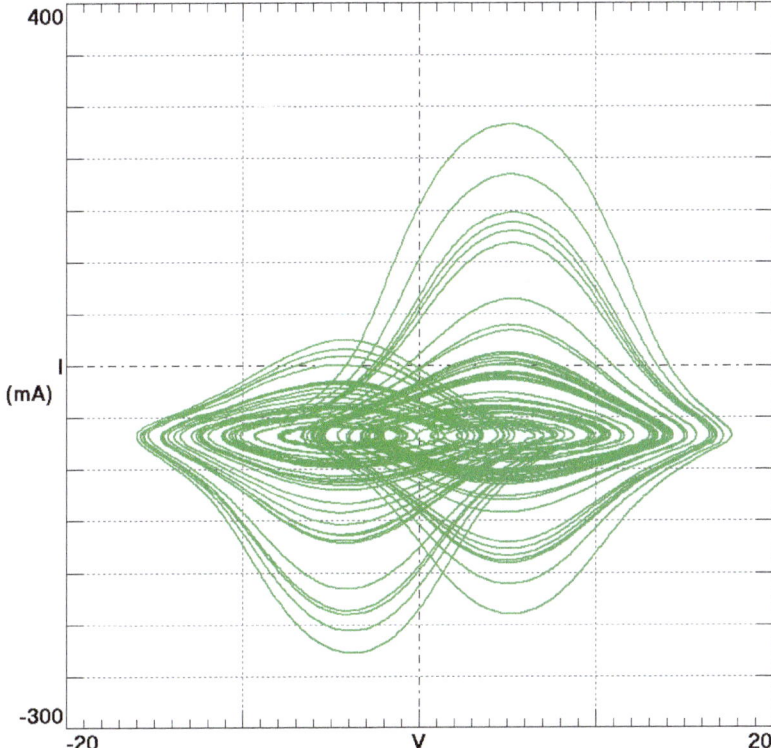

Fig. 8.10 Oscilloscope phase space plot of the forced ferroresonant circuit from Fig. 8.7.

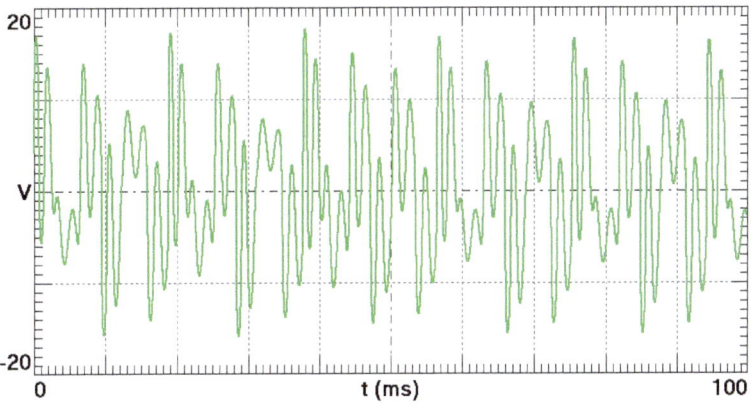

Fig. 8.11 Oscilloscope time plot of the forced ferroresonant circuit from Fig. 8.7.

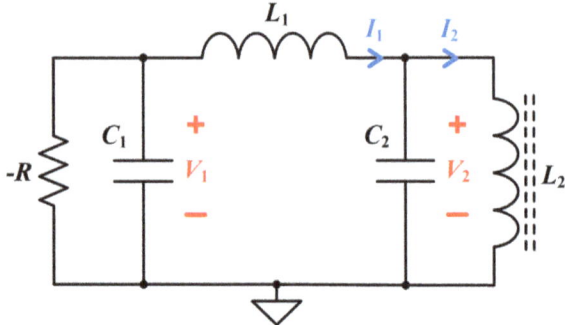

Fig. 8.12 Saito family inductor circuit described by Eq. (8.4).

Table 8.2 Component values for Fig. 8.12.

Component	Numerical	Simulated	Experimental
R	1 kΩ	1 kΩ	0.5 kΩ (adjust)
R_1	-	1 kΩ	0.5 kΩ (adjust)
R_2	-	1 kΩ	1 kΩ
R_3	-	1 kΩ	1 kΩ
R_{L1}	0.01 kΩ	0.01 kΩ	0.01 kΩ (1W, adjust)
R_{L2}	0.01 kΩ	0.01 kΩ	0.01 kΩ (1W, adjust)
C_1	2 μF	2.6 μF	2.6 μF (adjust)
C_2	8 μF	8 μF	8 μF
L_0	0.1 H	0.1 H	0.1 H
L_1	0.1 H	0.1 H	0.1 H
I_{sat}	20 mA	20 mA	20 mA
Op amp	-	Ideal	LM675

eigenvalues ($0.1300 \pm 2.5364i$, $0.0200 \pm 0.9743i$), rather than the more common saddle point.

The active resistor needs to deliver currents greater than I_{sat} in order to saturate the inductor and obtain chaos. For this reason we can no longer use the circuit from Fig. 2.42 since the AD844 will saturate at ± 10 mA, a nonlinearity previously used for the chaotic circuit in Section 6.3. Instead, we use the LM675 to implement a less elegant but more powerful version of the active resistor called a *negative impedance converter* shown in Fig. 8.13 that creates an effective negative resistance equal to $-R_1 R_3 / R_2$.

While testing this circuit, we noticed that the signal becomes noisy for certain component values such as $R_1 = 100$ Ω, $R_2 = 10$ kΩ, and $R_3 = 100$ kΩ. Be sure to check for problems such as this with your component

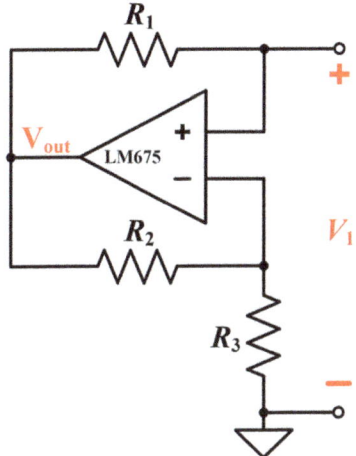

Fig. 8.13 Negative impedance converter.

values by measuring V versus $(V - V_{out})/R_1$, which should have a slope of the desired $-R$. We implemented a -0.5 kΩ resistor by using $R_1 = 0.5$ kΩ, $R_2 = 1$ kΩ, and $R_3 = 1$ kΩ as listed in Table 8.2.

The negative impedance converter works by using a non-inverting amplifier formed by R_2, R_3, and the operational amplifier that amplifies V_1 by a factor $K = 1 + R_2/R_3$. This allows the current in the active resistor to flow opposite to V_1 with a value of $V_1(1-K)/R_1$. However, the amplifier will saturate if K amplifies V_1 beyond the ± 30 V maximum supply voltage of the LM675. Since the signals are large in this circuit, you should routinely check V_{out} to avoid saturation whenever you obtain a stable oscillation.

One way is to avoid saturation is to reduce R_1 and increase R_3/R_2 while keeping R constant. If this method is insufficient, you can also scale down the voltages by reducing the resistances by the same factor while keeping R^2C/L constant. For this reason, the parameters are somewhat less elegant than they would be if saturation were not an issue.

In our circuit, the inductor L_1 has an internal DC resistance of 5.6 Ω while L_2 (our toroid) has a DC resistance of 3 Ω. Both inductors also had an additional series 10 Ω resistor rated for 1 W to limit the current and allow its measurement. The phase space plot of V_2 versus V_1 is shown in Fig. 8.16, and the time plot of $V_1(t)$ is shown in Fig. 8.17. This circuit has three resistors, two capacitors, two inductors, and one operational amplifier for a total of eight components.

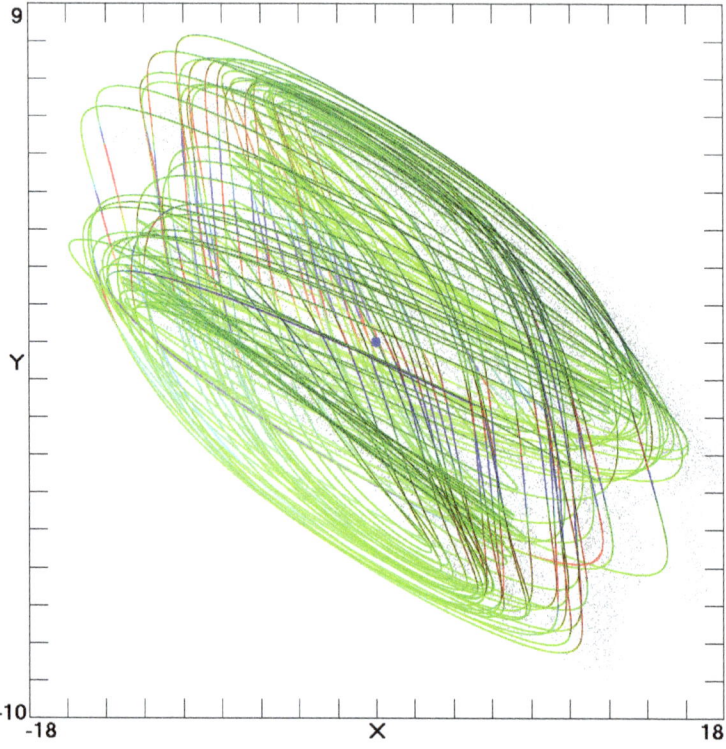

Fig. 8.14 Numerical solution of Eq. (8.5) with $a = 0.5$, $b = 0.125$, $c = 10$, $d = 0.01$, and $e = 0.0025$.

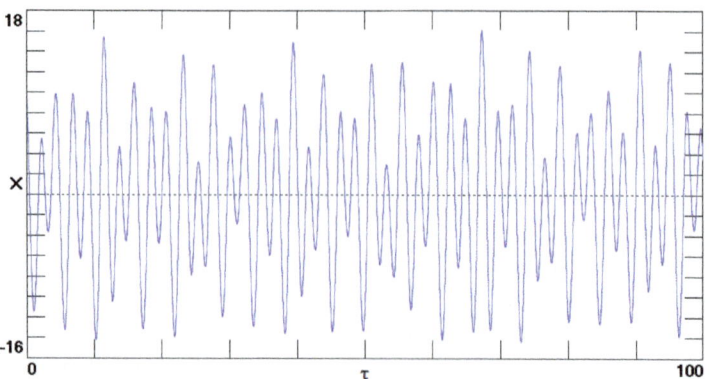

Fig. 8.15 Numerical waveform for $x(\tau)$ from Eq. (8.5) with $a = 0.5$, $b = 0.125$, $c = 10$, $d = 0.01$, and $e = 0.0025$.

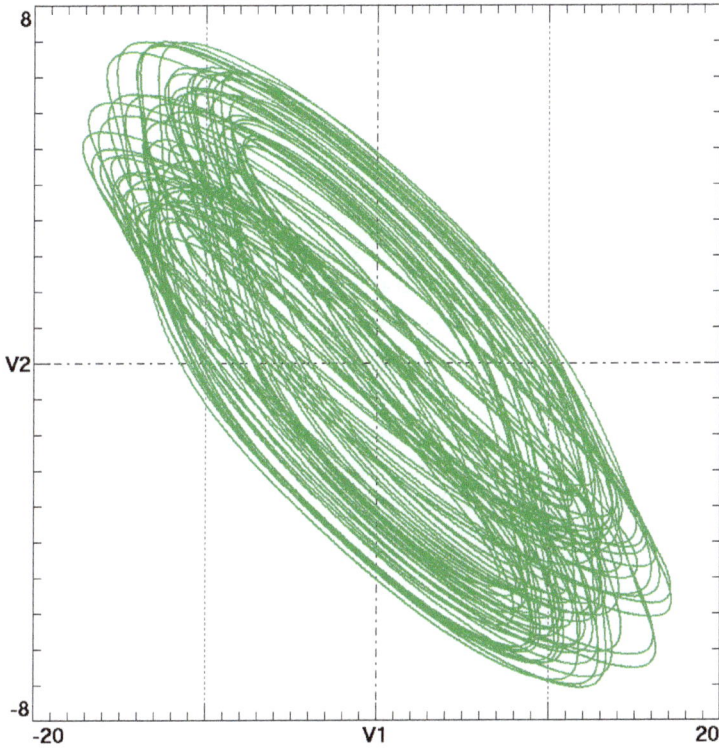

Fig. 8.16 Oscilloscope phase space plot of the Saito family inductor circuit from Fig. 8.12.

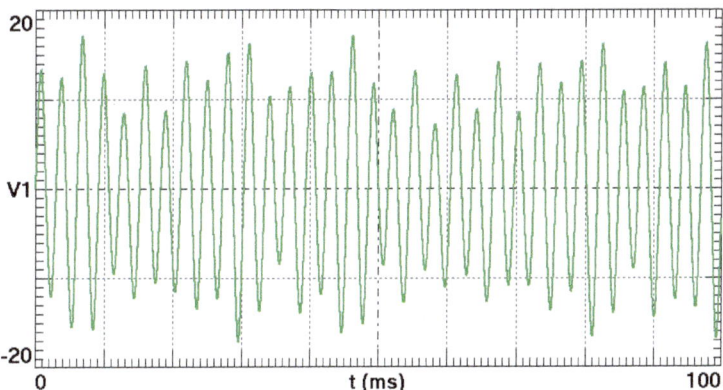

Fig. 8.17 Oscilloscope time plot of the Saito family inductor circuit from Fig. 8.12.

8.4 Minimal 3D Autonomous Inductor Circuit

We conclude this chapter with a new three-dimensional autonomous circuit in which the inductor is the only nonlinearity and provides the third dimension required for chaos without the need for an active resistor. We discovered it through a systematic search of all possible systems that involve two linear equations and one nonlinear equation governed by the saturating inductor that we constructed for this purpose.

The most general system contains nine terms (each of the three derivatives can depend linearly on each of the three variables), but we were able to eliminate three of the terms in a variety of combinations, one example of which is given by

$$\dot{V}_1 = -(V_1/R_1 + V_2/R_3)/C_1$$
$$\dot{V}_2 = -(V_1/R_2 + I)/C_2 \quad (8.6)$$
$$\dot{I} = (V_2 - IR_L)/L,$$

where $L = L_0/(1 + I^2/I_{sat}^2)$ from Eq. (8.1). The corresponding circuit is shown in Fig. 8.18.

Elegant circuit values that give chaos are $R_1 = 0.5$ kΩ, $R_2 = 0.3$ kΩ, $R_3 = 0.04$ kΩ, $R_L = 0.01$ kΩ, $C_1 = 8$ μF, $C_2 = 0.25$ μF, $L_0 = 0.1$ H, and $I_{sat} = 20$ mA. The parameters are less than ideal to limit the voltages. The dimensionless equations are

$$\dot{x} = -ax - by$$
$$\dot{y} = -cx - dz \quad (8.7)$$
$$\dot{z} = e(1 + fz^2)(y - gz).$$

The parameters $a = 0.25$, $b = 3.125$, $c = 20$, $d = 4$, $e = 10$, $f = 0.0025$, and $g = 0.01$ give the attractor shown in Fig. 8.19 and a time series as shown in Fig. 8.20. The Lyapunov exponents are (0.0586, 0, -0.5781), the Kaplan–Yorke dimension is $D_{ky} = 2.1014$, and the attractor has a global Class 1a basin of attraction. The system is inversion symmetric under the transformation $(x, y, z) \to (-x, -y, -z)$ with an unstable saddle node at the origin with eigenvalues (4.4800, 0.1675, -4.9974).

There are several considerations you should take into account due to the large power that this circuit dissipates. We had to scale down the voltages to prevent saturation by reducing the resistances by the same factor while keeping R^2C/L constant. The components R_2 and R_3 had to be further adjusted as shown in Table 8.3 to obtain chaos. The internal DC resistance of the inductor is 3 Ω, and no additional external resistor was required for chaos.

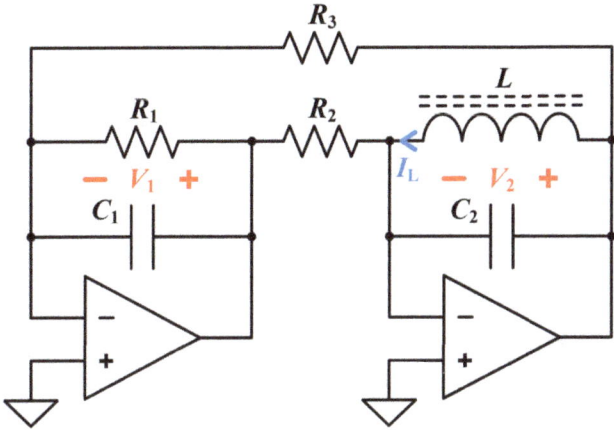

Fig. 8.18 Minimal 3D autonomous inductor circuit described by Eq. (8.6).

Table 8.3 Component values for Fig. 8.18.

Component	Numerical	Simulated	Experimental
R_1	0.5 kΩ	0.5 kΩ	0.5 kΩ (1W)
R_2	0.2 kΩ	0.5 kΩ	0.2 kΩ (10W, adjust)
R_3	0.04 kΩ	0.02 kΩ	0.04 kΩ (10W)
R_L	0.01 kΩ	1 Ω	0 Ω
C_1	8 μF	8 μF	8 μF (PP, ±3%)
C_2	0.25 μF	0.25 μF	0.25 μF (PP, ±3%)
L	0.1 H	0.1 H	0.1 H
I_{sat}	20 mA	20 mA	20 mA
Op amp	-	Ideal	LM675

Be sure that the resistors R_2 and R_3 have a sufficient power rating since the current through them can reach several hundred milliamperes. Additionally, the LM675 will become hot enough to activate its thermal shutdown function when it reaches around 110°C and then restart when it cools back down. You can prevent the LM675 from overheating by adding a large heat sink to the operational amplifiers.

There will be a short delay when the circuit is initially turned on because the initial conditions of the operational amplifiers start at the supply voltages (±30 V) which saturates and magnetizes the inductor. You can check this for yourself by measuring the magnetic flux through the secondary which likely will be initially offset in one polarity and become symmetric

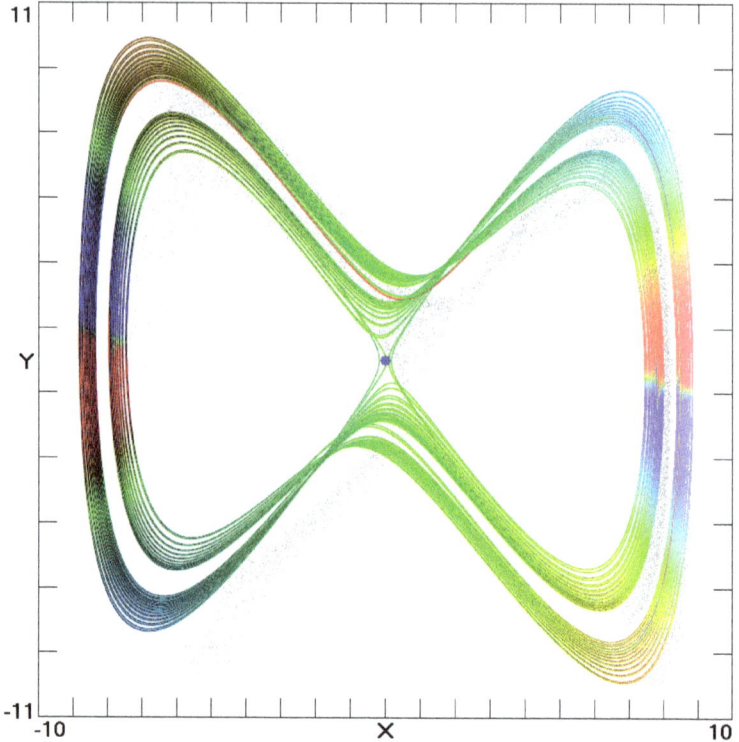

Fig. 8.19 Numerical solution of Eq. (8.7) with $a = 0.25$, $b = 3.125$, $c = 20$, $d = 4$, $e = 10$, $f = 0.0025$, and $g = 0.01$.

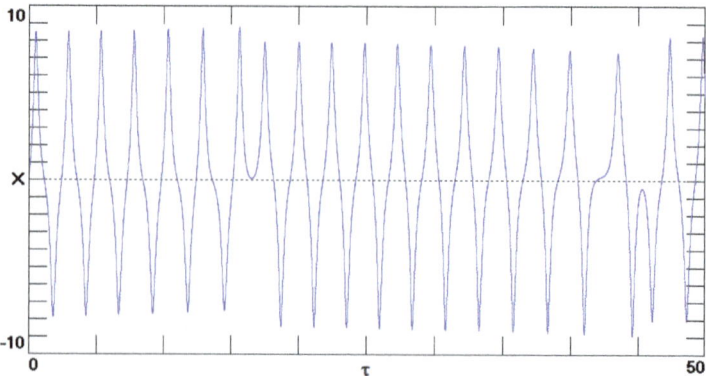

Fig. 8.20 Numerical waveform for $x(\tau)$ from Eq. (8.7) with $a = 0.25$, $b = 3.125$, $c = 20$, $d = 4$, $e = 10$, $f = 0.0025$, and $g = 0.01$.

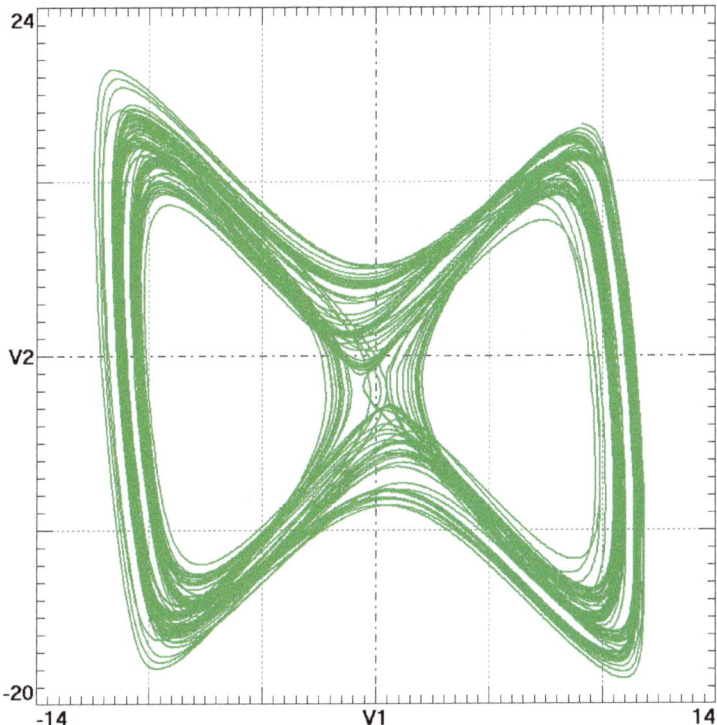

Fig. 8.21 Oscilloscope phase space plot of the minimal 3D autonomous inductor circuit from Fig. 8.18.

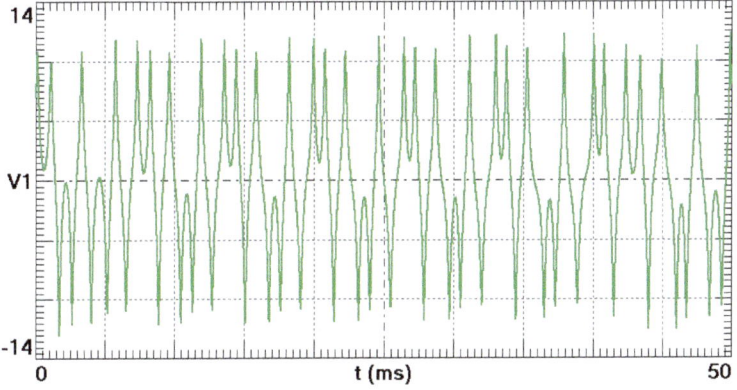

Fig. 8.22 Oscilloscope time plot of the minimal 3D autonomous inductor circuit from Fig. 8.18.

over time whereupon the chaos onsets. Alternately, you can avoid the delay by slowly turning up the power supply voltage until chaos onsets.

The system is delicate (a 0.55% change in parameters is likely to destroy the chaos), and may occasionally cause the circuit to get stuck in a limit cycle that resembles the attractor. If this happens, you can jar the circuit to obtain chaos again or use high-tolerance polypropylene capacitors for C_1 and C_2 to reduce the likelihood of this occurring. The phase space plot of V_2 versus V_1 is shown in Fig. 8.21, and the time plot of $V_1(t)$ is shown in Fig. 8.22. This circuit has three resistors, two capacitors, one inductor, and two operational amplifiers for a total of eight components.

Chapter 9

Memristor Circuits

The memristor is a two-terminal device that can be programmed to change its resistance when a voltage is applied to it, whereupon it will either increase or decrease its resistance depending on the polarity of the voltage. If this voltage is removed, the memristor retains its resistance value, making it useful as a computer memory element that can implement neural network algorithms. In this chapter, we use the resistance switching properties of a real solid-state memristor for several chaotic circuits that previously used memristor emulators.

9.1 Memristors

One of the important abilities of modern computers is to recognize useful patterns in data. For example, the computer in a self-driving car must identify road signs and signals from the images collected from its cameras. A small computer embedded in a medical device would need to accurately identify abnormal behavior such as seizures or arrhythmias from the recorded signals. Thus *neural network* algorithms were developed for computers to automatically learn useful trends in data just as a human would.

As computers evolved, several technologies were proposed to make them more efficient at running these algorithms. One of these technologies was the *memristor*, shown schematically in Fig. 9.1, which is a device that changes its resistance based on the polarity of its voltage or current. When a positive voltage is applied from anode to cathode, the memristor will *decrease* its resistance, whereas a negative voltage will *increase* the resistance. Furthermore, if the source is removed, the memristor retains its previous resistance.

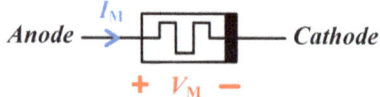

Fig. 9.1 Memristor schematic.

Memristors are also called *resistive random access memory* (RRAM) when they are used in computers with neural networks. Here the varying resistance of the memristors represents how well the neural network has learned a piece of data. This property eventually led to the first programmable memristor computer developed by Cai et al. (2019), that could identify Greek letters from a noisy image, find important features to reconstruct an image, and classify breast cancer tumors as malignant or benign. We now describe the fundamental characteristics of a memristor and how it is able to undergo the resistance change that is required for these applications.

9.1.1 *Memristor I-V characteristic*

The first model for a memristor was described by Chua (1971), who was attempting to find a framework to model the numerous new electronic devices that were being introduced at the time. His approach was to treat these new devices as a general form of the basic circuit elements: the resistor, capacitor, and inductor.

For example, a general form of a resistor following Ohm's law $V = IR$ would be $dV = R(I)dI$ which incorporates both linear and nonlinear resistors such as the current controlled diode in Eq. (2.4) and the tunnel diode (Chapter 4). This rule also applies to nonlinear capacitors (Section 2.1.5) given by $dq = C(V)dV$ and nonlinear inductors (Chapter 8) given by $d\Phi = L(I)dI$.

Since each of these circuit elements shares a variable with another element (for example V for the resistor and capacitor), they can be arranged in a diagram as shown in Fig. 9.2 with the diagonal lines representing the defining relationships between I and the charge q as well as V and the magnetic flux Φ. This diagram suggests a previously 'missing device' following the relation $d\Phi = M(q)dq$, which expressed as an I-V relationship gives

$$\begin{aligned} V_M &= M(q)I_M \\ \dot{q} &= I_M. \end{aligned} \quad (9.1)$$

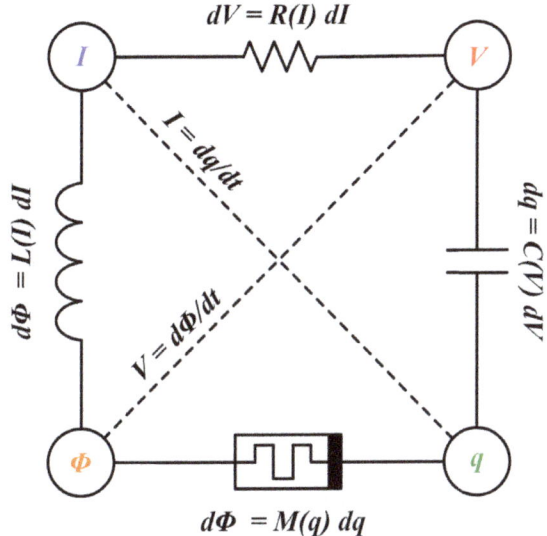

Fig. 9.2 Fundamental circuit elements.

A current I_M entering this device changes the charge q, which in turn changes the device resistance (measured in Ohms) represented by $M(q)$. For $I_M = 0$, the resistance no longer changes, and $M(q)$ retains the value determined by the previously applied I_M. Thus the device has a memory in the form of a resistance, which led to the term 'memristor.' A device that obeys Eq. (9.1) exactly is called an *ideal memristor* as defined by Chua (2015).

Just as how a diode can either be current or voltage-controlled, an ideal memristor can also be controlled by the magnetic flux, which can be modeled by

$$I_M = G(\Phi)V_M$$
$$\dot{\Phi} = V_M, \tag{9.2}$$

where $G(\Phi) = 1/M(\Phi)$ is the nonlinear conductance.

Chua and Kang (1976) introduced a more useful version of Eq. (9.1) which further generalizes memristors to *memristive systems* described by

$$V_M = M(s, I_M)I_M$$
$$\dot{s} = g(s, I_M), \tag{9.3}$$

where s is the 'internal state' and $g(s, I_M)$ is a function used to model the memristor behavior.

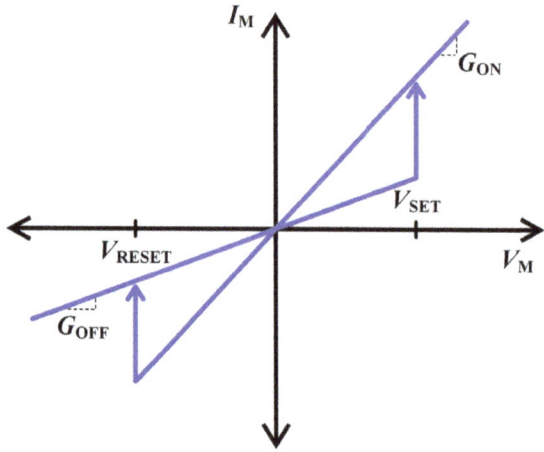

Fig. 9.3 Memristor I-V curve where $G_{ON} > G_{OFF}$.

When the device is controlled by a voltage, Eq. (9.3) becomes

$$I_M = G(s, V_M)V_M \\ \dot{s} = g(s, V_M). \qquad (9.4)$$

As an example, the I-V characteristic for one type of memristive system following Eq. (9.4) is shown in Fig. 9.3. In this particular system, there are two conductance levels where the higher conductance is termed G_{ON} since more current can flow through the memristor than when it is in the lower conductance state G_{OFF}. One important parameter of a memristor is the ON-OFF conductance ratio G_{ON}/G_{OFF}, which should be large so as to better differentiate the two states.

To change the memristor to a high conductance state, the voltage across the memristor must reach a positive threshold called the *SET* voltage (V_{SET}) which is defined as the minimum voltage needed to turn on the device. The device will stay at G_{ON} until an applied negative voltage reaches the minimum voltage needed to turn off the device called the *RESET* voltage threshold (V_{RESET}), whereupon it goes back to G_{OFF}.

Generally, any relation following Eq. (9.1) or Eq. (9.2) will form a figure eight in its I-V curve that Chua and Kang (1976) called a *pinched hysteresis loop* where the voltage and current are zero at the origin ('pinched') and the I-V curve has loops indicating the presence of hysteresis. There are no conditions related to the shape of the loops, and they will vary depending on the material used to make the memristor.

Although the resistance-changing behavior of memristors had been reported in the early 1960s by Hickmott (1962), memristors only gained attention much later when Strukov et al. (2008) identified their own devices to be the ones that Chua (1971) had described. Since then, many materials have been shown to behave like memristors, one of which we will describe in the next section and use in the chaotic circuits that follow.

9.1.2 Ag-chalcogenide memristor

Early research on memristor-based chaotic circuits focused on theoretical models called *emulators* that implemented Eq. (9.3) or Eq. (9.4) using analog computers (Chapter 7). Only a few studies were done during that time with solid-state memristors in chaotic circuits such as those by Kumar et al. (2017) and Minati et al. (2020).

In this chapter, we will use a particular memristor device (Knowm^{TM} M+SDC Chromium dopant) that was invented by Kozicki and West (1998) and later improved by Campbell (2017). An integrated circuit containing sixteen discrete memristors is shown in Fig. 9.4, where each memristor is individually laid out from anode (left) to cathode (right) in reference to our convention from Fig. 9.1.

Each memristor is made of chalcogenide glass, an amorphous material containing sulfur, selenium, and/or tellurium that is initially non-conducting (in the G_{OFF} state). Applying a SET voltage to the device repels Ag$^+$ ions from a Ag layer into the chalcogenide glass which increases the conductance to G_{ON}, whereas applying a RESET voltage attracts the ions out of the chalcogenide glass and reduces the conductance to G_{OFF}.

Before using this memristor in a chaotic circuit, it needs to be *formed* by increasing the voltage across the memristor so that the Ag$^+$ ions can easily enter the chalcogenide glass. You will need to protect the memristor during this process by using a series resistor R_1 and a transimpedance amplifier from Fig. 1.24 to set a current limit called the *current compliance* at $I_{max} = \pm V_{cc}/R_2$. Since the current through our device cannot safely exceed ± 2 mA, we set $R_2 = 7.5$ kΩ with $V_{cc} = 15$ V. We also experimentally determined the voltage limit to be approximately $+0.5$ V and -1 V, which you can avoid exceeding by monitoring the voltage across your device while forming.

Additionally, your AC voltage source will need to have a low frequency to properly view the pinched hysteresis loop when the device is formed. In fact, Chua and Kang (1976) theoretically predicted that increasing the

Fig. 9.4 Integrated circuit containing sixteen Ag-chalcogenide memristors. Left-side pins correspond to anodes of individual memristors, and right-side pins correspond to their cathodes.

frequency reduces the ability of the memristor to change conductance, eventually becoming a linear resistor. This behavior was later confirmed in experimental devices such as ours where we observe the pinched hysteresis loop to be large at 10 Hz but almost nonexistent at 1 kHz. There is also significantly less noise at 10 Hz, making the memristor easier to model and produce chaos since our circuits rely on deterministic behavior. For these reasons, we will operate our chaotic circuits at a low frequency, although doing so requires large inductance and capacitance values.

To form the device, start with a 10 Hz sine wave and slowly increase the amplitude from zero while observing the phase space plot of I versus V. At some point within ± 1 V, the current will begin to increase, and you should see a pinched hysteresis loop as shown in Fig. 9.6. You may occasionally

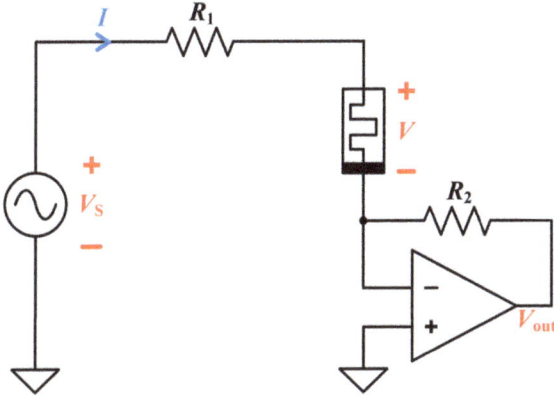

Fig. 9.5 Memristor forming and measurement circuit. Component values are $R_1 = 1$ kΩ and $R_2 = 7.5$ kΩ.

see some transient fluctuations in the waveform, but the phase space plot should be generally reproducible.

You may need to modify this procedure to form your device if you use a different material or type of memristor since there may be additional requirements to avoid damaging the device. However, you can still expect it to work in the circuits of this chapter provided the resistance-changing characteristics are similar.

9.1.3 *Memristor measurement and model*

Using the circuit in Fig. 9.5, we obtained the pinched hysteresis loop for the Ag-chalcogenide memristor. The data is plotted in Fig. 9.6 for 10 cycles, where the 10 Hz sine wave had an amplitude of 1.8 V and $R_1 = 1$ kΩ. The pinched hysteresis loop has some variance in V_{SET} and V_{RESET}, but it is sufficiently reproducible to allow modeling.

An empirical model for a similar memristor was developed by Pino *et al.* (2010) from which Yakopcic (2014) made a corresponding LTSpice model that you can use in your own simulations. The parameters of this model are the ON-OFF resistance values ($R_{ON} = 160$ Ω, $R_{OFF} = 2.5$ kΩ), the threshold voltages ($T_H = 0.3$ V, $T_L = -0.2$ V), and the fitting parameters ($K_{H1} = 5.5 \times 10^6$, $K_{H2} = -20$, $K_{L1} = 0.1 \times 10^6$, and $K_{L2} = 2.5$), which we adjusted to fit our data. This adjustment in parameters improved the agreement between the simulated and experimental results, although our simulations required an additional resistor in series with the memristor.

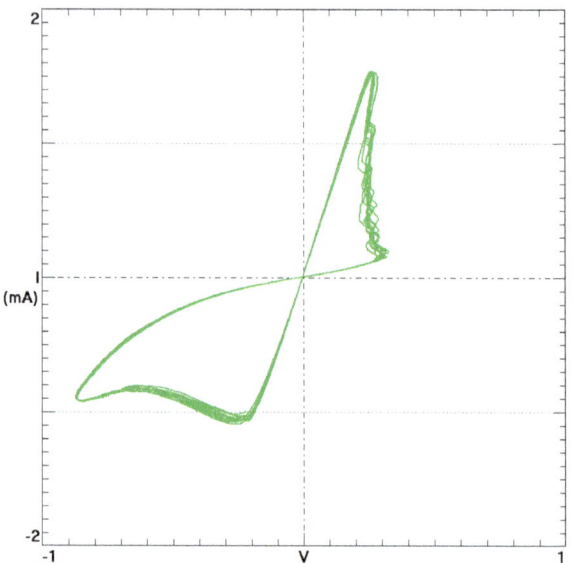

Fig. 9.6 Experimental memristor hysteresis plot when driven by a 10 Hz sine wave for 10 cycles.

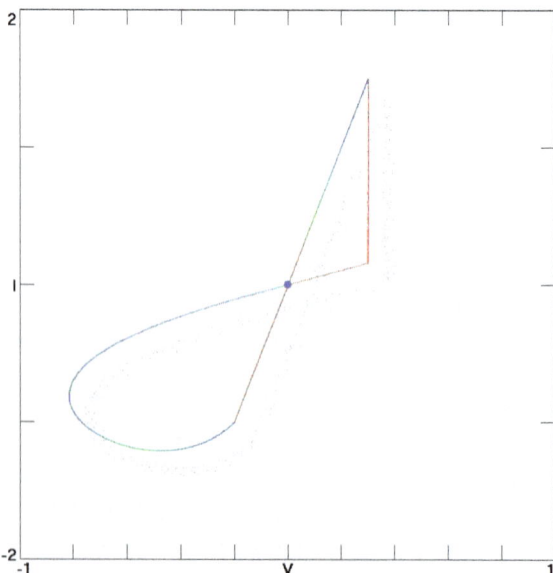

Fig. 9.7 Numerical memristor hysteresis plot from Eq. (9.5) with $\tau_0 = 10$ ms.

We explored about a dozen mathematical models for our memristor, some complicated and inelegant, and others simple and elegant. For example, $M(\Phi) = 0.6 - 0.4\tanh(\Phi/\Phi_0)$ gives a large resistance value of about 1 kΩ after V been negative long enough for its integral Φ to become negative, and another with a smaller value of about 0.2 kΩ after V has been positive long enough for its integral Φ to become positive. While such a model gives a plausible pinched hysteresis loop, it proved inadequate to model our chaotic circuits.

Instead, we used a more general model in the form of Eq. (9.4) given by

$$G(s, V_M) = 5.85s + 0.4$$

$$\dot{s} = \begin{cases} -s/\tau_0 & V_M < -0.2 \\ 0 & -0.2 \leq V_M \leq 0.3 \\ (1-s)/\tau_0 & V_M > 0.3, \end{cases} \quad (9.5)$$

where s is a state variable in the range of zero to one and τ_0 is a constant representing the response time of the memristor that controls the area of the I-V hysteresis loop and also determines the effective R_{OFF}/R_{ON} ratio since large values of τ_0 do not allow time for s to reach values near zero and one. Thus τ_0 is used as a free parameter in the model to account for variations among individual memristors and to best fit the data.

The other numerical values (5.85 and 0.4) correspond to $R_{OFF} = 2.5$ kΩ (when $s = 0$), $R_{ON} = 0.16$ kΩ (when $s = 1$), $V_{RESET} = -0.2$ V, and $V_{SET} = 0.3$ V. The model could be further generalized by using different values for τ_0 in the two regimes, but we did not find that necessary.

Although this model does not capture all the detail of the experimental plot, Fig. 9.7 with $\tau_0 = 10$ ms shows that it does produce an asymmetric pinched hysteresis loop of the proper size and shape and with two different resistances in the vicinity of the origin and a nonlinearity as required for chaos. This is the model that we will use in the circuits that follow.

9.2 Forced Memristor Circuit

One early memristor-based chaotic circuit was first proposed by Wang et al. (2009) who replaced the saturating amplifier in the Murali–Lakshmanan–Chua circuit (Section 6.3) with a memristor emulator as shown in Fig. 9.8. Since then, studies of this circuit have used memristor emulators that implement various forms of $\Phi(q)$ that were not related to physical devices,

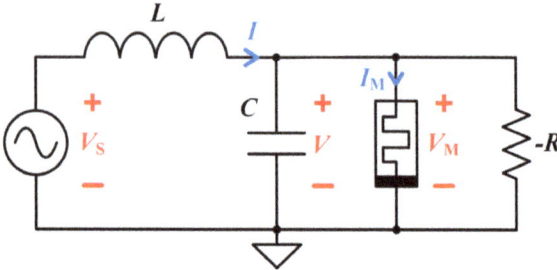

Fig. 9.8 Forced memristive circuit described by Eq. (9.6).

Table 9.1 Component values for Fig. 9.8.

Component	Numerical	Simulated	Experimental
R	1 kΩ	1 kΩ	1 kΩ
C	10 μF	10 μF	10 μF
L	10 H	10 H	10 H (gyrator)
V_0	0.2 V	0.4 V	0.15 V (adjust)
V_1	0.2 V	0.7 V	0.1 V (adjust)
ω	0.04π rad/ms	0.02π rad/ms	0.04π rad/ms (adjust)
τ_0	30 ms	-	-
Memristor	Eq. (9.5)	Pino et al. (2010)	Ag-chalcogenide

likely due to the lack of commercial memristors at that time. Here we implemented this circuit again with a Ag-chalcogenide memristor, although an additional active resistor is required to obtain chaos.

The equations describing the circuit in Fig. 9.8 are

$$\dot{V} = (I + V/R - GV)/C$$
$$\dot{I} = (V_S - V)/L$$
$$\dot{s} = \begin{cases} -s/\tau_0 & V < -0.2 \\ 0 & -0.2 \leq V \leq 0.3 \\ (1-s)/\tau_0 & V > 0.3, \end{cases} \quad (9.6)$$

where $V_S = V_1 \sin \omega t - V_0$ and $G = 5.85s + 0.4$ from Eq. (9.5).

Elegant circuit values that give chaos are $R = 1$ kΩ, $C = 10$ μF, $L = 10$ H, $V_0 = V_1 = 0.2$ V, $\omega = 0.04\pi$ rad/ms (20 Hz), and $\tau_0 = 30$ ms. The dimensionless equations are

$$\dot{x} = a[y + x - (5.85s + 0.4)x]$$
$$\dot{y} = b(c\sin\omega\tau - d - x)$$
$$\dot{s} = \begin{cases} -s/\tau_0 & x < -0.2 \\ 0 & -0.2 \leq x \leq 0.3 \\ (1-s)/\tau_0 & x > 0.3. \end{cases} \quad (9.7)$$

The parameters $a = b = 0.1$, $c = d = 0.2$, $\omega = 0.04\pi$, and $\tau_0 = 30$ give the attractor shown in Fig. 9.9 and a time series as shown in Fig. 9.10. The Lyapunov exponents are $(0.00348, 0, -0.0125, -0.0313)$, the Kaplan–Yorke dimension is $D_{ky} = 2.2696$, and the attractor has a Class 2 basin of attraction with $P_0 \approx 0.2$.

We will not use current compliance and voltage limiting for the circuits in this chapter since the required additional circuit elements are nonlinear and can be the source of the chaos rather than the memristor. This restriction makes the circuit more challenging to test because if you accidentally exceed the current limit (± 2 mA) or voltage limit (-1 V to $+0.5$ V) at any time while adjusting the parameters or due to some perturbation, you will immediately destroy the memristor. Prepare yourself with several replacements if you use our method.

The main risk to the memristor is the active resistor from Fig. 2.42 since the current feedback amplifier can saturate at a large voltage upon starting the circuit. To prevent this from happening, you should first leave the active resistor disconnected, turn on the frequency generator to output a sinusoidal waveform with a small amplitude, power the circuit on, and reconnect the active resistor. Be careful as you adjust the offset, amplitude, and frequency of the sine wave to find chaotic oscillations since this circuit is relatively delicate (a 3.6% variation in parameters is likely to destroy the chaos). We were only able to obtain chaos when the offset was adjusted so that V_S was mostly or completely in the negative region as shown in Table 9.1.

It might be helpful to observe the pinched hysteresis loop while adjusting parameters by measuring the memristor current using a transimpedance amplifier with a large I_{max} to verify that the memristor is still operational. Damaged memristors generally have more fluctuations in their measured currents and voltages as well as a much smaller G_{ON}/G_{OFF} ratio compared to Fig. 9.6 no matter how much you adjust the amplitude or frequency of V_S.

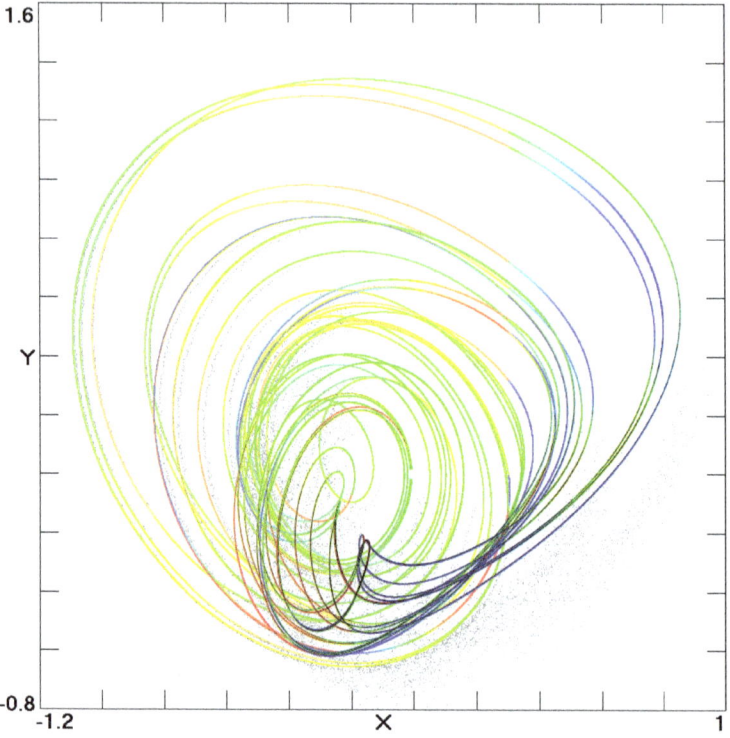

Fig. 9.9 Numerical solution of Eq. (9.7) with $a = b = 0.1$, $c = d = 0.2$, $\omega = 0.04\pi$, and $\tau_0 = 30$.

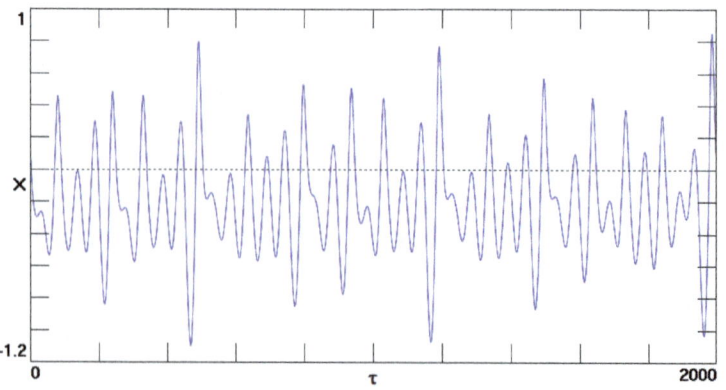

Fig. 9.10 Numerical waveform for $x(\tau)$ from Eq. (9.7) for $a = b = 0.1$, $c = d = 0.2$, $\omega = 0.04\pi$, and $\tau_0 = 30$.

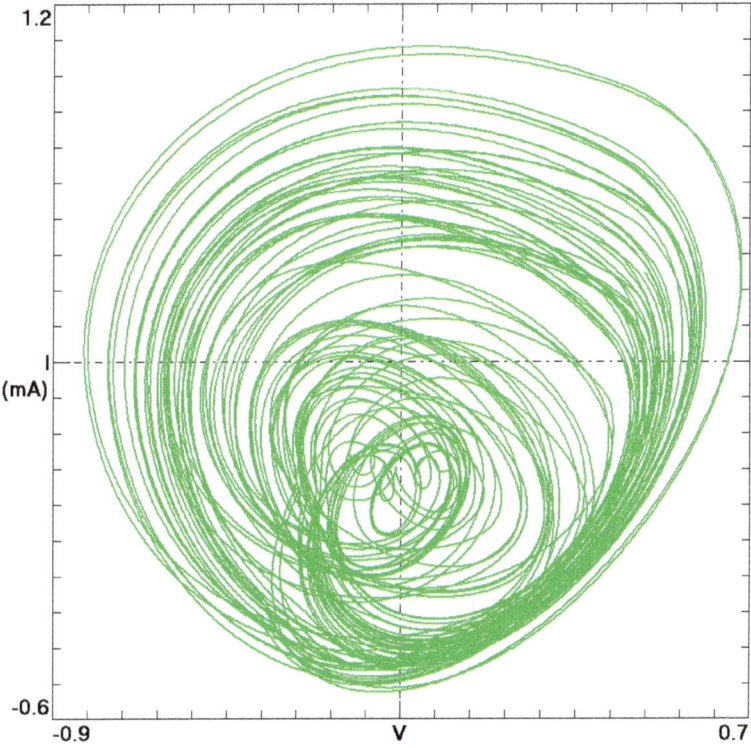

Fig. 9.11 Oscilloscope phase space plot of the forced memristive circuit from Fig. 9.8.

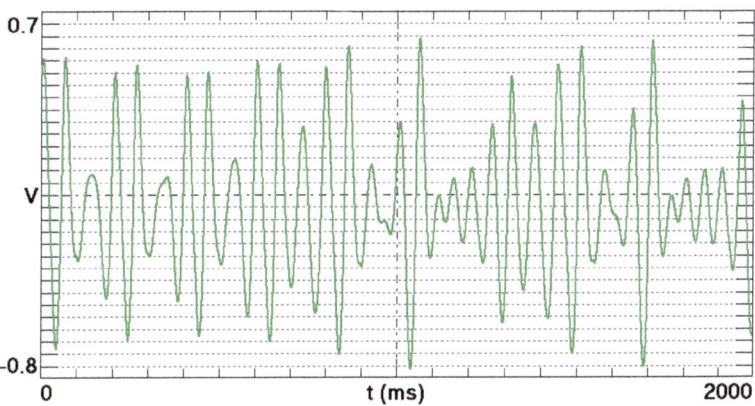

Fig. 9.12 Oscilloscope time plot of the forced memristive circuit from Fig. 9.8.

To make sure that the memristor is the nonlinearity responsible for chaos in the circuit, we checked the outputs of the operational amplifier of the gyrator and the TZ node of the active resistor to confirm that they were not saturating. We also determined the current saturation limits of the AD844 to be -11 mA and $+12$ mA when powered by a ± 15 V supply. If $R = 1$ kΩ, $V(t)$ would then saturate at -11 V or $+12$ V, which is not observed in the time plot in Fig. 9.12. Additionally, the circuit is still chaotic if the transimpedance amplifier used to measure I_M is removed.

A 600 Ω resistor in series with the memristor was also required to obtain chaos in the simulation, but not in the numerical or experimental circuit. Differences between the theoretical model and the experimental results presumably arise from the simple model used for the memristor, as well as variations in parameters among different memristors.

This circuit requires a large 10 H inductor with a small parasitic resistance since it is driven at a low frequency of 20 Hz. Since it is impractical to find or construct such an inductor, you will need to use the gyrator from Fig. 1.25 (with $R_1 = R_2 = R_3 = 1$ kΩ, $R_4 = 10$ kΩ, and $C = 1$ μF) with the grounded end of R_4 connected to the frequency generator instead.

Figure 9.11 shows the phase space plot of I versus V, where I can be calculated by probing across R_1 of the gyrator (with one end at V) and using Ohm's law. This circuit has one resistor, one capacitor, one inductor, one current feedback amplifier, and one memristor for a total of five components.

9.3 Saito Family Memristor Circuit

We have previously explored several variations of the chaotic circuit from Saito (1988) which contains a combination of active resistors, capacitors, inductors, and a device with hysteresis such as a thyristor (Section 5.7), an operational amplifier (Section 6.7), or a nonlinear inductor (Section 8.4).

Our final example of Saito's circuit was first explored by Itoh and Chua (2008) who used a theoretical memristor with a piecewise-linear equation for $M(\Phi)$. Figure 9.13 shows a version of their circuit that uses a Ag-chalcogenide memristor, where the hysteresis from state s of the memristor is the third minimal variable needed for chaos. We found that chaos only occurs in this circuit and for the other autonomous circuits in this chapter if a voltage source V_0 is added in series with the memristor so that it starts near V_{RESET} or in the G_{OFF} state when no signal is applied.

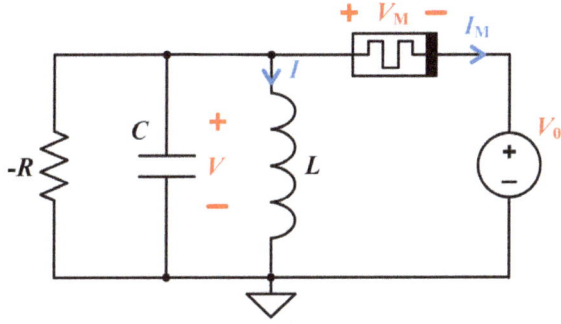

Fig. 9.13 Saito family memristor circuit described by Eq. (9.8).

Table 9.2 Component values for Fig. 9.13.

Component	Numerical	Simulated	Experimental
R	1 kΩ	1 kΩ	1 kΩ
C	10 μF	10 μF	10 μF
L	10 H	10 H	10 H (gyrator)
V_0	0.2 V	4.1 V	0.1 V (adjust)
τ_0	30 ms	-	-
Memristor	Eq. (9.5)	Pino et al. (2010)	Ag-chalcogenide

The equations describing the circuit in Fig. 9.13 are

$$\dot{V} = (V/R - I - GV_M)/C$$
$$\dot{I} = V/L$$
$$\dot{s} = \begin{cases} -s/\tau_0 & V_M < -0.2 \\ 0 & -0.2 \leq V_M \leq 0.3 \\ (1-s)/\tau_0 & V_M > 0.3, \end{cases} \quad (9.8)$$

where $V_M = V - V_0$ and $G = 5.85s + 0.4$ from Eq. (9.5).

Elegant circuit values that give chaos are $R = 1$ kΩ, $C = 10$ μF, $L = 10$ H, $V_0 = 0.2$ V, and $\tau_0 = 30$ ms. The dimensionless equations are

$$\dot{x} = a[x - y - (5.85s + 0.4)(x - c)]$$
$$\dot{y} = bx$$
$$\dot{s} = \begin{cases} -s/\tau_0 & x - c < -0.2 \\ 0 & -0.2 \leq x - c \leq 0.3 \\ (1-s)/\tau_0 & x - c > 0.3. \end{cases} \quad (9.9)$$

The parameters $a = b = 0.1$, $c = 0.2$, and $\tau_0 = 30$ give the attractor shown in Fig. 9.14 and a time series as shown in Fig. 9.15. The Lyapunov exponents are (0.0036, 0, -0.0313), and the Kaplan–Yorke dimension is $D_{ky} = 2.1143$. The attractor has a Class 3 basin of attraction with $P \approx 55/r^{1.7}$, and there is an unstable saddle node at (0, 1.25, 1) with eigenvalues (511.3, 0, -511.9).

Just as with the previous circuit, you will need to connect the memristor and supply power prior to connecting the active resistor to avoid saturating the amplifier. The circuit should begin to oscillate by increasing V_0 using the circuit in Fig. 1.29, but you are unlikely to obtain chaos because the amplitude at which the circuit starts oscillating exceeds the limit of what the memristor can handle and destroys it.

Instead of adding protection to the memristor, an alternative is to 'jump-start' the circuit by briefly applying a voltage to V to push it away from the origin. Then if V_0 is small, the circuit will oscillate at a small amplitude. We did this by setting V_0 to 100 mV as shown in Table 9.2 after reconnecting the active resistor and used another circuit from Fig. 1.29 to make a 'jump-start' voltage of 60 mV. Use a wire connected to this jump-start voltage to poke the node corresponding to V, and the circuit should begin to oscillate periodically. You can then carefully adjust V_0 to obtain chaos.

We picked these voltages because we found that the amplitude of the oscillation produced was within the memristor limits, but it may take some experimentation to find the right values of V_0 and jump-start voltage for your device without destroying it. One indication of a destroyed device is getting periodic oscillations but no chaos, which usually indicates that the memristor needs to be replaced.

We also checked that the TZ node of our current feedback amplifier will saturate at -10.8 mA and $+12$ mA corresponding to -10.8 V and $+12$ V for 1 kΩ, which was not observed in $V(t)$ in Fig. 9.17. You will also need a gyrator from Fig. 1.25 (with $R_1 = R_2 = R_3 = 1$ kΩ, $R_4 = 10$ kΩ, and $C = 1$ μF) to implement the 10 H inductor with a small parasitic resistance since it is difficult to construct a suitable passive inductor.

The phase space plot of I versus V is shown in Fig. 9.16, and the time plot of $V(t)$ is shown in Fig. 9.17. You can determine the current I by probing across R_1 of the gyrator (with one end at V) and using Ohm's law. This circuit has one resistor, one capacitor, one inductor, one current feedback amplifier, and one memristor for a total of five components.

9.4 Memristive Wien Bridge Oscillator

Another circuit that exhibits chaos with a variety of nonlinear devices is the Wien bridge oscillator, to which we added the Ag-chalcogenide memristor with a voltage source V_0 to replace the diode in the oscillator from Section 2.6. We found that this circuit could be converted into a potentially useful minimal-term chaotic oscillator by replacing the inductor with a resistor. However, the chaos in this circuit was delicate, and the circuit only oscillated transiently before returning to equilibrium. Coincidentally, Tamasevicius *et al.* (1996b) reported another Wien bridge oscillator using hysteresis through diodes that also had delicate behavior. Although such circuits can work in theory, they are challenging to construct and study, and so we instead describe the original version containing an inductor as shown in Fig. 9.18.

The equations describing the circuit in Fig. 9.18 are

$$\dot{V}_1 = (KV_1 - V_2 - V_1 - IR - GV_MR)/RC_1$$
$$\dot{V}_2 = (KV_1 - V_2 - V_1)/RC_2$$
$$\dot{I} = (V_1 - IR_L)/L \qquad (9.10)$$
$$\dot{s} = \begin{cases} -s/\tau_0 & V_M < -0.2 \\ 0 & -0.2 \leq V_M \leq 0.3 \\ (1-s)/\tau_0 & V_M > 0.3, \end{cases}$$

where $K = 1 + R_1/R_2$, $V_M = V_1 - V_0$, and $G = 5.85s + 0.4$ from Eq. (9.5).

Elegant circuit values that give chaos are $R = R_1 = R_2 = R_L = 1$ kΩ, $C_1 = 5$ μF, $C_2 = 10$ μF, $L = 10$ H, $V_0 = 1$ V, and $\tau_0 = 10$ ms. The dimensionless equations are

$$\dot{x} = a[x - y - z - (5.85s + 0.4)(x - d)]$$
$$\dot{y} = b(x - y)$$
$$\dot{z} = c(x - z) \qquad (9.11)$$
$$\dot{s} = \begin{cases} -s/\tau_0 & x - d < -0.2 \\ 0 & -0.2 \leq x - d \leq 0.3 \\ (1-s)/\tau_0 & x - d > 0.3. \end{cases}$$

The parameters $a = 0.2$, $b = c = 0.1$, $d = 1$, and $\tau_0 = 10$ give the attractor shown in Fig. 9.19 and a time series as shown in Fig. 9.20. The Lyapunov exponents are $(0.0033, 0, -0.0933, -0.1000)$, and the Kaplan–Yorke dimension is $D_{ky} = 2.0342$. The attractor has a Class 2 basin of attraction with $P_0 \approx 0.15$, and there is an unstable saddle focus at $(0.2857, 0.2857, 0.2857, 0)$ with eigenvalues $(0.0100 \pm 0.1670i, -1, -1)$.

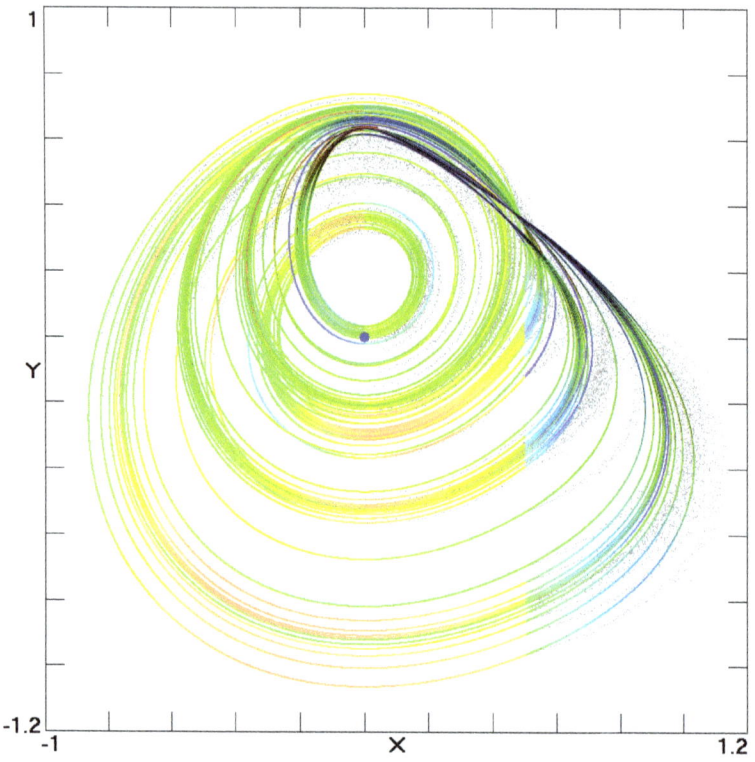

Fig. 9.14 Numerical solution of Eq. (9.9) with $a = b = 0.1$, $c = 0.2$, and $\tau_0 = 30$.

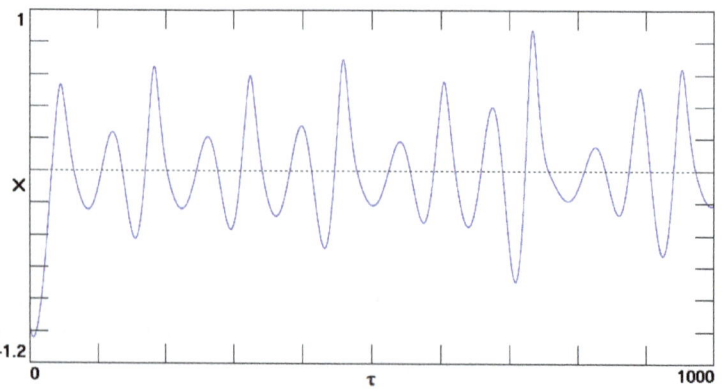

Fig. 9.15 Numerical waveform for $x(\tau)$ from Eq. (9.9) with $a = b = 0.1$, $c = 0.2$, and $\tau_0 = 30$.

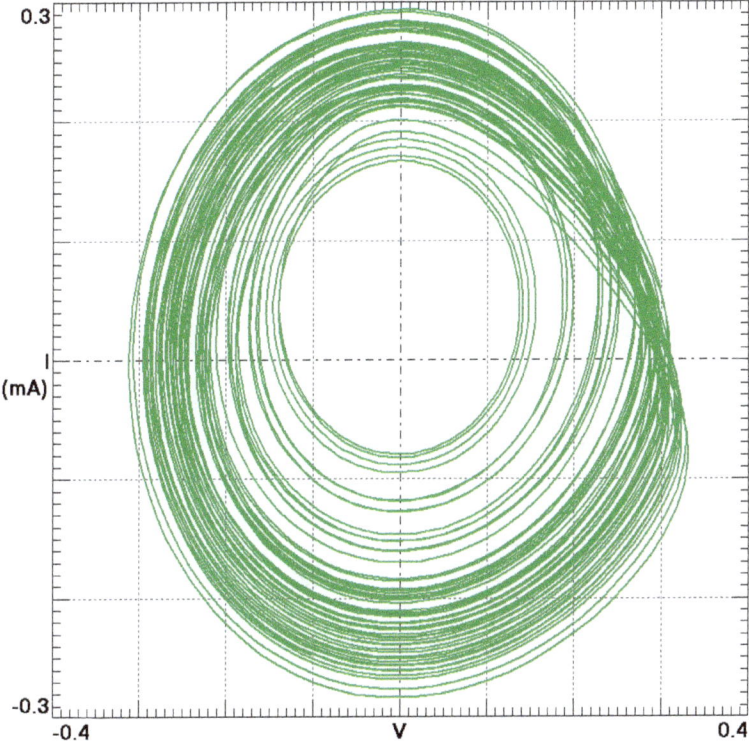

Fig. 9.16 Oscilloscope phase space plot of the Saito family memristor circuit from Fig. 9.13.

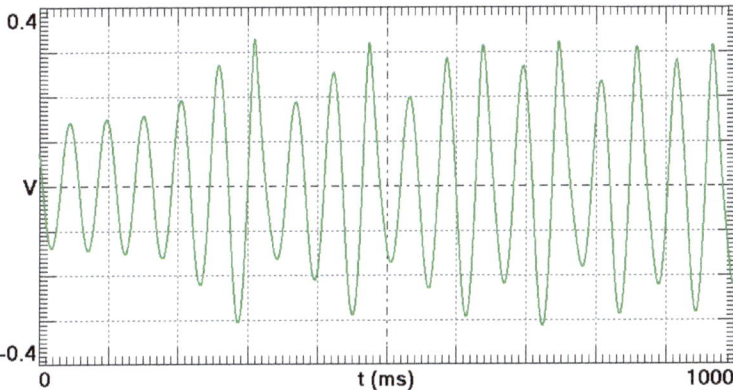

Fig. 9.17 Oscilloscope time plot of the Saito family memristor circuit from Fig. 9.13.

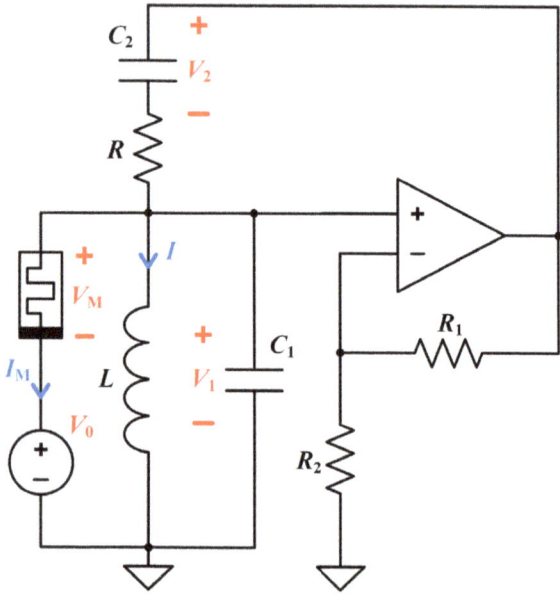

Fig. 9.18 Memristive Wien bridge oscillator described by Eq. (9.10).

Table 9.3 Component values for Fig. 9.18.

Component	Numerical	Simulated	Experimental
R	1 kΩ	1 kΩ	1 kΩ (±1%)
R_1	1 kΩ	1 kΩ	1 kΩ (±1%)
R_2	1 kΩ	1 kΩ	1 kΩ (±1%)
R_L	1 kΩ	1 kΩ	0 kΩ
C_1	5 μF	4 μF	4 μF
C_2	10 μF	10 μF	10 μF
L	10 H	10 H	10 H (gyrator)
V_0	1 V	2 V	0.11 V (adjust)
τ_0	10 ms	-	-
Memristor	Eq. (9.5)	Pino et al. (2010)	Ag-chalcogenide

We found that the best way to start this circuit without voltage or current limiting is to leave the memristor disconnected prior to powering on the circuit and short the output of the operational amplifier to ground, which prevents the memristor from being damaged by any voltage spikes that arise from turning on the power supply or from the operational amplifier saturating at the power supply voltage. It should be safe to disconnect the short circuit after powering on the circuit and reconnecting the memristor.

You will then need to increase V_0 from zero until oscillations begin and then carefully reduce it until you see chaos. You may have to repeat this several times since the chaos is delicate (a 5.3% variation in parameters is likely to destroy the chaos), and oscillations cease if V_0 is lowered too much.

We found it easier to obtain chaos when we used $\pm 1\%$ tolerance resistors for those listed in Table 9.3, whereas the capacitors and inductors can have $\pm 10\%$ tolerance. The experimental circuit did not need an additional resistor in series with the inductor, although this resistor was required in the numerical and simulated results.

The phase space plot of V_2 versus V_1 is shown in Fig. 9.21, and the time plot of $V_1(t)$ is shown in Fig. 9.22. You can measure V_2 by probing below the capacitor which has a voltage $KV_1 - V_2$ and calculate V_2 using V_1, or interchange the positions of C_2 and R_2 and take the voltage across C_2. This circuit has three resistors, two capacitors, one inductor, one operational amplifier, and one memristor for a total of eight components.

9.5 Elwakil–Kennedy Memristor Oscillator

Chaos also occurs in the Elwakil–Kennedy oscillator from Section 2.7 when the diode is replaced with a memristor. We encountered a similar situation where we found a minimal-term chaotic oscillator version of this circuit by replacing the inductor with a resistor that was difficult to operate because of its delicate behavior. Here we describe the original version containing an inductor as shown in Fig. 9.23.

The equations describing the circuit in Fig. 9.23 are

$$\dot{V}_1 = (V_1/R - V_2/R - I - GV_M)/C_1$$
$$\dot{V}_2 = (V_1 - V_2)/RC_2$$
$$\dot{I} = V_1/L$$
$$\dot{s} = \begin{cases} -s/\tau_0 & V_M < -0.2 \\ 0 & -0.2 \leq V_M \leq 0.3 \\ (1-s)/\tau_0 & V_M > 0.3, \end{cases}$$

(9.12)

where $V_M = V_1 - V_0$ and $G = 5.85s + 0.4$ from Eq. (9.5).

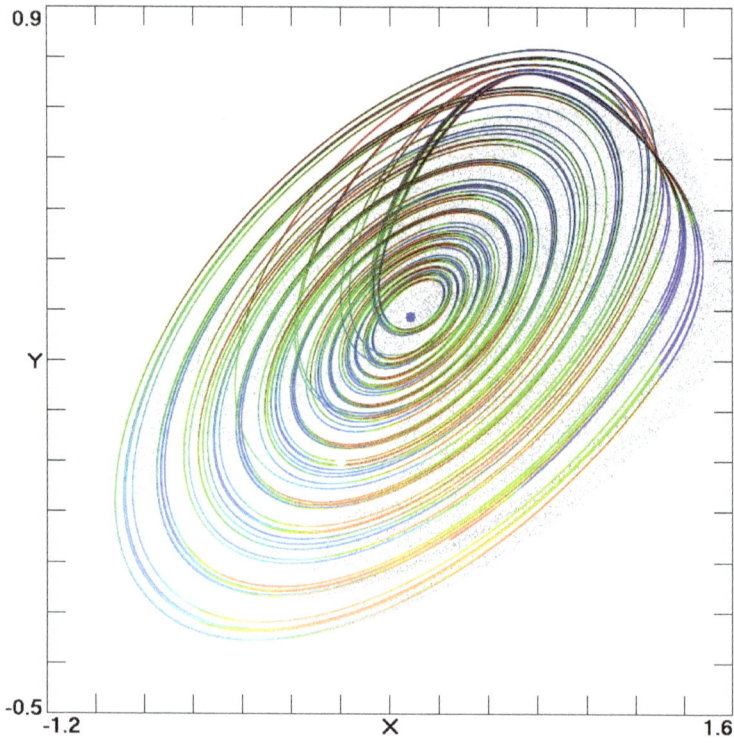

Fig. 9.19 Numerical solution of Eq. (9.11) with $a = 0.2$, $b = c = 0.1$, $d = 1$, and $\tau_0 = 10$.

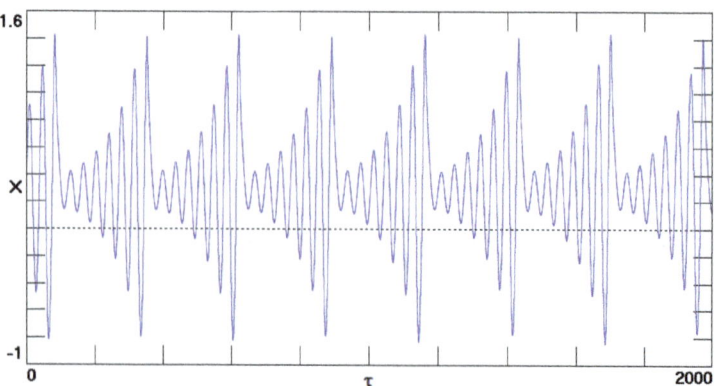

Fig. 9.20 Numerical waveform for $x(\tau)$ from Eq. (9.11) with $a = 0.2$, $b = c = 0.1$, $d = 1$, and $\tau_0 = 10$.

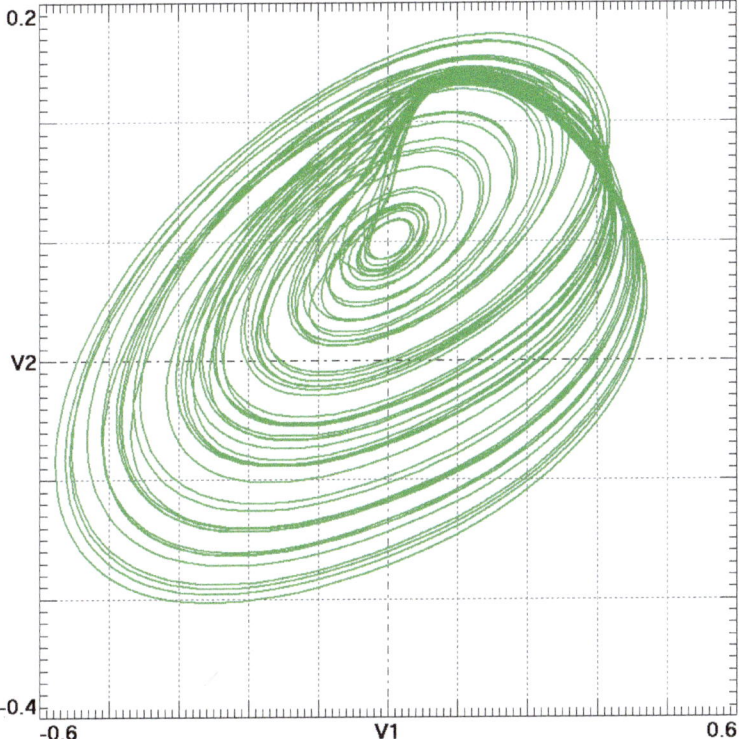

Fig. 9.21 Oscilloscope phase space plot of the memristive Wien bridge oscillator from Fig. 9.18.

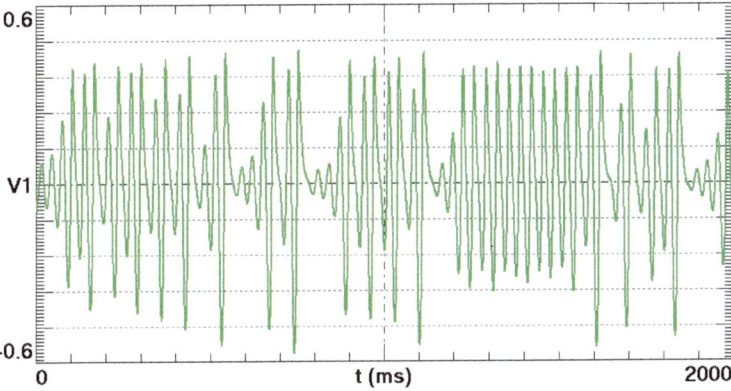

Fig. 9.22 Oscilloscope time plot of the memristive Wien bridge oscillator from Fig. 9.18.

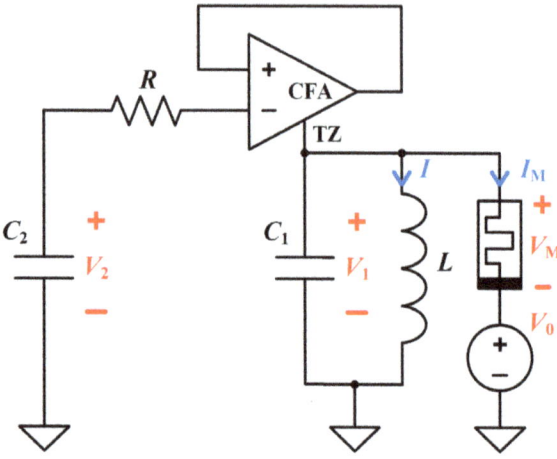

Fig. 9.23 Elwakil–Kennedy memristor oscillator described by Eq. (9.12).

Table 9.4 Component values for Fig. 9.23.

Component	Numerical	Simulated	Experimental
R	1 kΩ	1 kΩ	1 kΩ
C_1	4 µF	4 µF	4 µF
C_2	10 µF	10 µF	10 µF
L	10 H	10 H	10 H (gyrator)
V_0	0.34 V	2.5 V	0.15 V (adjust)
τ_0	13 ms	-	-
Memristor	Eq. (9.5)	Pino et al. (2010)	Ag-chalcogenide

Elegant circuit values that give chaos are $R = 1$ kΩ, $C_1 = 4$ µF, $C_2 = 10$ µF, $L = 10$ H, $V_0 = 0.34$ V, and $\tau_0 = 13$ ms. The dimensionless equations are

$$\dot{x} = a[x - y - z - (5.85s + 0.4)(x - d)]$$
$$\dot{y} = b(x - y)$$
$$\dot{z} = cx$$
$$\dot{s} = \begin{cases} -s/\tau_0 & x - d < -0.2 \\ 0 & -0.2 \leq x - d \leq 0.3 \\ (1 - s)/\tau_0 & x - d > 0.3 \end{cases} \quad (9.13)$$

The parameters $a = 0.25$, $b = c = 0.1$, $d = 0.34$, and $\tau_0 = 13$ give the attractor shown in Fig. 9.24 and a time series as shown in Fig. 9.25. The

Lyapunov exponents are (0.0084, 0, −0.0541, −0.0847), and the Kaplan–Yorke dimension is $D_{ky} = 2.1548$. The attractor has a Class 3 basin of attraction with $P \approx 10/r^{0.75}$, and there is an unstable saddle focus at (0, 0, 0.136, 0) with eigenvalues (0.0550 ± 0.1964i, −0.0601, −0.0769).

The main parameter to adjust for chaos is V_0, which should start at zero to prevent the circuit from oscillating and potentially destroying the memristor. This circuit is less delicate than the others (a 15% variation in parameters is likely to destroy the chaos), and it is much easier to start since you can turn on the circuit with the memristor connected.

Carefully increase V_0 until oscillations occur, and then reduce it until you observe chaos, repeating whenever the oscillations stop. We found chaos at $V_0 = 0.15$ V as shown in Table 9.4, but you may need a different value due to the variation in memristor devices. The phase space plot of V_2 versus V_1 is in Fig. 9.26, and the time plot of $V_1(t)$ is shown in Fig. 9.27. This circuit has one resistor, two capacitors, one inductor, one current feedback amplifier, and one memristor for a total of six components.

9.6 Senani–Singh Memristor Oscillator

The final circuit we will consider is based on a periodic oscillator proposed by Senani and Singh (1996) that uses two current feedback amplifiers. We followed the conjecture by Elwakil and Kennedy (2000a, 2001) and converted this circuit into a chaotic one by using a Ag-chalcogenide memristor as the nonlinear device. Since chaos occurs when the memristor is in parallel with either capacitor, we decided to use C_2 as shown in Fig. 9.28. The resulting circuit has the advantage of not requiring an inductor and also has the minimum number of variables needed for chaos, one of which is the state s of the memristor.

The equations describing the circuit in Fig. 9.28 are

$$\dot{V}_1 = [(V_1 - V_2)/R_3 - V_1/R_1]/C_1$$
$$\dot{V}_2 = (V_1/R_2 - GV_M)/C_2$$
$$\dot{s} = \begin{cases} -s/\tau_0 & V_M < -0.2 \\ 0 & -0.2 \leq V_M \leq 0.3 \\ (1-s)/\tau_0 & V_M > 0.3, \end{cases} \quad (9.14)$$

where $V_M = V_2 - V_0$ and $G = 5.85s + 0.4$ from Eq. (9.5).

Elegant circuit values that give chaos are $R_1 = 2$ kΩ, $R_2 = R_3 = 1$ kΩ, $C_1 = C_2 = 10$ μF, $V_0 = 0.3$ V, and $\tau_0 = 20$ ms. The dimensionless

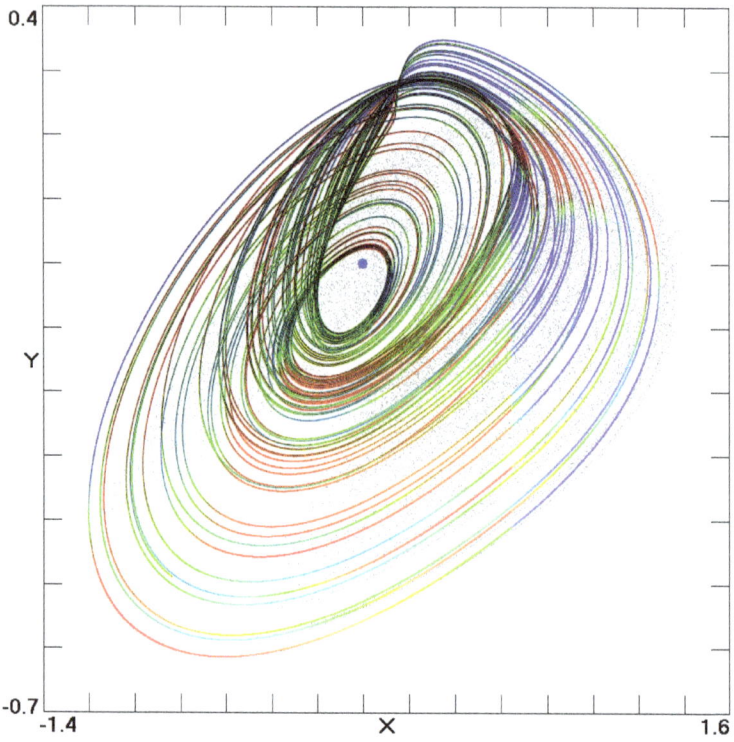

Fig. 9.24 Numerical solution of Eq. (9.13) with $a = 0.25$, $b = c = 0.1$, $d = 0.34$, and $\tau_0 = 13$.

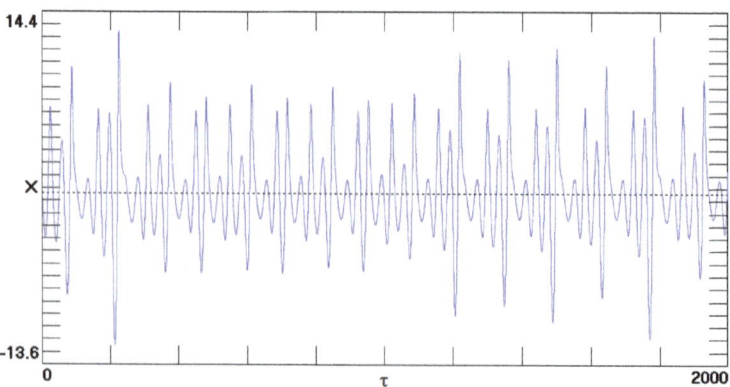

Fig. 9.25 Numerical waveform for $x(\tau)$ from Eq. (9.13) with $a = 0.25$, $b = c = 0.1$, $d = 0.34$, and $\tau_0 = 13$.

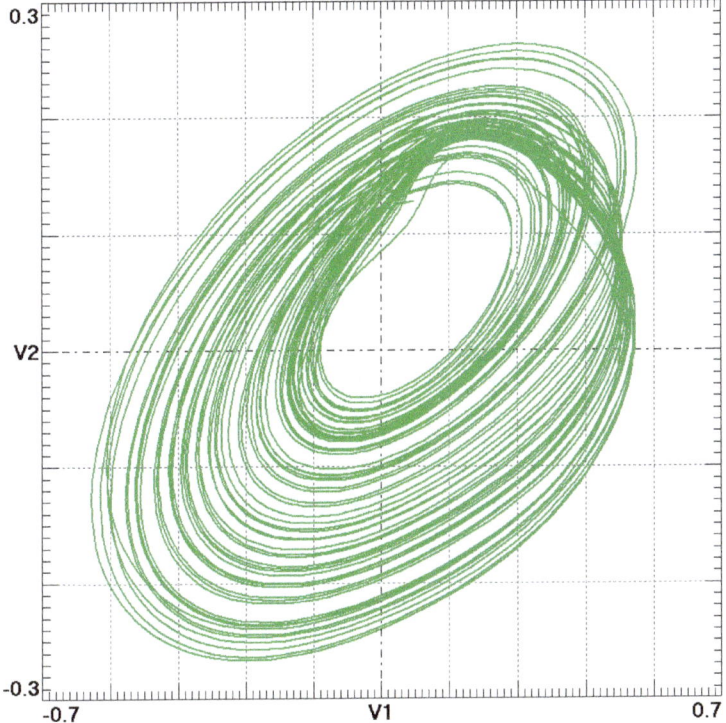

Fig. 9.26 Oscilloscope phase space plot of the Elwakil–Kennedy memristor oscillator from Fig. 9.23.

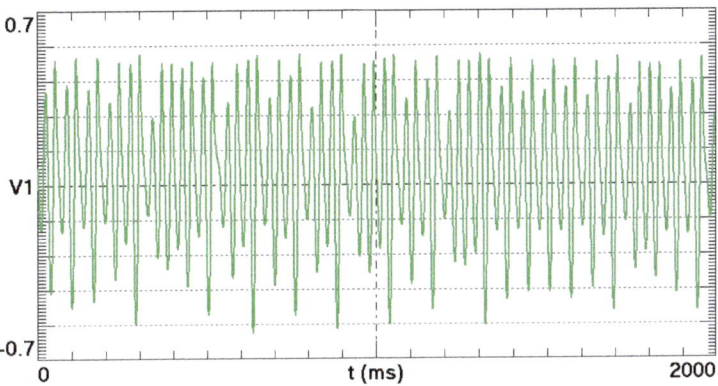

Fig. 9.27 Oscilloscope time plot of the Elwakil–Kennedy memristor oscillator from Fig. 9.23.

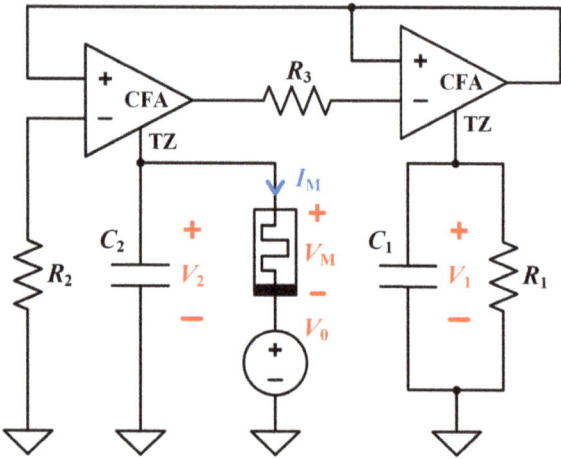

Fig. 9.28 Senani–Singh memristor oscillator described by Eq. (9.14).

Table 9.5 Component values for Fig. 9.28.

Component	Numerical	Simulated	Experimental
R_1	2 kΩ	2 kΩ	4 kΩ (adjust)
R_2	1 kΩ	1 kΩ	1 kΩ
R_3	1 kΩ	1 kΩ	1 kΩ
C_1	10 μF	10 μF	10 μF
C_2	10 μF	10 μF	10 μF
V_0	0.3 V	1 V	0.05 V (adjust)
τ_0	20 ms	-	-
Memristor	Eq. (9.5)	Pino et al. (2010)	Ag-chalcogenide

equations are

$$\dot{x} = a(x/2 - y)$$
$$\dot{y} = b[x - (5.85s + 0.4)(y - c)]$$
$$\dot{s} = \begin{cases} -s/\tau_0 & y - c < -0.2 \\ 0 & -0.2 \leq y - c \leq 0.3 \\ (1-s)/\tau_0 & y - c > 0.3. \end{cases} \quad (9.15)$$

The parameters $a = b = 0.1$, $c = 0.3$, and $\tau_0 = 20$ give the attractor shown in Fig. 9.29 and a time series as shown in Fig. 9.30. The Lyapunov exponents are (0.0032, 0, −0.0658) and the Kaplan–Yorke dimension is

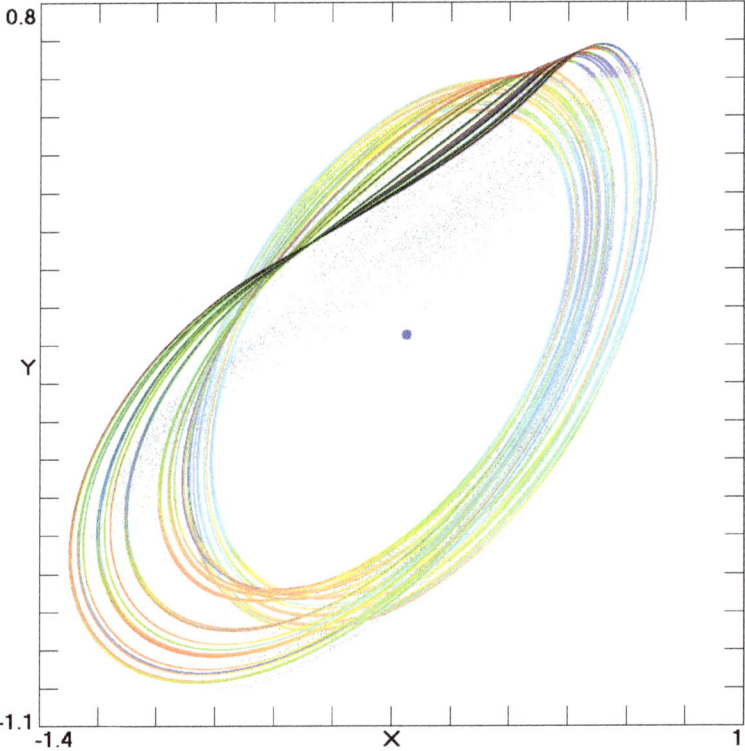

Fig. 9.29 Numerical solution of Eq. (9.15) with $a = b = 0.1$, $c = 0.3$, and $\tau_0 = 20$.

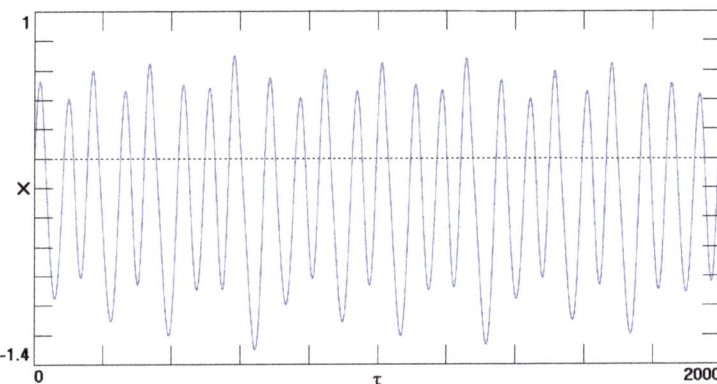

Fig. 9.30 Numerical waveform for $x(\tau)$ from Eq. (9.15) with $a = b = 0.1$, $c = 0.3$, and $\tau_0 = 20$.

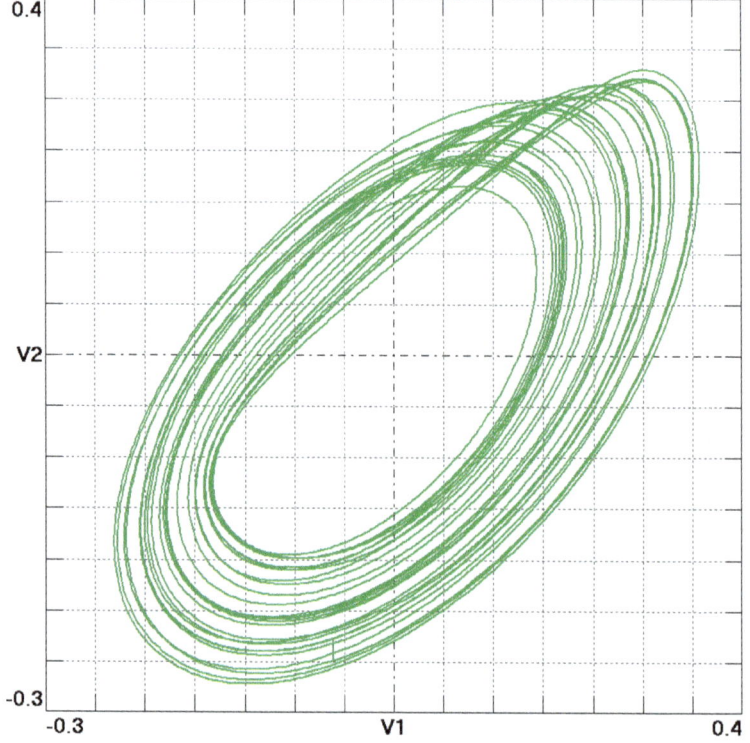

Fig. 9.31   Oscilloscope phase space plot of the Senani–Singh memristor oscillator from Fig. 9.28.

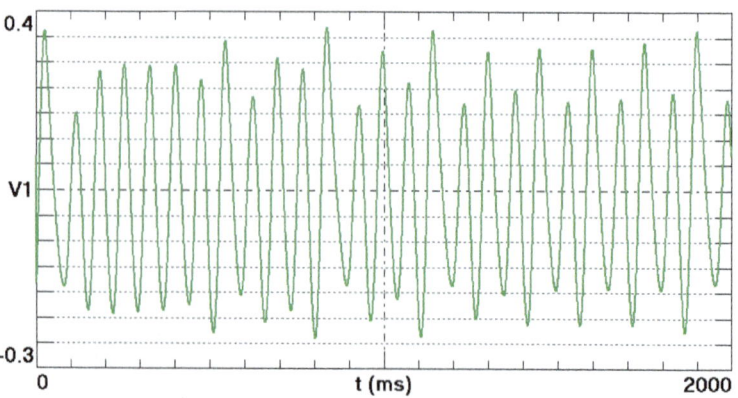

Fig. 9.32   Oscilloscope time plot of the Senani–Singh memristor oscillator from Fig. 9.28.

$D_{ky} = 2.0494$. The attractor has a Class 3 basin of attraction with $P \approx 2/r^2$, and there is an unstable saddle focus at $(-0.15, -0.075, 0)$ with eigenvalues $(0.0050 \pm 0.0893i, -0.0500)$.

You can start the circuit with the memristor connected, but make sure that V_0 and R_1 are both initially zero. Then carefully increase these parameters together until you see oscillation in V_1 and V_2, and then reduce V_0 until you see bifurcation and eventually chaos. You may need to increase R_1 slightly if you are having difficulty, but be cautious because that can cause the amplifiers to saturate and destroy the memristor.

We found chaos at $R_1 = 4$ kΩ and $V_0 = 0.05$ V as shown in Table 9.5, but you will likely require different values due to the variation among memristors. The phase space plot of V_2 versus V_1 is shown in Fig. 9.31, and the time plot of $V_1(t)$ is shown in Fig. 9.32. This circuit has three resistors, two capacitors, two current feedback amplifiers, and one memristor for a total of eight components.

Parting Comment

Like stars in the sky, there are countless chaotic systems and circuits, only a tiny sample of which have we included in this book. We have provided you with the fundamentals for understanding the most common chaotic circuits and hopefully piqued your interest in discovering new examples, maybe using devices not yet invented. If so, we encourage you to explore them yourself. Perhaps you will find a chaotic circuit more elegant than those in this book or an application for them that will improve the lives of others or extend humanity's knowledge of our incredible universe.

Bibliography

Abdomerovic, I., Lozowski, A. G., and Aronhime, P. B. (2000). High-frequency Chua's circuit, *Proc. 43rd IEEE Midwest Symp. on Circuits and Systems*, pp. 1026–1028.
Abraham, R. and Ueda, Y. (2000). *The Chaos Avant-Garde: Memories of the Early Days of Chaos Theory* (World Scientific, Singapore).
Antoniou, A. (1969). Realisation of gyrators using operational amplifiers, and their use in RC-active-network synthesis, *Proceedings of the Institution of Electrical Engineers* **116**, pp. 1838–1850.
Aoki, K., and Yamamoto, K. (1989). Nonlinear response and chaos in semiconductors induced by impact ionization, *Appl. Phys. A-Mater.* **48**, pp. 111–125.
Barboza, R. and Chua, L. O. (2008). The four-element Chua's circuit, *Int. J. Bifurcat. Chaos* **18**, pp. 943–955.
Bernhardt, P. A. (1991). The autonomous chaotic relaxation oscillator: An electrical analogue to the dripping faucet, *Phys. Nonlinear Phenom.* **52**, pp. 489–427.
Bhushan, M. and Newcomb, R. W. (1967). Grounding of capacitors in integrated circuits, *Electron. Lett.* **3**, pp. 148–149.
Blakely, J. N., Eskridge, M. B., and Corron, N. J. (2007). A simple Lorenz circuit and its radio frequency implementation, *Chaos* **17**, pp. 023112-1–5.
Buscarino, A., Fortuna, L., Frasca, M., and Sciuto, G. (2014). A Concise Guide to Chaotic Electronic Circuits (Springer, Heidelberg).
Cai, F., Correll, J. M., Lee, S. H., Lim, Y., Bothra, V., Zhang, Z., Michael, P. F., and Lu, W. D. (2019). A fully integrated reprogrammable memristor–CMOS system for efficient multiply–accumulate operations. *Nature Electronics* **2**, pp. 290–299.
Campbell, K. A. (2017). Tunable variable resistance memory device, U.S. patent number 9,583,703 B2.
Chua, L. (1971). Memristor — The missing circuit element, *IEEE T. Circuits Syst.* **18**, pp. 507–519.
Chua, L. (1992). The genesis of Chua's circuit, *Archiv fur Elektronik und Ubertragungstechnik* **46**, pp. 250–257.
Chua, L. (2015). Everything you wish to know about memristors but are afraid to ask. *Radioengineering* **24**, pp. 319–368.

Chua, L. and Kang, S. M. (1976). Memristive devices and systems, *Proc. IEEE* **64**, pp. 209–223.

Chua, L. O., Yu, J., and Yu, Y. (1983). Negative resistance devices. *IEEE T. Circuits-II* **11**, pp. 161–186.

Clapp, J. K. (1948). An inductance-capacitance oscillator of unusual frequency stability, *Proc. IRE* **367**, pp. 356–358.

Cuomo, K. M. and Oppenheim, A. V. (1993). Circuit implementation of synchronized chaos, with applications to communications, *Phys. Rev. Lett.* **71**, pp. 65–68.

Cuomo, K. M., Oppenheim, A. V., and Strogatz, S. H. (1993). Synchronization of Lorenz-based chaotic circuits with applications to communications, *IEEE T. Circuits-II* **40**, pp. 626–633.

Deane, J. H. B. (1994). Modeling the dynamics of nonlinear inductor circuits, *IEEE Trans. Magn.* **30**, pp. 2795–2801.

Deane, J. H. B. and Hamill, D. C. (1990). Instability, subharmonics, and chaos in power electronic systems, *IEEE Trans. Power Electron.* **5**, pp. 260–268.

Doike, Y., Uwate, Y., Nishio, Y., and Xin, J. (2013). Investigation of chaos-generating circuit composed of RL circuits, *2013 Shikoku-Section Joint Convention Record of the Institutes of Electrical and Related Engineers (Tokushima)*, p. 14.

Early, J. M. (1952). Effects of space-charge layer widening in junction transistors, *Proceedings of the IRE* **40**, pp. 1401–1406.

Ebers, J. J. and Moll, J. L. (1954). Large-signal behavior of junction transistors, *Proceedings of the IRE* **42**, pp. 1761–1772.

Eichhorn, R., Linz, S. J., and Hänggi, P. (1998). Transformations of nonlinear dynamical systems to jerky motion and its application to minimal chaotic flows, *Phys. Rev. E* **58**, pp. 7151–7164.

Eichhorn, R., Linz, S. J., and Hänggi, P. (2002). Simple polynomial classes of chaotic jerky dynamics, *Chaos, Solitons & Fractals* **13**, pp. 1–15.

Elwakil, A. S. (2000). Low-voltage relaxation oscillator, *Electron. Lett.* **36**, pp. 1256–1257.

Elwakil, A. S. and Soliman, A. M. (1997a). A family of Wien-type oscillators modified for chaos, *Int. J. Circ. Theor. Appl* **25**, pp. 561–579.

Elwakil, A. S. and Soliman, A. M. (1997). Chaos from a family of minimum-component oscillators, *Chaos, Soliton Fract.* **8**, pp. 335–356.

Elwakil, A. S. and Kennedy, M. P. (1998). High frequency Wien-type chaotic oscillator, *Electron. Lett.* **34**, pp. 1161–1162.

Elwakil, A. S. and Kennedy, M. P. (1999a). A family of Colpitts-like chaotic oscillators, *J. Franklin Inst.* **336**, pp. 687–700.

Elwakil, A. S. and Kennedy, M. P. (1999b). Chaotic oscillators derived from Saito's double-screw hysteresis oscillator, *IEICE Trans. Commun.* **82**, pp. 1769–1775.

Elwakil, A. S. and Kennedy, M. P. (2000a). A semi-systematic procedure for producing chaos from sinusoidal oscillators using diode-inductor and FET-capacitor composites, *IEEE T. Circuits-I* **47**, pp. 582–590.

Elwakil, A. S. and Kennedy, M. P. (2000b). Chaotic oscillators derived from sinusoidal oscillators based on the current feedback op amp, *Analog Integr. Circ. S.* **24**, pp. 239–251.

Elwakil, A. S. and Kennedy, M. P. (2000c). A low-voltage, low-power, chaotic oscillator, derived from a relaxation oscillator, *Microelectron. J.* **31**, pp. 459–468.

Elwakil, A. S. and Kennedy, M. P. (2001). Construction of classes of circuit-independent chaotic oscillators using passive-only nonlinear devices, *IEEE T. Circuits-I* **48**, pp. 289–307.

Elwakil, A. S. and Ozoguz, S. (2003). Chaos in pulse-excited resonator with self feedback, *Electron. Lett.* **39**, pp. 831–833.

Freire, E., Franquelo, L. and Aracil, J. (1984). Periodicity and chaos in an autonomous electronic system, *IEEE T. Circuits* **31**, pp. 237–247.

Gleick, J. (1987). Chaos: Making a New Science (Viking, New York).

Gollub, J. P., Brunner, T. O., and Danly, B. G. (1978). Periodicity and chaos in coupled nonlinear oscillators, *Science* **200**(7), pp. 48–50.

Gollub, J. P., Romer, E. J., and Socolar, J. E. (1980). Trajectory divergence for coupled relaxation oscillators: Measurements and Models, *J. Stat. Phys.* **23**, pp. 321–333.

González-López, R. (1998). Estudio de Osciladores Electrónicos Autónomos: Aplicación a un Oscilador de Puente de Wien Modificado, Ph.D. dissertation, Dep. Syst. Eng., Univ. de Cádiz, Cádiz, Spain (in Spanish).

Hamill, D. C. (1993). Learning about chaotic circuits with SPICE, *IEEE T. Educ.* **36**, pp. 28–35.

Heidel, J. and Zhang, F. (1999). Nonchaotic behaviour in three-dimensional quadratic systems II. The conservative case, *Nonlinearity* **12**, pp. 617–633.

Hickmott, T. W. (1962). Low-frequency negative resistance in thin anodic oxide films, *J. Appl. Phys.* **33**, pp. 2669–2682.

Hindmarsh, J. L. and Rose, R. M. (1984). A model of neuronal bursting using three coupled first order differential equations, *Proc. R. Soc. London, Ser. B* **221**, pp. 87–102.

Hirsch, M. W., Smale, S., and Devaney, R. L. (2004). *Dynamical Systems, and an Introduction to Chaos* (2nd edn) (Elsevier/Academic Press, Amsterdam).

Hoh, K. and Yasuda, Y. (1994). Electronic chaos in silicon thyristor, *Jpn. J. Appl. Phys. 1* **33**, pp. 594–598.

Hoover, W. G. (1995). Canonical dynamics: Equilibrium phase-space distributions, *Phys. Rev. A* **31**, pp. 1695–1697.

Hunt, E. R. (1982). Comment on a driven nonlinear oscillator, *Phys. Rev. Lett.* **49**, p. 1054.

Inaba, N. and Mori, S. (1989). Chaotic phenomena in four circuits with an ideal diode due to the change of the oscillation frequency, *Circuits and Systems, 1989 International Conference on* **3**, pp. 2147–2150.

Inaba, N. and Saito, T. (1991). Chaotic Circuit family with a diode due to the change of the oscillation frequency. In *Circuits and Systems, IEEE International Symposium on*, DOI: 10.1109/ISCAS.1991.176145.

Inaba, N., Saito, T., and Mori, S. (1987). Chaotic phenomena in a circuit with

a negative resistance and an ideal switch of diodes, *IEICE Trans. (1976–1990)* **70**, pp. 744–754.
Irita, T., Tsujita, T., Fujishima, M., and Hoh, K. (1995). Physical mechanism of chaos in thyristors and coupled-transistor structures, *Jpn. J. Appl. Phys.* **34**, pp. 1409–1412.
Irita, T., Tsujita, T., Fujishima, M., and Hoh, K. (1998). A simple chaos-generator for neuron element utilizing capacitance–npn-transistor pair, *Computers & Electrical Engineering* **24**, pp. 43–61.
Itoh, M. and Chua, L. O. (2008). Memristor oscillators, *Int. J. Bifurcat. Chaos* **18**, pp. 3183–3206.
Ivanov, Y. V. (1991). Unstable and chaotic modes in switching resonant invertors, *Izvestiya Vysshikh Uchebnykh Zavedenii Radioelektronika* **34**, pp. 3–6.
Izhikevich, E. M. (2003). Simple model of spiking neurons, *IEEE T. Neural Networ.* **14**, pp. 1569–1572.
Kaplan, J. and Yorke, J. (1979). Chaotic behavior of multidimensional difference equations. In *Functional Differential Equations and Approximation of Fixed Points, Lecture Notes in Mathematics*, Vol. 730 (ed. H.-O. Peitgen and H.-O. Walther), pp. 228–237 (Springer, Berlin).
Karadzinov, L. V., Arsov, G. L., Dzekov, T. A., and Jefferies, D. J. (1996). Charge-control piecewise-linear bipolar junction transistor model. In *Proceedings of IEEE International Symposium on Industrial Electronics* **2**, pp. 561–566.
Kennedy, M. P. (1992). Robust op amp realization of Chua's circuit, *Frequenz* **46**, pp. 66–80.
Kennedy, M. P. (1994). Chaos in the Colpitts oscillator, *IEEE T. Circuits-I* **41**, pp. 771–774.
Ketthong, P., San-Um, W., Srisuchinwong, B., and Tachibana, M. (2017) A simple current-reversible chaotic jerk circuit using inherent tanh(x) of an opamp, *IEICE Electron. Expr.* **14**, pp. 20170192-1–17.
Keuninckx, L., Van der Sande, G., and Danckaert, J. (2014). A simple two-transistor chaos generator based on a resistor-capacitor phase shift oscillator, *International Symposium on Nonlinear Theory and its Applications NOLTA2014, Luzern, Switzerland, September 14–18, 2014*, pp. 490–493.
Kozicki, M. N. and West, W. C. (1998). Programmable metallization cell structure and method of making same, U.S. patent number 5,761,115.
Kumar, S., Strachan, J. P., and Williams, R. S. (2017). Chaotic dynamics in nanoscale NBO_2 Mott memristors for analogue computing, *Nature* **548**, pp. 318–321.
Kuznetsov, S. P. (2016). Parametric chaos generator operating on a varactor diode with the instability limitation decay mechanism, *Tech. Phys.* **61**, pp. 436–445.
Kvarda, P. (2002). Chaos in Hartley's oscillator, *Int. J. Bifurcat. Chaos* **12**, pp. 2229–2232.
Lacy, J. G. (1996). A simple piecewise-linear non-autonomous circuit with chaotic behavior, *Int. J. Bifurcat. Chaos* **6**, pp. 2097–2100.
Leonov, G. A. and Kuznetsov, N. V. (2013). Hidden attractors in dynamical

systems. From hidden oscillations in Hilbert–Kolmogorov, Aizerman, and Kalman problems to hidden chaotic attractor in Chua circuits, *Int. J. Bifurcat. Chaos* **23**, pp. 1330002-1–69.

Li, C. and Sprott, J. C. (2013). Amplitude control approach for chaotic signals, *Nonlinear Dynam.* **73**, pp. 1335–1341.

Lindberg, E., Murali, K., and Tamasevicius, A. (2005). The smallest transistor-based nonautonomous chaotic circuit, *IEEE T. Circuits-II* **52**, pp. 661–664.

Linsay, P. S. (1981). Period doubling and chaotic behavior in a driven anharmonic oscillator, *Phys. Rev. Lett.* **47**, pp. 1349–1352.

Linz, S. J. and Sprott, J. C. (1999). Elementary chaotic flow, *Phys. Lett. A* **259**, pp. 240–245.

Lorenz, E. N. (1963). Deterministic nonperiodic flow, *J. Atmos. Sci.* **20**, pp. 130–141.

Lorenz, E. N. (1993). *The Essence of Chaos* (University of Washington Press, Seattle).

Louodop, P., Fotsin, H., Kountchou, M., Ngouonkadi, E. B. M., Cerdeira, H. A., and Bowong, S. (2014). Finite-time synchronization of tunnel-diode-based chaotic oscillators, *Phys. Rev. E* **89**, pp. 032921-1–11.

Mandelbrot, B. B. (1982). *The Fractal Geometry of Nature* (Freeman, New York).

Marsden, J. E. and McCracken, M. (1976). *The Hopf Bifurcation and its Applications* (Springer, New York).

Martienssen, V. O. (1910). Über neue, resonanzerscheinungen in wechselstromkreisen, *Physik Zeitschrift – Leipzig* **11**, pp. 448–460.

Masuda, S., Uchitani, Y. and Nishio, Y. (2009). Simple chaotic oscillator using two RC circuits, *Proceedings of RISP International Workshop on Nonlinear Circuits and Signal Processing* (NCSP'09), pp. 89–92.

Matsumoto, T. (1984). A chaotic attractor from Chua's circuit, *IEEE T. Circuits Syst.* **31**, pp. 1055–1058.

Matsumoto, T., Chua, L., and Tokumasu, K. (1986). Double scroll via a two-transistor circuit, *IEEE T. Circuits Syst.* **33**, pp. 828–835.

Minati, L. (2013). Emergence of chaos in transistor circuits evolved towards maximization of approximate signal entropy. In *International Symposium on Image and Signal Processing and Analysis*, ISPA, pp. 755–760. [06703838] IEEE Computer Society.

Minati, L. (2014). Experimental dynamical characterization of five autonomous chaotic oscillators with tunable series resistance, *Chaos* **24**, pp. 033110-1–12.

Minati, L., Frasca, M., Oświęcimka, P., Faes, L., and Brozdz, S. (2017). Atypical transistor-based chaotic oscillators: Design, realization, and diversity, *Chaos* **27**, pp. 073113-1–13.

Minati, L., Frasca, M., Giustolisi, G, Oświęcimka, P., Drożdż, S., and Ricci, L. (2018). High-dimensional dynamics in a single-transistor oscillator containing Feynman–Sierpiński resonators: Effect of fractal depth and irregularity, *Chaos* **28**, pp. 093112-1–15.

Minati, L., Gambuzza, L. V., Thio, W. J., Sprott, J. C., and Frasca, M. (2020). A chaotic circuit based on a physical memristor, *Chaos, Solitons & Fractals* **138**, pp. 109990-1–9.

Morgul, O. (1995). Inductorless realisation of Chua oscillator, *Electron. Lett.* **31**, pp. 1403–1404.

Munakata, T., Sinha, S., and Ditto, W. L. (2002). Chaos computing: implementation of fundamental logical gates by chaotic elements, *IEEE T. Circuits-I* **49**, pp. 1629–1633.

Munmuangsaen, B. and Srisuchinwong, B. (2013). Chaos in modified CFOA-based inductorless sinusoidal oscillators using a diode, *Chaotic Modeling and Simulation (CMSIM)* **1**, pp. 179–185.

Murali, K., Lakshmanan, M., and Chua, L. O. (1994). The simplest dissipative nonautonomous chaotic circuit, *IEEE T. Circuits-I* **41**, pp. 462–463.

Murali, K., Lindberg, E., and Leung, H. (2002). Design principles of hyperchaotic circuits, *AIP Conf. Proc.* **622**, pp. 15–26.

Muthuswamy, B. and Banerjee, S. (2015). *A Route to Chaos Using FPGAs* (Springer, Heidelberg).

Mykolaitis, G., Tamaševičius, A., Bumelienė, S., Namajūnas, A., Pyragas, K., and Pyragas, V. (2005). Application of ultrafast Schottky diodes to high megahertz chaotic oscillators, *Acta Phys. Pol. A. Part II, Proceedings of the 12th International Symposiums on Ultrafast Phenomena in Semiconductors (12-UFPS), Vilnius, Lithuania, August 22-25, 2004*. Warszawa: Polish Academy of Sciences **107**, pp. 365–368.

Nakano, H. and Saito, T. (2002). Basic dynamics from a pulse-coupled network of autonomous integrate-and-fire chaotic circuits, *IEEE T. Neural Networ.* **13**, pp. 92–100.

Namajunas, A. and Tamasevicius, A. (1995). Modified Wien-bridge oscillator for chaos, *Electron. Lett.* **31**, pp. 335–336.

Namajunas, A. and Tamasevicius, A. (1996). Simple RC chaotic oscillator, *Electron. Lett.* **32**, pp. 945–946.

Nguyen, N. M. (1991). Monolithic microwave oscillators and amplifiers, *EECS Department University of California, Berkeley, Technical Report No. UCB/ERL M91/36*.

Nienhaus, H. A., Bowers, J. C., and Herren, P. C. (1976). Computer model for a high power SCR, *1976 IEEE Power Electronics Specialists Conference*, pp. 56–61.

Nishio, Y. and Mori, S. (1991). Chaotic phenomena in an LCR oscillator with a hysteresis inductor, *IEEE International Sympoisun on Circuits and Systems*, pp. 2873–2876.

Nishio, Y., Inaba, N., Mori, S., and Saito T. (1990a). Rigorous analyses of windows in a symmetric circuit, *IEEE T. Circuits* **37**, pp. 473–487.

Nishio, Y., Inaba, N., and Mori, S. (1990b). Chaotic phenomena in an autonomous circuit with nonlinear inductor, *IEEE International Symposium on Circuits and Systems*, pp. 942–945.

Nishio, Y., Mori, S., and Inaba, N. (1992). Chaotic phenomena in an autonomous circuit with a nonlinear negative inductor, *Electr. Commun. Jpn. 3* **75**, pp. 72–81.

Nosé, S. (1984). A unified formulation of the constant temperature molecular dynamics methods, *J. Chem. Phys.* **81**, pp. 511–519.

Petrzela, J. and Polak, L. (2019). Minimal realizations of autonomous chaotic oscillators based on trans-immittance filters, *IEEE Access* **7**, pp. 17561–17577.
Pham, C., Korehisa, M., and Tanaka, M. (1996). Chaotic behavior and synchronization phenomena in a novel chaotic transistors circuit, *IEEE T. Circuits-I* **43**, pp. 1006–1011.
Pikovsky, A. S., and Rabinovich, M. I. (1981). Stochastic oscillations in dissipative systems, *Phys. Nonlinear Phenom.* **2**, pp. 8–24.
Pino, R. E., Bohl, J. W., McDonald, N., Wysocki, B., Rozwood, P., Campbell, K. A., Oblea, A. and Timilsina, A. (2010). Compact method for modeling and simulation of memristor devices: Ion conductor chalcogenide-based memristor devices, In *2010 IEEE/ACM International Symposium on Nanoscale Architectures*, pp. 1–4.
Piper, J. and Sprott, J. C. (2010). Simple autonomous chaotic circuits, *IEEE T. Circuits-II* **57**, pp. 730–734.
Press, W. H., Teukolsky, S. A., Vetterling, W. T., and Flannery, B. P. (2007). *Numerical Recipes: The Art of Scientific Computing* (3rd edn) (Cambridge University Press, Cambridge).
Raadhakrishnan, P., Yagle, A. E., Rao, B. V., and Dorband, E. (1992). On the upper bounds of the equivalent oscillator and notch-filter circuits: a non-commutative group theoretic approach, *IEEE T. Circuits-I* **39**, pp. 756–759.
Rodriguez-Vazquez, A., Huertas, J., and Chua, L. (1985). Chaos in switched-capacitor circuit, *IEEE T. Circuits* **32**, pp. 1083–1085.
Rodriguez-Vazquez, A., Huertas, J., and Chua, L. (1987). Chaos from switched-capacitor circuits: Discrete maps, *Proc. IEEE* **75**, pp. 1090–1106.
Rollins, R. W. and Hunt, E. R. (1982). Exactly solvable model of a physical system exhibiting universal chaotic behavior, *Phys. Rev. Lett.* **49**, pp. 1295–1298.
Rössler, O. E. (1976). An equation for continuous chaos, *Ann. New York Acad. Sci.* **57**, pp. 397–398.
Rössler, O. E. (1979). Continuous chaos – four prototype equations, *Phys. Lett. A* **316**, pp. 376–392.
Saito, T. (1985). Hysteresis chaos generator, *Electron. Comm. Jpn. 1* **68**, pp. 49–54.
Saito, T. (1988). The hysteresis chaos generator family, *IEEE International Symposium on Circuits and Systems, IEEE, 1988*.
Saito, T. (1989). A chaotic circuit family including one diode, *Electr. Commun. Jpn. 3* **72**, pp. 52–59.
Saito, T. and Fujita, H. (1981). Chaos in a manifold piecewise linear system, *Electron. Comm. Jpn. 1* **64**, pp. 9–17.
San-Um, W., Suksiri, B., and Ketthong, P. (2014). A simple RLCC-diode-opamp chaotic oscillator, *Int. J. Bifurcat. Chaos* **24**, pp. 1450155-1–8.
Schot, S. H. (1978). The time rate of change of acceleration, *Am. J. Phys.* **46**, pp. 1090–1094.
Sedra, A. S., and Smith, K. C. (1970). A second-generation current conveyor and its applications, *IEEE T. Circuits Syst.* **17**, pp. 132–134.

Sedra, A. S. and Smith, K. C. (2015). *Microelectronic Circuits* 7th ed. (Oxford, New York).
Senani, R. and Singh, V. K. (1996). Novel single-resistance-controlled-oscillator configuration using current feedback amplifiers, *IEEE T. Circuits-I* **43**, pp. 698–700.
Shaw, R. (1984). *The Dripping Faucet as a Model Chaotic System* (Aerial Press, Santa Cruz, CA).
Shi, Z., and Ran, L. (2004). Tunnel diode based Chua's circuit. In *Emerging Technologies: Frontiers of Mobile and Wireless Communication, 2004. Proceedings of the IEEE 6th Circuits and Systems Symposium on* **1**, pp. 217–220.
Shinriki, M., Yamamoto, M., and Mori, S. (1981). Multimode oscillations in a modified van der Pol oscillator containing a positive nonlinear conductance, *P. IEEE* **69**, pp. 394–395.
Sprott, J. C. (1981). *Introduction to Modern Electronics* (Wiley, New York).
Sprott, J. C. (1994). Some simple chaotic flows, *Phys. Rev. E* **50**, pp. R647–R650.
Sprott, J. C. (1997a). Simplest dissipative chaotic flow, *Phys. Lett. A* **228**, pp. 271–274.
Sprott, J. C. (1997b). Some simple chaotic jerk functions, *Am. J. Phys.* **65**, pp. 537–543.
Sprott, J. C. (2000a). Simple chaotic systems and circuits, *Am. J. Phys.* **68**, pp. 758–763.
Sprott, J. C. (2000b). A new class of chaotic circuit, *Phys. Lett. A* **266**, pp. 19–23.
Sprott, J. C. (2010). *Elegant Chaos: Algebraically Simple Chaotic Flows* (World Scientific, Singapore).
Sprott, J. C. (2011). A new chaotic jerk circuit, *IEEE T. Circuits-II* **58**, pp. 240–443.
Sprott, J. C. (2018). Ergodicity of one-dimensional oscillators with a signum thermostat, *Computational Methods in Science and Technology* **24**, pp. 169–176.
Sprott, J. C. (2019). *Elegant Fractals: Automated Generation of Computer Art* (World Scientific, Singapore).
Sprott, J. C. and Thio, W. J. (2020). A chaotic circuit for producing Gaussian random numbers, *Int. J. Bifurcat. Chaos* **30**, pp. 2050116-1–8.
Sprott, J. C. and Xiong, A. (2015). Classifying and quantifying basins of attraction, *Chaos* **25**, pp. 083101.
Srisuchinwong, B. and Munmuangsaen, B. (2012). Four current-tunable chaotic oscillators in set of two diode reversible pairs, *Electron. Lett.* **48**, pp. 1051–1053.
Srisuchinwong, B. and Treetanakorn, R. (2014). Current-tunable chaotic jerk circuit based on only one unity-gain amplifier, *Electron. Lett.* **24**, pp. 1815–1817.
Srisuchinwong, B., Munmuangsaen, B., Ahmad, I., and Suibkitwanchai, K. (2019). On a simple single-transistor-based chaotic snap circuit: A maximized attractor dimension at minimized damping and a stable equilibrium, *IEEE Access* **7**, pp. 116643–116660.

Streetman, B. G. and Banerjee S. K. (2016). *Solid State Electronic Devices* (Pearson Education, Boston).
Strogatz, S. H. (1994). *Nonlinear Dynamics and Chaos with Applications to Physics, Biology, Chemistry, and Engineering* (Westview Press, Boulder, CO).
Strukov, D. B., Snider, G. S., Stewart, D. R., and Williams, R. S. (2008). The missing memristor found, *Nature* **453**, pp. 80–83.
Svoboda, J. A. (1989). Current conveyors, operational amplifiers and nullors, *IEE Proc-G.* **136**, pp. 317–322.
Sze, S. M. and Kwok, K. N. (2006). *Physics of Semiconductor Devices* (John Wiley & Sons, New York).
Tamasevicius, A., Mykolaitis, G., Pyragas, V., and Pyragas, K. (2004). A simple chaotic oscillator for educational purposes, *Eur. J. Phys.* **26**, pp. 61–63.
Tamasevicius, A., Namajunas, A., Cenys, A. (1996a). Simple 4D chaotic oscillator, *Electron. Lett.* **32**, pp. 957–958.
Tamasevicius, A., Mykolaitis, G., and Namajunas, A. (1996b). Double scroll in a simple '2D' chaotic oscillator, *Electron. Lett.* **32**, pp. 1250–1251.
Tchitnga, R., Fotsin, H. B., Nana, B., Fotso, P. H. L., and Woafo, P. (2012). Hartley's oscillator: The simplest chaotic two-component circuit, *Chaos, Solitons & Fractals* **45**, pp. 306–313.
Tchitnga, R., Nguazon, T., Fotso, P. H. L., and Gallas, J. A. C. (2016). Chaos in a single op-amp-based jerk circuit: Experiments and simulations, *IEEE T. Circuits-II* **63**, pp. 239–243.
Tchitnga, R., Nanfa'a, R., Pelap, F., Louodop, P., and Woafo, P. (2017). A novel high-frequency interpretation of a general purpose Op-Amp-based negative resistance for chaotic vibrations in a simple a priori nonchaotic circuit, *J. Vib. Control* **23**, pp. 744–751.
Tekam, R. B. W., Kengne, J., and Kenmoe, G. D. (2019). High frequency Colpitts' oscillator: A simple configuration for chaos generation, *Chaos, Solitons & Fractals* **126**, pp. 351–360.
Testa, J., Pérez, J. and Jeffries, C. (1982). Evidence for universal chaotic behavior of a driven nonlinear oscillator, *Phys. Rev. Lett.* **48**, pp. 714–717.
Thompson, J. M. T. and Stewart, H. B. (1986). *Nonlinear Dynamics and Chaos* (Wiley, New York).
Ueda, Y. and Akamatsu, N. (1981). Chaotically transitional phenomena in the forced negative-resistance oscillator, *IEEE T. Circuits* **28**, pp. 217–224.
Ueda, Y., Akamatsu, N., and Hayashi, C. (1973). Computer simulation of differential equations and non-periodic oscillations (in Japanese), *Trans. IEE Japan* **98-A**, pp. 167–173.
van der Pol, B. (1920). A theory of the amplitude of free and forced triode vibrations, *Radio Review* **1**, pp. 701–710 and 754–762.
van der Pol, B. (1926). On relaxation-oscillations, *The London, Edinburgh, and Dublin Philos. Mag.* **2**, pp. 978–992.
van der Pol, B. and van der Mark, J. (1927). Frequency demultiplication, *Nature* **120**, pp. 363–364.

Wang, D., Zhao, H., and Yu, J. (2009). Chaos in memristor based Murali-Lakshmanan-Chua circuit, *Proceedings of the International Conference on Communications, Circuits and Systems (ICCCAS '09)*, pp. 958–960.

Wang, N., Zhang, G., and Bao, H. (2020). A simple autonomous chaotic circuit with dead-zone nonlinearity, in *IEEE T. Circuits-II*, doi: 10.1109/TCSII.2020.3005726.

Wegener, C. and Kennedy, M. P. (1995). Chaotic Colpitts oscillator, *Proc. NDES* **95**, pp. 255–258.

Wijekoon, J. H. B. and Dudek, P. (2008). Compact silicon neuron circuit with spiking and bursting behaviour, *Neural Networks* **21**, pp. 524–534.

Wolf, A., Swift, J. B., Swinney, H. L., and Vastano, J. A. (1985). Determining Lyapunov exponents from a time series, *Phys. Nonlinear Phenom.* **16**, pp. 285–317.

Yakopcic, C. (2014). Memristor device modeling and circuit design for read out integrated circuits, memory architectures, and neuromorphic systems. University of Dayton, Ph.D. thesis.

Yim, G., Ryu, J., Park, Y., Rim, S., Lee, S., Kye, W., and Kim, C. (2004). Chaotic behaviors of operational amplifiers, *Phys. Rev. E* **69**, pp. 045201.

Zhang, F. and Heidel, J. (1997). Non-chaotic behaviour in three-dimensional quadratic systems, *Nonlinearity* **10**, pp. 1289–1303.

Zhang, F. and Heidel, J. (1999). Erratum: Non-chaotic behaviour in three-dimensional quadratic systems, *Nonlinearity* **12**, p. 738.

Index

action potential, 111
active resistor, 77
amplifier
  buffer, 57
  current feedback, 73
  high gain, 205
  instrumentation, 32
  operational, 23, 205
  saturating, 205
  transimpedance, 33
amplitude parameter, 25
analog computer, 241
analog multiplier, 241
analog switch, 229
analog-to-digital converter, 146
antidamping, 8
aperiodicity, 12
attractor, 8
  hidden, 17
  limit cycle, 8
  multiple, 17
  multiscroll, 108
  point, 17
  self-excited, 17
  strange, 10, 12
  stretching and folding, 14
  torus, 10
  two-lobe, 208
autonomous system, 10

B-H curve, 276
band-pass filter, 67

bang-bang controller, 272
barrier voltage, 44
basin of attraction, 17
bifurcation parameter, 25
bipolar junction transistor (BJT), 89
boundedness, 14
breakover voltage, 173
buffer amplifier, 57
bursting oscillation, 114
butterfly effect, 13

chaos, 13
chaotic sea, 272
charge control model, 121
choke, 29
Chua's circuit, 21, 108, 212
Chua, Leon, vii, 20
Clapp oscillator, 99
clipper circuit, 34
coercivity, 276
Colpitts oscillator, 97
Colpitts, Edwin, 97
comparator, 4, 23, 207
constant-current diode, 133
corner frequency, 281
correlation dimension, 184
criticality, 114
current conveyor, 73
current feedback amplifier (CFA), 73
current source, 56
current-to-voltage converter, 33

damping, 2
depletion region, 44
deterministic system, 10
diffusion current, 44
dimension
 attractor, 16
 correlation, 184
 fractal, 12
 Kaplan–Yorke, 16
diode
 capacitance, 48
 constant-current, 133
 conventional, 41
 I-V characteristic, 45
 ideal, 41
 light emitting (LED), 42
 PN junction, 42
 real, 42
 Schottky, 51
 tunnel, 149
 varactor, 51
 Zener, 175
drift current, 44
dripping faucet, 158

Early effect, 92
Ebers–Moll model, 93
eddy current, 24, 277
electromotive force, 277
electronic filter, 61
elegance, 38
emulator, 39, 152, 301
entrainment, 12
entropy, 16
ergodicity, 272
Esaki, Leo, 149
Euler's number, 2

ferrite, 275
field effect transistor (FET), 94
field programmable analog array (FPAA), 141, 242
field programmable gate array (FPGA), 242

filter, 61
 active and passive, 64
 band-pass, 67
 high-pass, 67
 low-pass, 33
 resonant, 64
forced diode resonator, 49
forward bias, 42, 44
four-quadrant multiplier, 243
fractal, 12
frequency, 2
frequency locking, 12
function generator, 30

Gaussian distribution, 272
ground, 22
gyrator, 33

harmonic oscillator, 8, 21, 272
Hartley oscillator, 2, 99, 130
Hartley, Ralph, 130
heat bath, 272
hidden attractor, 17
high-gain amplifier, 205
high-pass filter, 67
Hindmarsh model, 114
Hopf bifurcation, 26, 245
hyperbolic tangent, 206
hyperchaoticity, 17
hysteresis
 comparator, 7
 inductor, 24, 276
 memristor, 300
 thyristor, 173
 tunnel diode, 151

I-V converter, 33
incommensurate frequencies, 10
initial condition, 38
instrumentation amplifier, 32
integrator, 64
inversion layer, 95
inverter, 152

Jacobian matrix, 16
jerk circuit, 84
jerk system, 253
junction capacitance, 48
junction field effect transistor, 94

Kaplan–Yorke dimension, 16
Kirchhoff's law, 21

laminations, 277
Landauer, Rolf, 191
light emitting diode (LED), 42
limit cycle, 8, 25
Lorenz system, 243
Lorenz, Edward, vii, 20
low-pass filter, 33
LTSpice, 35
Lyapunov exponent, 14

magnetic flux, 277
magnetization curve, 276
Mandelbrot, Benoit, vii
Matsumoto, Takashi, 20
memristor, 297
metal oxide semiconductor field effect transistor (MOSFET), 95
mixing, 14
multiscroll attractor, 108
multistability, 17
multivibrator, 6

negative impedance converter, 288
negative resistance, 77, 150, 177
neon lamp, 19, 184
neural network, 297
nonautonomous system, 10
nonlinear inductor, 275
Nosé–Hoover system, 261

operational amplifier, 23, 205
oscillator, 1
    Clapp, 99
    Colpitts, 2, 97
    harmonic, 8, 21, 272
    Hartley, 2, 99, 130

relaxation, 4, 153, 187
sinusoidal, 25
van der Pol, 8, 19, 184
Vilnius, 51
Wien bridge, 67, 133, 208
oscilloscope, 31

parameter
    amplitude, 25
    bifurcation, 25
    elegant, 26
    scaling, 26
parasitic effect, 24
permanent magnet, 276
permeability, 275
phase space plot, 2
pinch-off voltage, 95
pinched hysteresis loop, 300
PN junction, 42
Poincaré section, 181
Poincaré–Bendixson theorem, 16
point attractor, 17
power supply, 30
predictability, 13
probability distribution function, 272

quality factor (Q), 2, 25
quasiperiodicity, 10
quiescent (Q) point, 54

Rössler prototype-4 system, 245
Rössler, Otto, 245
rail splitter, 30
random number generator, 272
rectification, 42
relaxation oscillator, 4, 153, 187
resistive random access memory (RRAM), 298
resonant filter, 64
retentivity, 276
reverse bias, 42, 45
reverse recovery time, 51
ripple, 42
robustness, 17

Saito family circuit
    diode, 77
    hysteresis, 224
    inductor, 285
    memristor, 310
    switch, 228
    thyristor, 199
sample-and-hold circuit, 146
saturating amplifier, 205
saturation, 275, 277
Schmitt trigger, 224
Schottky diode, 51
self-excited attractor, 17
semiconductor, 42
set of measure zero, 14
signum function, 23, 207
signum thermostat, 268
silicon bilateral switch (SBS), 175
silicon controlled rectifier (SCR), 173
sinusoidal oscillator, 25
skin effect, 24
slew rate, 7
spontaneous symmetry breaking, 209
state space, 10
stiff system, 8
strange attractor, 10, 12
switched capacitor, 140
symmetry, 208, 285, 292

thermostat, 261
thyristor, 125, 173
toroid, 278
torus, 10
transformer, 278

transient, 15
transimpedance amplifier, 33
transistor, 89
    bipolar junction, 89
    field effect, 94
    junction field effect, 94
    metal oxide semiconductor field
        effect, 95
    model, 90
    modes, 89
    switch, 117
transistor-transistor logic (TTL), 31
transmission gate, 147
tunnel diode, 149
tunneling, 45, 149
two-lobe attractor, 208

Ueda system, 249
Ueda, Yoshisuke, vii, 20, 249

van der Pol oscillator, 8, 19, 184
    forced, 10
van der Pol, Balthasar, vii, 4
varactor diode, 51
Vilnius oscillator, 51

Wien bridge oscillator
    comparator, 208
    diode, 67
    memristor, 313
    transistor, 133

Zener diode, 175

About the Authors

JULIEN CLINTON SPROTT is Emeritus Professor of Physics at the University of Wisconsin–Madison. After a twenty-five year career in plasma physics and controlled nuclear fusion, he became interested in chaos in 1988 after hearing a lecture on the subject by George Rowlands. He is author of about five hundred technical papers and a dozen books including *Introduction to Modern Electronics* (Wiley, 1981), *Chaos and Time-Series Analysis* (Oxford, 2003), and *Elegant Chaos* (World Scientific, 2010). He has produced thirty-eight hour-long videos of his popular public presentation of *The Wonders of Physics* and four commercial software packages. His award-winning website is at http://sprott.physics.wisc.edu/.

WESLEY JOO-CHEN THIO obtained his B.S. in Electrical and Computer Engineering at The Ohio State University in 2018 and his M.S. in Electrical and Computer Engineering at the University of Michigan in 2022. His undergraduate research studies at Ohio State centered on batteries and self-powered wearable technologies. During that time, he developed an interest in chaotic circuits through *Elegant Chaos* and Chua's circuit. His work in electronics and batteries has resulted in no less than a dozen publications as well as several patents, and he was awarded Ohio State's Next-Generation Innovator Award in 2019. His personal website is at http://wesleythio.com/.

The authors discussing a chaotic circuit at the University of Wisconsin in 2014.

CPSIA information can be obtained
at www.ICGtesting.com
Printed in the USA
BVHW061108300122
627168BV00002B/5